CELL BIOLOGY AND GENETICS

C E C I E   S T A R R

BELMONT, CALIFORNIA

R A L P H   T A G G A R T

MICHIGAN STATE UNIVERSITY

BIOLOGY:

THE UNITY AND DIVERSITY OF LIFE

SEVENTH EDITION

WADSWORTH PUBLISHING COMPANY

I(T)P™ AN INTERNATIONAL THOMSON PUBLISHING COMPANY

Belmont • Albany • Bonn • Boston
Cincinnati • Detroit • London • Madrid
Melbourne • Mexico City • New York
Paris • San Francisco • Singapore
Tokyo • Toronto • Washington

BIOLOGY PUBLISHER: Jack C. Carey

ASSISTANT EDITOR: Kristin Milotich

EDITORIAL ASSISTANT: Kerri Abdinoor

PRINT BUYER: Randy Hurst

PRODUCTION SERVICES COORDINATOR: Sandra Craig

PRODUCTION: Mary Douglas, Rogue Valley Publications

TEXT AND COVER DESIGN, ART DIRECTION: Gary Head,
Gary Head Design

EDITORIAL PRODUCTION: Myrna Engler-Forkner, Melissa Andrews,
Rosaleen Bertolino, Marilyn Evenson, Susan Gall, Ed Serdziak,
Karen Stough

ARTISTS: Raychel Ciemma, Robert Demarest, Hans & Cassady, Inc.
(Hans Neuhart), Darwen Hennings, Vally Hennings, Betsy
Palay, Precision Graphics (Jan Flessner), Nadine Sokol, Kevin
Somerville, Lloyd Townsend

PHOTO RESEARCH AND PERMISSIONS: Marion Hansen

COVER PHOTOGRAPH: © Frans Lanting/Minden Pictures

COMPOSITION: American Composition & Graphics, Inc.
(Jim Jeschke, Jody Ward, and Valerie Norris)

COLOR PROCESSING: H & S Graphics, Inc. (Tom Anderson,
Nancy Dean, and John Deady)

PRINTING AND BINDING: R. R. Donnelley & Sons Company/Willard

BOOKS IN THE WADSWORTH BIOLOGY SERIES

For more information, contact Wadsworth Publishing Company:

Wadsworth Publishing Company
10 Davis Drive, Belmont, California 94002, USA

International Thomson Publishing Europe
Berkshire House 168-173, High Holborn
London, WC1V 7AA, England

Thomas Nelson Australia
102 Dodds Street
South Melbourne 3205, Victoria, Australia

Nelson Canada
1120 Birchmount Road
Scarborough, Ontario, Canada M1K 5G4

International Thomson Editores
Campos Eliseos 385, Piso 7
Col. Polanco, 11560 México D.F. México

International Thomson Publishing GmbH
Königswinterer Strasse 418
53227 Bonn, Germany

International Thomson Publishing Asia
221 Henderson Road, #05-10 Henderson Building
Singapore 0315

International Thomson Publishing Japan
Hirakawacho Kyowa Building, 3F
2-2-1 Hirakawacho, Chiyoda-ku, Tokyo 102, Japan

CONTENTS IN BRIEF

Highlighted chapters are included in CELL BIOLOGY AND GENETICS.

DETAILED CONTENTS

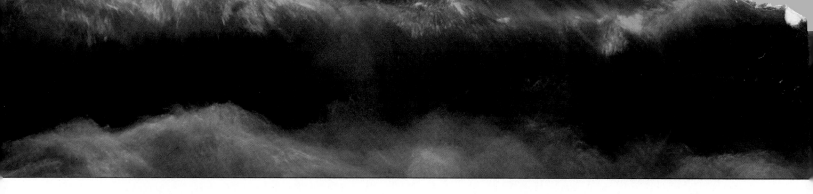

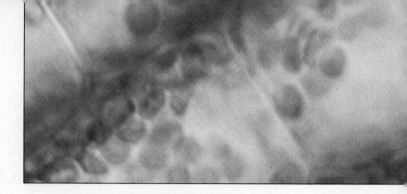

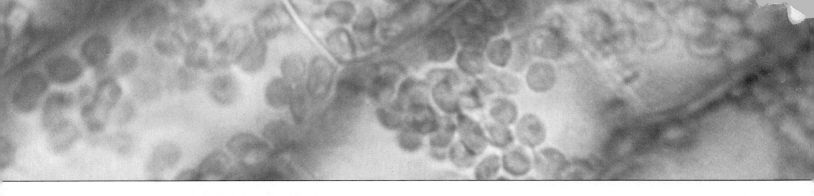

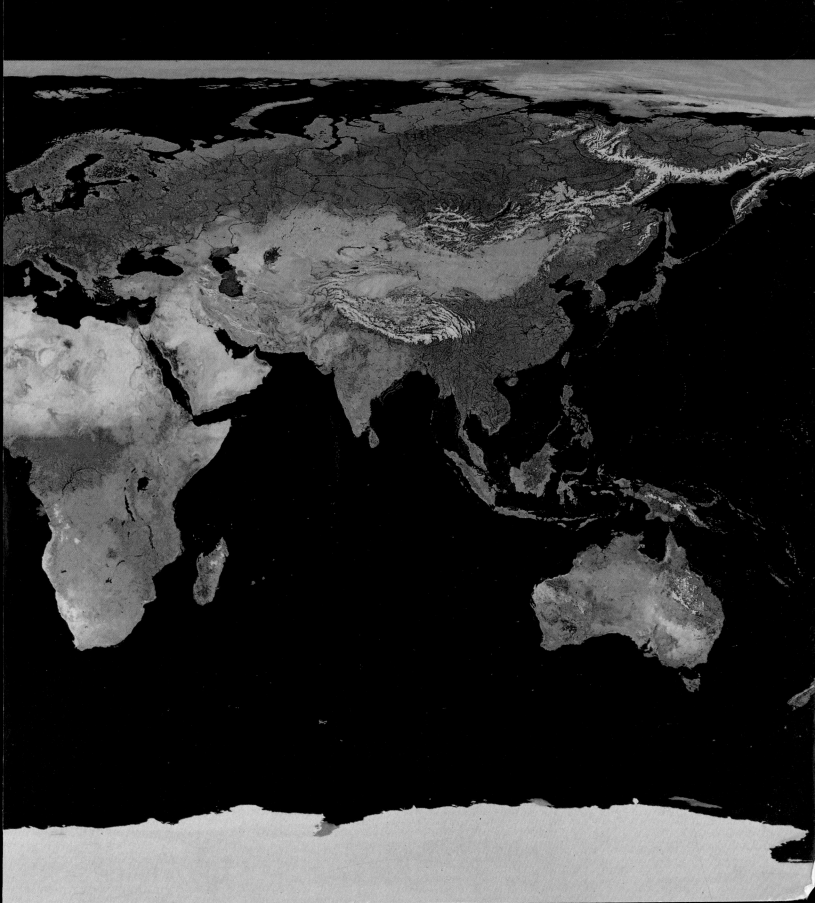

1 METHODS AND CONCEPTS IN BIOLOGY

Biology Revisited

Buried somewhere in that mass of tissue just above and behind your eyes are memories of your first encounters with the living world. Still in residence are memories of discovering your hands and feet, your family, friends, the change of seasons, the smell of rain-drenched earth and grass. In that brain are memories of early introductions to a great disorganized parade of insects, spiders, flowers, frogs, and furred things—mostly living, sometimes dead. There are memories of questions—*"What is life?"* and, inevitably, *"What is death?"* There are memories of answers, some satisfying, others less so.

Figure 1.1 Think back on all you have known and seen. This is a foundation for your deeper probes into the world of life.

By observing, asking questions, and accumulating answers, you have built up a store of knowledge about the world of life. Experience and education have been refining your questions, and no doubt some answers are difficult to come by. Think of a young man whose brain is functionally dead as a result of a motorcycle accident. If his breathing and other basic functions proceed only as long as he remains hooked up to mechanical support systems, is he no longer "alive"? Think of a recently fertilized egg growing inside a pregnant woman, but currently no more than a cluster of a few dozen tiny cells. At what point in its development is it a definably "human" life? If questions like this have crossed your mind, your thoughts about life obviously run deep.

The point is, this book isn't your introduction to biology—"the study of life"—for you have been studying life ever since information began penetrating your brain. This book simply is biology *revisited*, in ways that may help carry your thoughts to more organized levels of understanding.

Return to the question, *What is life?* Offhandedly, you might reply that you know it when you see it. To biologists, however, the question opens up a story that has been unfolding in countless directions for several billion years! "Life" is an outcome of ancient events by which nonliving materials became assembled into the first living cells. "Life" is a way of capturing and using energy and raw materials. "Life" is a way of sensing and responding to specific changes in the environment. "Life" is a capacity to reproduce, grow, and develop. And "life" evolves, meaning that details in the body plan and functions of organisms can change through successive generations.

Yet this short description only hints at the meaning of life. Deeper insight requires wide-ranging study of life's characteristics.

Throughout this book you will come across many examples of how organisms are constructed, how they function, where they live, and what they do. The examples support certain concepts which, taken together, will give you a sense of what "life" is. This chapter provides an overview of the basic concepts. As you continue reading the book, you may find it useful to return to this overview to reinforce your grasp of details.

1. There is unity in the living world, for all organisms are alike in key respects. Their structural organization and functions depend on properties of matter and energy. They obtain and use energy and materials from the environment. They make controlled responses to changing conditions. They grow and reproduce, based on instructions contained in DNA.

2. There is diversity in the living world, for organisms vary immensely in body plans, body functions, and behavior. Evolutionary theories explain this diversity.

3. Biology, like other branches of science, is based on systematic observations, hypotheses, predictions, and relentless tests. The external world, not internal conviction, is the testing ground for scientific theories.

1.1 SHARED CHARACTERISTICS OF LIFE

Energy, DNA, and Life

Picture a frog on a rock, busily croaking. Without even thinking about it, you know the frog is alive and the rock is not. At a much deeper level, however, the difference between them blurs. They and all other things are composed of the same particles (protons, electrons, and neutrons). The particles are organized as atoms, according to the same physical laws. At the heart of those laws is something called **energy**—a capacity to make things happen, to do work. Energetic interactions bind atom to atom in predictable patterns, giving rise to the structured bits of matter we call molecules. Energetic interactions among molecules hold a rock together—and they hold a frog together.

It takes a special molecule called deoxyribonucleic acid, or **DNA**, to set living things apart from the nonliving world. No chunk of granite or quartz has it. DNA molecules contain instructions for assembling new organisms from "lifeless" molecules that contain carbon and a few other kinds of atoms. By analogy, with proper instructions and a little effort, you can turn a disordered heap of ceramic tiles—even just two kinds of tiles—into ordered patterns such as these:

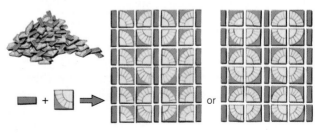

Similarly, life emerges from lifeless matter with DNA "directions," raw materials, and energy inputs.

Levels of Organization in Nature

Look carefully at Figure 1.2, which outlines the levels of organization in nature. The properties of life emerge at the level of cells. A **cell** is an organized unit that can survive and reproduce on its own, given DNA instructions and sources of energy and raw materials. In other words, the cell is the basic *living* unit. This definition obviously fits a free-living, single-celled organism such as an amoeba. Does it fit a **multicelled organism**, that has specialized cells organized into tissues and organs? Yes. You may find this a strange answer. After all, your own cells could never live all by themselves in nature. They must be bathed by fluids inside your body. Yet even human cells can be isolated and kept alive under controlled laboratory conditions. Researchers around the world routinely maintain human cells for use in important experiments, including cancer studies.

Referring to Figure 1.2, we find a more inclusive level of organization—the **population**. This is a group of single-celled or multicelled organisms of the same kind, such as a breeding colony of Emperor penguins in Antarctica. Next is the **community**, which includes all populations of all species (penguins, whales, seals, fishes, and so on) living in the same area. The next level, the **ecosystem**, includes the community and its physical and chemical environment. The most inclusive level of organization in nature is the **biosphere**. The word refers to all regions of the earth's waters, crust, and atmosphere in which organisms live.

Within the hierarchy of organization in nature, the properties of life emerge at the level of cells. Cells emerge through a convergence of raw materials, sources of energy, and instructions contained in DNA molecules.

Metabolism: Life's Energy Transfers

You never, ever will find a rock engaged in metabolic activities. Only living cells can do this. **Metabolism** refers to the cell's capacity to (1) extract and convert energy from its surroundings and (2) use energy and so maintain itself, grow, and reproduce. Simply put, metabolism means *energy transfers* within cells.

Think of a rice plant. Many of its cells engage in **photosynthesis**. In the first stage of this process, cells trap sunlight energy, then convert it to another form of energy. In the second stage, cells use the chemical energy to build sugars, starch, and other substances. As part of the process of photosynthesis, molecules of **ATP**, an "energy carrier," are put together. ATP transfers energy to other molecules that function as metabolic workers (enzymes), building blocks, or energy reserves.

Biosphere
Those regions of the earth's waters, crust, and atmosphere in which organisms can exist

Ecosystem
A community and its physical environment

Community
The populations of *all* species occupying the same area

Population
A group of individuals of the same kind (that is, the same species) occupying a given area at the same time

Multicellular Organism
An individual composed of specialized, interdependent cells arrayed in tissues, organs, and often organ systems

Organ System
Two or more organs interacting chemically, physically, or both in ways that contribute to the survival of the whole organism

Organ
A structural unit in which tissues are combined in specific amounts and patterns that allow them to perform a common task

Tissue
A group of cells and surrounding substances, functioning together in a specialized activity

Cell
Smallest *living* unit; may live independently or may be part of a multicellular organism

Organelle
Sacs or other compartments that separate different activities inside the cell

Molecule
A unit of two or more atoms of the same or different elements bonded together

Atom
Smallest unit of an element that still retains the properties of that element

Subatomic Particle
An electron, proton, or neutron; one of the three major particles of which atoms are composed

Figure 1.2 Levels of organization in nature. Cells represent the first level at which the properties of life emerge.

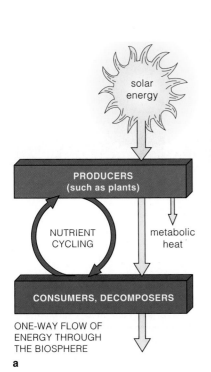

solar energy

PRODUCERS
(such as plants)

NUTRIENT CYCLING

metabolic heat

CONSUMERS, DECOMPOSERS

ONE-WAY FLOW OF
ENERGY THROUGH
THE BIOSPHERE

a

b

c

d

Figure 1.3 (**a**) Direction of energy flow and the cycling of materials through the biosphere.

(**b**) Example of interdependency through nutrient cycling, although this cast of characters may seem a bit improbable at first. In the warm, dry grassland called the African savanna, we come across an adult male elephant. It eats huge quantities of plants to maintain its eight-ton self, and it produces huge piles of solid wastes—dung—that still contain some unused nutrients. Thus, although most organisms would not recognize it as such, elephant dung *is* an exploitable food source.

(**c**) And so we next have little dung beetles rushing to the scene almost simultaneously with the uplifting of an elephant tail. Working rapidly, they carve fragments of moist dung into round balls, which they roll off and bury in burrows. In these balls the beetles lay eggs, a reproductive behavior that assures forthcoming offspring (**d**) of a compact food supply.

Thanks to beetles, dung does not pile up and dry out into rock-hard mounds in the intense heat of the day. Instead, the surface of the land is tidied up, beetle offspring are fed, and leftover dung accumulates in burrows—there to enrich the soil that nourishes the plants that sustain (among others) the elephants.

In rice plants, some of the stored energy becomes concentrated in starchy seeds—rice grains. Energy reserves in countless trillions of rice grains provide energy for billions of rice-eating humans around the world. How? In humans, as in most animals and plants, stored energy is released and transferred to ATP by way of **aerobic respiration**, another metabolic process.

Living things show metabolic activity. Their cells acquire and use energy to stockpile, tear down, build, and eliminate materials in ways that promote survival and reproduction.

Interdependency Among Organisms

With few exceptions, a flow of energy from the sun maintains the great pattern of organization in nature. Plants and some other photosynthetic organisms are the entry point for this flow. They are the food **producers**. Animals are **consumers**. Directly or indirectly, they feed on energy stored in plant parts. Thus zebras tap directly into the stored energy when they nibble on grass, and lions tap into it indirectly when they chomp on zebras. Many kinds of bacteria and fungi are **decomposers**. When they feed on tissues or remains of other organisms, they break down sugars and other biological molecules to simple raw materials—which can be cycled back to producers.

And so we have interdependency among organisms, based on a one-way flow of energy *through* them and a cycling of materials *among* them (Figure 1.3).

Such interactions among organisms influence populations, communities, and ecosystems. They even influence the global environment. Understand the extent of the interactions and you will gain insight into amplification of the greenhouse effect, acid rain, and many other modern-day problems.

Webs of organization connect all organisms in nature, in that organisms depend directly or indirectly on one another for energy and raw materials.

a   b   c   d

Figure 1.4 "The insect"—a continuous series of stages in development. Different adaptive properties emerge at each stage. Shown here, a silkworm moth, from egg (**a**) to larval stage (**b**), to pupal form (**c**), to the splendid adult form (**d**,**e**).

Sensing and Responding to Change

It is often said that only organisms "respond" to the environment. Yet a rock also "responds" to the environment, as when it yields to gravity and tumbles downhill or changes shape slowly under the battering of wind, rain, or tides. The real difference is this: *Organisms have the cellular means to sense changes in the environment and make controlled responses to them.* They do so with the help of **receptors**, which are molecules and structures that can detect specific information about the environment. When cells receive signals from receptors, they adjust their activities in ways that bring about an appropriate response.

Your body, for example, can withstand only so much heat or cold. It must rid itself of harmful substances. Certain foods must be available, in certain amounts. Yet temperatures shift, harmful substances may be encountered, and food is sometimes plentiful or scarce.

Think about what happens after you eat and simple sugar molecules enter your bloodstream. Blood is part of the body's "internal environment" (the other part is the tissue fluid bathing your cells). When the sugar level in blood rises, cells of the pancreas step up their secretion of insulin. Most cells in your body have receptors for insulin, a hormone that prods the cells into taking up sugar molecules. With so many cells taking up sugar, the blood sugar level returns to normal.

Suppose you skip breakfast, then lunch, and the blood sugar level falls. Now a different hormone prods liver cells to dig into their stores of energy-rich molecules. Those molecules are broken down to simple sugars, which are released into the bloodstream—and again the blood sugar level returns to normal.

Usually, the internal environment of a multicelled organism is kept fairly constant. When conditions in the internal environment are being maintained within tolerable limits, we call this a state of **homeostasis**.

Organisms have the means to sense and respond to changes in their environment. The responses help maintain favorable operating conditions inside the cell or multicelled body.

Reproduction

We humans tend to think we enter the world rather abruptly and leave it the same way. Yet we and all other organisms are more than this. *We are part of an immense, ongoing journey that began billions of years ago.* Think of the first cell produced when a human sperm penetrates an egg. The cell would not even exist if the sperm and egg had not formed earlier, according to DNA instructions passed down through countless generations. With those time-tested instructions, a new human body develops in ways that will prepare it, ultimately, for helping to produce individuals of the next generation. With **reproduction**—that is, the production of offspring—life's journey continues.

Or think of a moth. Do you simply picture a winged insect? What of the tiny fertilized egg deposited on a branch by a female moth (Figure 1.4)? The egg contains the instructions necessary to become an adult. By those instructions, the egg develops into a caterpillar, a larval stage adapted for rapid feeding and growth. The caterpillar eats and grows until an internal "alarm clock" goes off. Then its body enters a so-called pupal stage of development, which involves wholesale remodeling. Some cells die, and others multiply and become organized in different patterns. In time an adult moth emerges. It has organs that contain eggs or sperm. Its wings are brightly colored and flutter at a frequency appropriate for attracting a mate. In short, the adult stage is adapted for reproduction.

None of these stages is "the insect." The insect is a series of organized stages from one fertilized egg to the

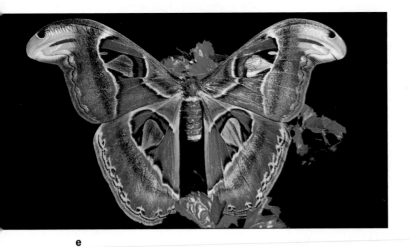

e

a

b

Figure 1.5 An example of how two different forms of the same trait (coloration of moths) are each adaptive under different environmental conditions.

next. Each stage is vital for the ultimate production of new moths. The instructions for each stage were written into moth DNA long before each moment of reproduction—and so the ancient moth story continues.

Each organism arises through reproduction.

Each organism is part of a reproductive continuum that extends back through countless generations.

Mutation: Source of Variations in Heritable Traits

Reproduction involves **inheritance**. The word means that parents transmit DNA instructions for duplicating their traits, such as body form, to offspring.

DNA has two striking qualities. Its instructions assure that offspring will resemble parents—and they also permit *variations* in the details of traits. For example, having five fingers on each hand is a human trait. Yet some humans are born with six fingers on each hand instead of five! Variations in traits arise through **mutations**, which are abnormal, heritable changes in the structure of DNA molecules.

Many mutations are harmful. A change in even a bit of DNA may be enough to sabotage the steps necessary to produce a vital trait. In *hemophilia A*, for example, a tiny mutation leads to an impaired ability to clot blood. Bleeding continues for an abnormally long time after even a small cut or bruise.

Yet some mutations are harmless, even beneficial, under prevailing conditions. A classic example is a mutation in light-colored moths that leads to dark-colored offspring. Moths fly by night and rest during the day, when birds that eat them are active. What happens when a light moth rests on a light-colored tree trunk (Figure 1.5)? Birds simply don't see it. Suppose, as a result of heavy industry, light trunks in a forested region become soot covered—and dark. The dark moths are less conspicuous, so they have a better chance of living long enough to reproduce. Under sooty conditions, the dark form of the trait is more adaptive.

An **adaptive trait** simply is one that helps an organism survive and reproduce under a given set of environmental conditions.

DNA is the molecule of inheritance in organisms. Its instructions for reproducing traits are passed on from parents to offspring.

Mutations introduce variations in heritable traits.

Although many mutations are harmful, some give rise to variations in form, function, or behavior that turn out to be adaptive under prevailing conditions.

1.2 LIFE'S DIVERSITY

So Much Unity, Yet So Many Species

Until now, we have focused on the *unity* of life—on characteristics shared by all organisms. Superimposed on the shared heritage is immense *diversity*. Many millions of different kinds of organisms, or **species**, inhabit the earth. Many millions more lived in the past and became extinct. Attempts to make sense of diversity led to a classification scheme in which each species is assigned a two-part name. The first part designates the **genus** (plural, genera). A genus encompasses all species related by descent from a common ancestor. The second part designates a particular species within that genus. For instance, *Quercus alba* is the scientific name of the white oak. *Q. rubra* is the name of the red oak. (Once the genus name is spelled out in a document, subsequent uses of it can be abbreviated.)

Life's diversity is further classified by assigning species to groups at more encompassing levels. Genera that share a common ancestor are placed in the same *family*, related families are placed in the same *order*, then related orders are placed in the same *class*. Related classes are placed in a *phylum* (plural, phyla) or *division*, which is assigned to one of five *kingdoms*:

Monerans   Bacteria (singular, bacterium). Single cells, all prokaryotic (their DNA is not enclosed in a membrane-bound compartment called a nucleus). Producers, consumers, decomposers. Kingdom of greatest metabolic diversity.

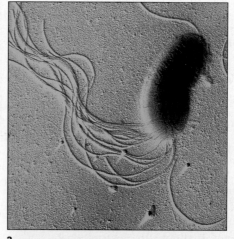
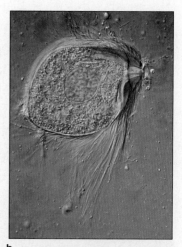

a

b

c

Figure 1.6 Representatives of life's diversity.

Kingdom Monera. (**a**) A bacterium, a microscopically small single cell. Bacteria live nearly everywhere, including in or on other organisms. The ones in your gut and on your skin outnumber the cells of your body.

Kingdom Protista. (**b**) A trichomonad, living as a parasite in a termite's gut. Most protistans are single celled, but they generally are much larger and have much greater internal complexity than bacteria.

Kingdom Plantae. (**c**) A grove of California coast redwoods. Like nearly all members of the plant kingdom, they produce their own food through photosynthesis. (**e**) From a plant called a composite, a flower having a pattern that guides bees to nectar. The bees get food, the plants get help in reproducing. Many organisms are locked in mutually helpful interactions.

Kingdom Fungi. (**d**) A stinkhorn fungus. The kingdom of fungi includes many major decomposers, which break down the remains and wastes of organisms. Without decomposers, communities would gradually become buried in their own garbage.

Kingdom Animalia. (**f**) Male bighorn sheep competing for females. Like all members of the animal kingdom, they cannot produce their own food; they depend on other organisms for it. They generally move about far more than other kinds of organisms.

Protistans Mostly single cells, larger than bacteria. Eukaryotic (DNA enclosed in a nucleus). Producers, consumers. Diverse life-styles.

Fungi Mostly multicelled. Eukaryotic. Decomposers, consumers. Their secretions digest food outside the fungal body; their cells absorb the breakdown products.

Plants Mostly multicelled. Eukaryotic. Nearly all producers that rely on photosynthesis.

Animals Multicelled. Eukaryotic. Consumers, typically motile, with diverse life-styles.

Figure 1.6 shows a few members of these kingdoms. Table 1.1 summarizes the characteristics of life that have been described so far. With very few exceptions, the living organisms in all five kingdoms display these characteristics.

Table 1.1 Characteristics of Living Organisms
1. Complex structural organization based on instructions contained in DNA molecules.
2. Directly or indirectly, dependence on other organisms for energy and material resources.
3. Metabolic activity by the single cell or multiple cells composing the body.
4. Use of homeostatic controls that maintain favorable operating conditions in the body within tolerable limits.
5. Reproductive capacity, by which the instructions for heritable traits are passed from parents to offspring.
6. Diversity in body form, in the functions of various body parts, and in behavior. Such traits are adaptations to prevailing conditions in the environment.
7. The capacity to evolve, based ultimately on variations in traits that arise through mutations in DNA.

d

e

f

a   b   c

Figure 1.7 Representatives of the more than 300 varieties of domesticated pigeons resulting from artificial selection practices. Pigeon breeders started with variant forms of traits among captive populations of wild rock doves (**a**).

d   e

f   g

An Evolutionary View of Diversity

Given that organisms are so much alike, what could account for their diversity? One key explanation is called evolution by means of natural selection. A few simple examples will be enough to introduce you to the main premises of this explanation.

Evolution Defined Suppose a DNA mutation gives rise to a different form of a trait in a few members of a population. We can use those dark moths in a sooty forest as an example. Birds see and eat many of their light-colored relatives, but dark moths escape detection and live to reproduce. So do their dark offspring—and so do *their* offspring. The dark version of the trait is popping up with greater frequency. In time it may become the more common form—and we might end up referring to "the dark moth population." **Evolution** is taking place—meaning that the features that characterize a population are changing through successive generations. Because heritable changes in the DNA are the source of variation in those features, they are the starting point for evolution.

Natural Selection Defined Long ago, Charles Darwin used pigeons to explain the link between variation in traits and evolution. Domesticated pigeons vary greatly in size, feather color, and other traits (Figure 1.7). As Darwin knew, pigeon breeders "select" certain traits. Suppose the desired traits are black tail feathers with curly edges. Only those pigeons having the most black and the most curl in their tail feathers are allowed to mate. Over time, those forms of the traits become most common as others are eliminated from the pigeon population.

Because pigeon breeders do their selecting in an artificial environment rather than in nature, the practice is an example of *artificial* selection. Even so, Darwin recognized it as a way to explain evolution by *natural* selection—that is, selection of adaptive traits in nature. Later in the book, we will consider the mechanisms by which organisms evolve. In the meantime, keep the following key points of Darwin's explanation in mind. They are absolutely central to biological inquiry, and we

will be using them to explain topics throughout the book. We express them here in modern terms:

---

1. Members of a population vary in form, function, and behavior. Much of this variation is heritable.

2. Some forms of heritable traits are more adaptive than others—they improve chances of surviving and reproducing. As a result, those forms become more common in the population.

3. **Natural selection** is simply the result of differences in survival and reproduction that have occurred among individuals that differ in one or more traits.

4. Any population *evolves* when some forms of traits become more or less common, or even disappear, over the generations. In this manner, variations have accumulated in different lines of organisms. Life's diversity is the sum total of those variations.

1.3 THE NATURE OF BIOLOGICAL INQUIRY

On Scientific Methods

Biology, like science generally, is an ongoing, methodical search for information that helps reveal the secrets of the natural world. Since Darwin's time, the searchers have branched out and established a great number of specialized subdivisions. Biologists now pursue topics that range from the molecular structure of the virus that causes AIDS to the ozone hole in the stratosphere. There is still so much to be learned about each topic that few now claim the whole of nature as their research interest. And no one claims that one method alone can be used to study its complexity.

Despite the specialization, scientists everywhere still have methods in common. *They ask questions, make educated guesses about possible answers, then devise ways to test predictions that will hold true if their guesses are good ones.* The following list is a more formal description of how scientists generally proceed with an investigation:

---

1. Ask a question (or identify a problem) about some aspect of nature, then develop one or more **hypotheses**, or educated guesses, about what the answer (or solution) might be. This might involve sorting through existing information about related phenomena.

2. Using hypotheses as a guide, make a **prediction**— that is, a statement of what you should be able to observe in nature, if you were to go looking for it. This is often called the "if-then" process. (*If* gravity pulls objects toward the earth, *then* it should be possible to observe apples falling down, not up, from a tree.)

3. Devise ways to test the accuracy of predictions. You might do so by making observations, developing models, and doing experiments.

4. If the tests do not turn out as expected, check to see what might have gone wrong. (Maybe you overlooked something that influenced the results. Or maybe the hypothesis isn't a good one.)

5. Repeat the tests or devise new ones—the more the better. (Hypotheses supported by many different tests are more likely to be correct.) Then objectively report the test results and conclusions drawn from them.

In broad outline, a scientific approach to studying nature is that simple. You yourself can use this approach to advantage. You can use it to satisfy curiosity about mammoths or moth wings. You can use it to pick your way logically through environmental, medical, and social land mines of the sort described later in the book. And you can use it to understand the past and predict possible futures for life on this planet.

About the Word "Theory"

How does a hypothesis differ from a theory? By way of example, go back to Darwin's ideas about evolution. When Darwin proposed these ideas more than a century ago, he ushered in one of the most dramatic of all scientific revolutions. The core of his thinking became popularly known as "the theory of evolution."

In science, a **theory** is a related set of hypotheses that, taken together, form a broad-ranging, testable explanation about some fundamental aspect of the natural world. A scientific theory differs from a scientific hypothesis in its *breadth of application*. Darwin's theory fits this description—it is a big, encompassing "Aha!" explanation that, in a few intellectual strokes, makes sense of a huge number of observable phenomena. Think of it. Darwin's theory explains how most of the diversity among many millions of different living things came about! The *Focus* essay on the next page is but one example of how biologists use the theory to formulate and test ideas about the natural world.

Yet is any theory an "absolute truth" in science? Ultimately, no. Why? It would be impossible to perform the infinite number of tests required to show that a theory holds true under all possible conditions! Objective scientists say only that they are *relatively* certain that a theory is (or is not) correct.

Even so, "relative certainty" can be impressive. Especially after exhaustive tests by many scientists, a theory may be as close to the truth as we can get with the evidence at hand. After more than a century's worth of thousands of different tests, Darwin's theory still stands, with only minor modification. Most biologists accept the modified theory—although they still keep their eyes open for contradictory evidence.

Scientists must keep asking themselves: "Will some other evidence show my idea to be incorrect?" They are expected to put aside pride or bias by testing ideas, even in ways that might prove them wrong. If an individual doesn't (or won't) do this, *others will*—for science proceeds as a community that is both cooperative and competitive. Ideas are shared, with the understanding that it is just as important to expose errors as it is to applaud insights. Individuals can change their mind when presented with new evidence—and this is a strength of science, not a weakness.

A scientific theory is a testable explanation about the cause or causes of a broad range of related phenomena. As is true of hypotheses, theories are open to tests, revision, and tentative acceptance or rejection.

Darwin's Theory and Doing Science

A time-tested theory serves as a general frame of reference for studying nature. Consider Charles Darwin's theory of evolution by natural selection. How might you use it to explain the patterned wings of the moth shown in Figure 1.4? According to this theory, the traits of moths and all other organisms exist because they have contributed to reproductive success. So your question might be this: "I wonder how the wing pattern helps the moth leave descendants." Then you hypothesize about the answer.

"Maybe a wing pattern is a mating flag that helps males and females of the same species identify each other." This may be correct, but you don't limit yourself to one hypothesis. Why? Nearly always, there's more than one possible answer to a question about some aspect of nature. So you also come up with an alternative: "Maybe the pattern camouflages moths during the day." In science, *alternative hypotheses are the rule, not the exception.*

Testing Hypotheses   Of any number of alternative hypotheses, how do you identify the most plausible one? The trick is to let each one guide you in making testable predictions. If wing patterns help moths identify mates (the hypothesis), then it follows that moths should mate only when the patterns are visible (the prediction). Moths mate at night. If wing pattern is a mating flag, then on moonless nights, moths shouldn't be able to see it and you won't see moths mating. To test the prediction, you watch moths on a moonlit and then on a moonless night. You notice they mate with or without help from the light of the moon. Here is evidence that the prediction—and, by extension, the hypothesis—might be wrong. Then again, maybe you overlooked something important. For example, maybe moths (like cats) see better than you do in the dark.

The Role of Experiments   Now you decide to test the same hypothesis by experimentation. An **experiment** is a test in which nature is manipulated to reveal one of its secrets. It requires careful design of a set of controls to evaluate possible side effects of the manipulation.

If wing pattern is a mating flag, then moths with altered color patterns might have a tough time attracting a mate. To test this new prediction, you capture new moths, paint an altered pattern on their wings (Figure *a*), then put them in a cage with unaltered moths to see what happens.

You also set up a **control group** to evaluate possible side effects of a test involving an experimental group. Ideally, members of a control group should be identical to those of an experimental group in all respects—except for the key **variable** (the factor being investigated). The number of individuals in both groups also must be large enough so results won't be due to chance alone.

Besides wing pattern, what other variables between the two groups might affect the outcome of your experiment? Maybe paint fumes are as repulsive as a painted-on pattern to a potential mate. Maybe when you paint the moths you somehow rough them up, making them less desirable than those in the control group. Maybe the paint weighs enough to change the flutter frequency of the wings.

So you decide the control group also must be painted with the same kind of paint, using the same brushes, and they must be handled the same way. But for this group, you *duplicate* the natural wing color pattern as you paint. Now your experimental and control groups are identical except for the variable under study. If only those moths with altered wing patterns turn out to be unlucky in love, then the greater reproductive success of your control group will help substantiate your hypothesis.

| 1.4 | **THE LIMITS OF SCIENCE** |

The call for objective testing strengthens the theories that emerge from scientific studies. Yet it also puts limits on the kinds of studies that can be carried out. Beyond the realm of scientific analysis, some events remain unexplained. Why do we exist, for what purpose? Why does any one of us have to die at a particular moment and not another? Answers to such questions are *subjective*. This means they come from within us, as an outcome of all the experiences and mental connections that shape our consciousness. Because people differ so enormously in this regard, subjective answers do not readily lend themselves to scientific analysis.

This is not to say that subjective answers are without value. No human society can function without a shared commitment to standards for making judgments, even if the judgments are subjective. Moral, aesthetic, economic, and philosophical standards vary from one society to the next. But all guide their mem-

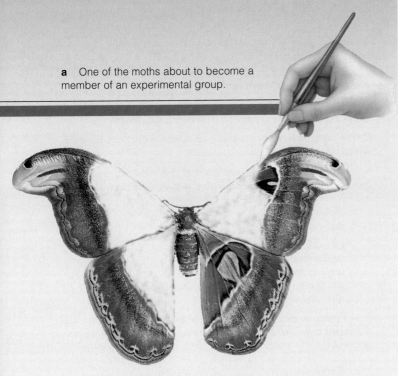

Generally, experiments are devised to disprove a hypothesis. Why? It would be impossible to prove beyond a shadow of a doubt that a hypothesis is correct. It would take an infinite number of experiments to demonstrate that it holds under all possible conditions.

Have you been thinking that painting moths is a rather fanciful example of a scientific approach? As reported in *Nature* in 1993, Karen Marchetti, a graduate student of the University of California at Davis, wielded a paintbrush on birds in a forest in Kashmir, India. She found evidence for her hypothesis that bright feather color, not patterning, gives male yellow-browed leaf warblers an edge in mating. In early tests, Marchetti put a patternless dab of yellow paint on head feathers. In later tests, she painted larger-than-normal yellow bands on wings of one group and painted out part of the bands of another group. She used transparent paint for a third group. (Can you guess why?) As Marchetti discovered, color-enhanced birds secured larger territories and produced more offspring than toned-down birds.

bers in deciding what is important and good, and what is not. All attempt to give meaning to what we do.

Every so often, scientists stir up controversy when they explain part of the world that was considered beyond natural explanation—that is, belonging to the "supernatural." This is sometimes true when moral codes are interwoven with religious narratives. Exploring some longstanding view of the world from a scientific perspective may be misinterpreted as questioning morality, even though the two are not at all the same thing.

For example, centuries ago Nicolaus Copernicus studied the movements of planets and stated that the earth circles the sun. Today the statement seems obvious. Back then, it was heresy. The prevailing belief was that the Creator had made the earth (and, by extension, humanity) the immovable center of the universe! Not long afterward a respected professor, Galileo Galilei, studied the Copernican model of the solar system. He thought it was a good one and said so. He was forced to retract his statement publicly, on his knees, and to put the earth back as the fixed center of things. (Word has it that when he stood up he muttered, "Even so, it does move.")

Today, as then, society has its sets of standards. Today, as then, those standards may be called into question when a new, natural explanation runs counter to supernatural belief. This doesn't mean the scientists who raise the questions are less moral, less lawful, less sensitive, or less caring than anyone else. It simply means one more standard guides their work: *The external world, not internal conviction, must be the testing ground for scientific beliefs.*

Systematic observations, hypotheses, predictions, tests—in all these ways, science differs from systems of belief that are based on faith, force, or simple consensus.

SUMMARY

1. All organisms share the following characteristics:
   a. Their structure, organization, and interactions arise from basic properties of matter and energy.
   b. They use metabolic and homeostatic processes.
   c. They have the capacity for growth, development, and reproduction, based on instructions contained in their DNA molecules.

2. There are many millions of different organisms. Each kind of organism is a species. In classification schemes, species are placed in increasingly inclusive groupings, from genus on up through family, order, class, phylum (or division), and kingdom.

3. Diversity among organisms arises through mutation. Mutations introduce changes in the DNA. The changes may lead to variation in heritable traits (traits that parents transmit to offspring, including most details of the body's form and functioning).

4. Different versions of the same heritable trait occur among individuals of a population. These influence the ability to survive and reproduce. Under prevailing conditions, some may be more adaptive than others and

will become more common ("selected") in subsequent generations. Others will become less common and may disappear. Thus the population changes over time; it evolves. These points are central to the theory of evolution by natural selection.

5. There are many specialized scientific methods, corresponding to many different fields of inquiry. The following terms are important to all of them:

a. Theory: An explanation of a broad range of phenomena that has successfully withstood intensive testing.

b. Hypothesis: A possible explanation of a specific phenomenon. Sometimes called an educated guess.

c. Prediction: A claim about what you can expect to see in nature if a theory or hypothesis is correct.

d. Test: An attempt to produce actual observations that match predicted or expected observations.

e. Conclusion: A statement about whether a theory or hypothesis should be accepted, rejected, or modified, based on tests of the predictions derived from it.

6. Scientific theories are based on systematic observations, hypothesizing, predictions, and tests. The external world, not internal conviction, is the testing ground for those theories.

Review Questions

1. Why is it difficult to give a simple definition of life? *3* (For this and subsequent chapters, *italic numbers* following review questions indicate the pages on which the answers may be found.)

2. What characteristics do all organisms have in common? *3*

3. What is energy? What is DNA? *3*

4. Study Figure 1.2. Then, on your own, arrange and define the levels of biological organization. *4*

5. Define metabolic activity. *4–5*

6. Make a sketch of the one-way flow of energy and the cycling of materials through the biosphere. *5*

7. What is mutation? How is it related to the diversity of life? *7*

8. Witnesses in a court of law are asked to "swear to tell the truth, the whole truth, and nothing but the truth." What are some problems inherent in the question? Can you think of a better alternative?

9. Design a test to support or refute the following hypothesis: Body fat appears yellow in certain rabbits—but only when those rabbits also eat leafy plants that contain a yellow pigment molecule called xanthophyll.

Self-Quiz *(Answers in Appendix IV)*

1. The _____ is the smallest unit of life.

2. _____ is the ability of cells to extract and transform energy from the environment and use it to maintain themselves, grow, and reproduce.

3. _____ is a state in which the body's internal environment is being maintained within tolerable limits.

4. If a form of a trait improves chances for surviving and reproducing in a particular environment, it is a(n) _____ trait.

5. The capacity to evolve is based on variations in heritable traits, which originally arise through _____ .

6. You have some number of traits that also were present in your great-great-great-great-grandmothers and -grandfathers. This is an example of _____ .
   a. metabolism   c. a control group
   b. homeostasis   d. inheritance

7. DNA molecules _____ .
   a. contain instructions for traits
   b. undergo mutation
   c. are transmitted from parents to offspring
   d. all of the above

8. For many years in a row, a dairy farmer allowed his best milk-producing cows but not the poor producers to mate. Over the generations, milk production increased. This outcome is an example of _____ .
   a. natural selection   c. evolution
   b. artificial selection d. both b and c

9. A related set of hypotheses that explains some aspect of the natural world is a scientific _____ .
   a. prediction   c. theory
   b. test   d. observation

Selected Key Terms

For this and subsequent chapters, these are the **boldface** terms that occur in the text on the pages indicated by *italic* numbers. Make a list of these terms, write a definition next to each, then check it against the one in the text. (You will be using these terms later on.)

adaptive trait *7*
aerobic respiration *5*
animal *9*
ATP *4*
biosphere *4*
cell *4*
community *4*
consumer *5*
control group *12*
decomposer *5*
DNA *3*
ecosystem *4*

energy *3*
evolution *10*
experiment *12*
fungus *9*
genus *8*
homeostasis *6*
hypothesis *11*
inheritance *7*
metabolism *4*
moneran *8*
multicelled organism *4*
mutation *7*

natural selection *10*
photosynthesis *4*
plant *9*
population *4*
prediction *11*
producer *5*
protistan *9*
receptor *6*
reproduction *6*
species *8*
theory *11*
variable *12*

Readings

Committee on the Conduct of Science. 1989. *On Being a Scientist.* Washington, D.C.: National Academy of Sciences. Paperback.

Larkin, T. June 1985. "Evidence vs. Nonsense: A Guide to the Scientific Method." *FDA Consumer* 19:26–29.

FACING PAGE: *Living cells of a green plant* (Elodea), *as seen with the aid of a microscope. Each rectangular cell contains efficient chemical factories called chloroplasts (the green spheres).*

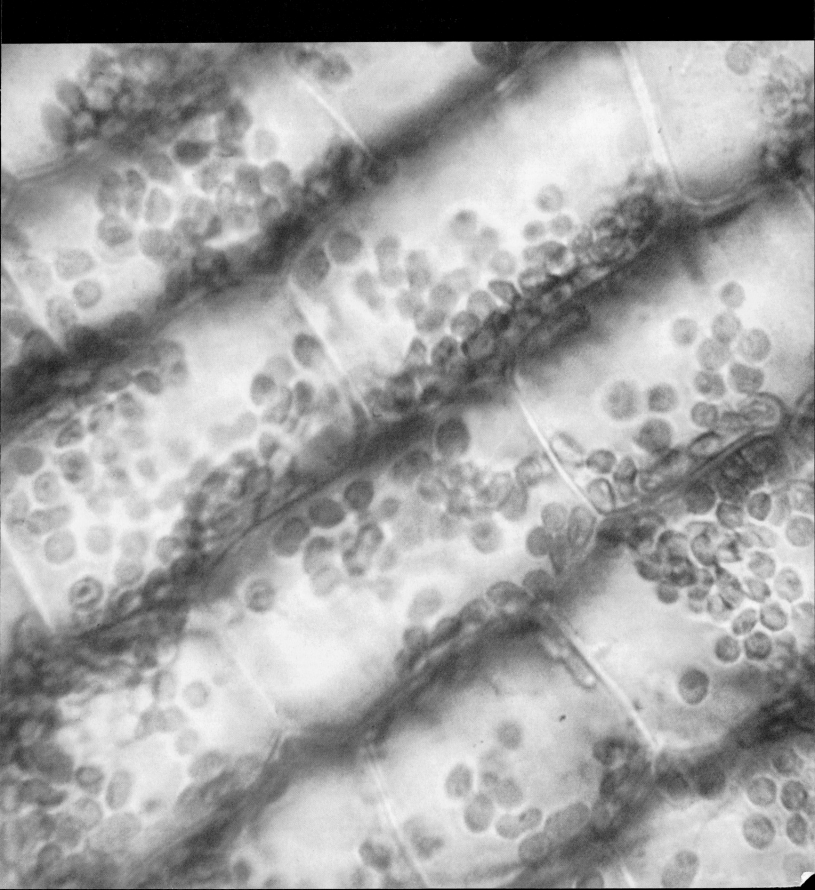

2 CHEMICAL FOUNDATIONS FOR CELLS

The Chemistry In and Around You

Right now you are breathing in oxygen. You would die without it. But where did the oxygen come from? Whether it's noon or midnight, countless plants on the sunlit side of our planet are busily converting energy from the sun to forms of energy they can use—and they release oxygen during the conversions. Aquatic algae have been doing this for more than 900 million years. Some types of bacteria have been doing the same thing for more than 3 *billion* years! All of that released oxygen adds up.

In the past two centuries (which is a mere blip of evolutionary time), we managed to discover what substances are made of and how they can be converted into different forms. With this amazing knowledge of chemistry, we developed such products as fertilizers, nylons, vaccines, lipsticks, antibiotics, and plastic

parts of refrigerators, computers, television sets, jet planes, and cars.

Our chemical "magic" brings benefits *and* problems. Without synthetic fertilizers to help grow crops, more humans than you might imagine would starve to death. But weeds don't know that fertilizers are for crop plants, and animal pests don't know that crops aren't being raised for them. Each year pests ruin or gobble up nearly half of what we grow. In 1945 we began using synthetic pesticides. Among other things, pesticides kill weeds, moths, worms, and rats that threaten our food supplies, health, pets, and ornamental plants. In 1988 alone, Americans spread more than a billion pounds of pesticides through homes, gardens, offices, industries, and farmlands (Figure 2.1).

Among the pesticides, we find the carbamates, organophosphates (including malathion), and halogenated compounds (including chlordane). Most are toxins that block vital communication signals between cells of the nervous system. Some remain active for days, others for weeks or years. Besides pests, they kill great numbers of pest-eaters, including dragonflies and birds. Besides this, pest populations build up resistance to pesticides, for reasons that will become apparent in later chapters.

We, too, inhale pesticides, ingest them with food, or absorb them through skin. Many pesticides cause headaches, rashes, and asthma in susceptible people. Some trigger hives, joint pain, even life-threatening allergic reactions in millions of people in the United States alone.

Maintaining our crops, industries, and health depends on chemistry. So does our chance of reducing harmful side effects of the application of chemistry. You owe it to yourself and others to gain understanding of chemical substances. By demystifying chemistry's "magic," you will be better equipped to assess its benefits and risks.

KEY CONCEPTS

1. All atoms of each element have the same number of protons and electrons, but they may differ slightly in their number of neutrons. These variant forms of atoms are called isotopes.

2. Atoms give up, acquire, or share electrons with other atoms in specific ways. These interactions are the basis for the structural organization and activities of all organisms.

3. Chemical bonds are unions between the electron structures of different atoms. The number and arrangement of electrons in the atoms of different elements give rise to differences in their bonding behavior. Some atoms tend to give up, gain, or share one or more electrons with another atom. Carbon, hydrogen, nitrogen, and oxygen—the main building blocks of life—are like this.

4. In biological molecules, the most common bonds are ionic bonds, hydrogen bonds, and covalent bonds.

5. An ion is an atom or molecule that has gained or lost one or more electrons, and so has acquired an overall positive or negative charge.

6. Life depends on the properties of water, including its temperature-stabilizing effects, cohesiveness, and capacity to dissolve many substances. Hydrophilic substances dissolve in water; hydrophobic ones are repelled by it.

7. Life depends on the controlled formation, use, and disposal of hydrogen ions (H^+). The pH scale is a measure of the concentration of these ions in different solutions.

Figure 2.1 A low-flying cropduster, with its rain of pesticides.

ORGANIZATION OF MATTER

"Matter" is anything that occupies space and has mass. All solids, liquids, and gases within and around you are forms of matter, and each consists of one or more kinds of elements. An **element** is a fundamental substance that cannot be broken down to a different substance, at least by ordinary means. About ninety-two elements occur naturally on earth. It takes only four kinds—oxygen, carbon, hydrogen, and nitrogen—to make up most of the human body.

You and all other organisms also require seemingly insignificant amounts of other elements. Collectively, these so-called *trace* elements make up less than 0.01 percent of any organism, yet normal body functioning depends on them. Copper is an example. Carefully dry out and analyze tissues of a maple tree and you will find they are about 0.006 percent copper. If that tree has a copper deficiency, its leaf buds will die and its growth will suffer.

Take a look at Figure 2.2. It shows the proportions of elements that make up a human and the fruit of pumpkin plants, and it compares them to the proportions of elements in the "nonliving" materials of the earth's crust. In what respects are they similar? In what respects do they differ?

By international agreement, a one- or two-letter chemical symbol stands for each element, regardless of the element's name in different languages. What we call nitrogen is *azoto* in Italian and *stickstoff* in German—but the symbol remains N. Similarly, the symbol for sodium is Na (from the Latin *natrium*). Table 2.1 lists the chemical symbols of elements that are common in living things.

Figure 2.2 Proportions of different elements in the earth's crust, the human body, and a pumpkin (the fruit of a pumpkin plant).

EARTH'S CRUST		HUMAN		PUMPKIN	
Oxygen	46.6	Oxygen	65	Oxygen	85
Silicon	27.7	Carbon	18	Hydrogen	10.7
Aluminum	8.1	Hydrogen	10	Carbon	3.3
Iron	5.0	Nitrogen	3	Potassium	0.34
Calcium	3.6	Calcium	2	Nitrogen	0.16
Sodium	2.8	Phosphorus	1.1	Phosphorus	0.05
Potassium	2.6	Potassium	0.35	Calcium	0.02
Magnesium	2.1	Sulfur	0.25	Magnesium	0.01
Other		Sodium	0.15	Iron	0.008
elements:	1.5	Chlorine	0.15	Sodium	0.001
		Magnesium	0.05	Zinc	0.0002
		Iron	0.004	Copper	0.0001
		Iodine	0.0004	Other:	0.00005

Table 2.1	Atomic Number and Mass Number of Elements Common in Living Things		
Element	Symbol	Atomic Number	Most Common Mass Number
Hydrogen	H	1	1
Carbon	C	6	12
Nitrogen	N	7	14
Oxygen	O	8	16
Sodium	Na	11	23
Magnesium	Mg	12	24
Phosphorus	P	15	31
Sulfur	S	16	32
Chlorine	Cl	17	35
Potassium	K	19	39
Calcium	Ca	20	40
Iron	Fe	26	56
Iodine	I	53	127

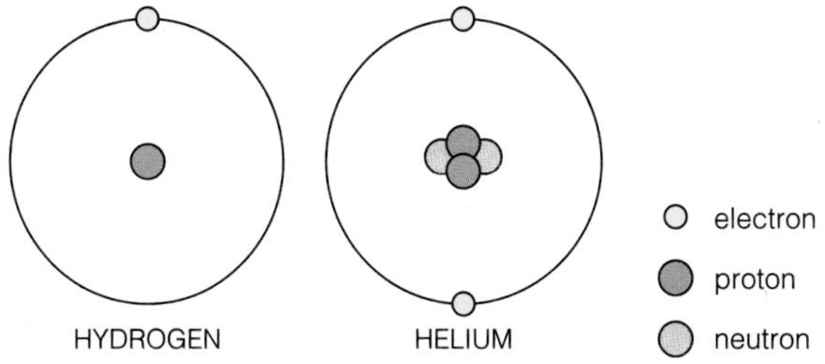

Figure 2.3 Simplified model of atomic structure, using hydrogen and helium as examples. At this scale, the nucleus actually would be an invisible speck at the atom's center.

The Structure of Atoms

Each kind of **atom** is the smallest unit of matter that is unique to a particular element. A **molecule** is two or more joined-together atoms of the same or different elements. Pure water (if there still is such a thing) consists of a stupendous number of water molecules, each composed of one oxygen and two hydrogen atoms.

Water, incidentally, is a compound. A **compound** is a substance in which the relative proportions of two or more elements never vary. Water in rainclouds or a Siberian lake or your bathtub always has twice as many hydrogen atoms as oxygen atoms, period.

Atomic Building Blocks The organization and activities of living things start with charged particles called **protons** and **electrons**. Together with **neutrons**, which have no charge, these are building blocks of atoms (Figure 2.3). An atom's core region, or nucleus, consists of some number of protons and (except for hydrogen) neutrons. Protons carry a positive charge (p^+). Electrons move rapidly around the nucleus, they occupy most of the atom's volume, and they carry a negative charge (e^-). Because an atom has just as many electrons as protons, it has no *net* charge, overall.

Atomic Number and Mass Number The number of protons in the nucleus is called the **atomic number**. It differs for each element. For example, that number is 1 for the hydrogen atom (with one proton) and 6 for the carbon atom (with six protons). Table 2.1 lists other examples. The **mass number** is the number of protons *and* neutrons in the nucleus. A carbon atom with six protons and six neutrons has a mass number of 12.

The relative masses of atoms are also called atomic weights. This term is not precise—mass is not quite the same thing as weight—but its use continues.

As you will see, knowing the atomic numbers and mass numbers gives us an idea of whether certain atoms can lose, gain, or share electrons. *Such electron activity is the basis for the organization of materials and the flow of energy through the living world.*

Isotopes: Variant Forms of Atoms

All the atoms of an element have the same number of protons and electrons, but they can differ in the number of neutrons. Such atoms are **isotopes**. For example, "a carbon atom" might be carbon 12 (six protons, six neutrons), carbon 13 (six protons, seven neutrons), or carbon 14 (six protons, eight neutrons). You can write these as ^{12}C, ^{13}C, and ^{14}C. All isotopes of an element interact with other atoms the same way. Thus cells can use any carbon isotope for a metabolic reaction.

You have probably heard of radioactive isotopes, or **radioisotopes**. These are unstable and tend to break apart (decay) into more stable atoms. The *Focus* essay on the pages that follow describes some uses of radioisotopes in research, medicine, and studies of the history of life.

Each kind of atom is the smallest unit of matter that is unique to a given element.

All atoms of an element have the same number of protons and electrons, but they can vary slightly in the number of neutrons.

Variant atoms of the same element are isotopes.

Using Radioisotopes to Date Fossils, Track Chemicals, and Save Lives

In the winter of 1896, the physicist Henri Becquerel tucked a heavily wrapped rock of uranium into a desk drawer, on top of an unexposed photographic plate. A few days later, he opened the drawer and discovered a faint image of the rock on the plate—apparently caused by energy emitted from the rock. His coworker, Marie Curie, named the phenomenon "radioactivity."

As we now know, radioisotopes are unstable atoms with dissimilar numbers of protons and neutrons. They emit electrons and energy. In this spontaneous process (radioactive decay), an isotope changes to a new, stable one that is not radioactive.

Radioactive Dating Each type of radioisotope has a certain number of protons and neutrons, and it decays spontaneously at a certain rate into a particular isotope of a different element. "Half-life" is the time it takes for half the nuclei in any given amount of a radioactive element to decay into another element. Half-life can't be modified by temperature, pressure, chemical reactions, or any other environmental factor. That's why radioactive dating is a reliable way to discern the age of different rock layers in the earth—and of fossils contained in the layers. To discover a rock's age, we can compare the amount of one of its radioisotopes with the amount of the decay product for that isotope.

For example, ^{40}potassium has a half-life of 1.3 billion years and decays to ^{40}argon, a stable isotope. The age of anything that contains ^{40}potassium can be determined by measuring the ratio of ^{40}argon to ^{40}potassium. In such ways, researchers have dated ancient fossils. Figure *a* shows two examples. Similarly, researchers who used ^{238}uranium (with a half-life of 4.5 billion years) discovered that the earth formed more than 4.6 billion years ago. The following table lists useful ranges of some radioisotopes that are employed in dating methods:

Main Radioisotopes Used in Dating

Radioisotope (unstable)		Stable Product	Half-Life (years)	Useful Range (years)
^{87}rubidium	→	^{87}strontium	49 billion	100 million
^{232}thorium	→	^{208}lead	14 billion	200 million
^{238}uranium	→	^{206}lead	4.5 billion	100 million
^{40}potassium	→	^{40}argon	1.3 billion	100 million
^{235}uranium	→	^{207}lead	704 million	100,000
^{14}carbon	→	^{14}nitrogen	5,730	0–60,000

a (*Left*) A fossilized frond of a tree fern, one of many species that lived more than 250 million years ago. (*Right*) A fossilized sycamore leaf that dropped 50 million years ago.

Tracking Chemicals Scintillation counters and other devices can detect emissions from radioisotopes. Thus radioisotopes can be used as **tracers**. Tracers reveal a pathway or destination of a substance that has entered a cell, the human body, an ecosystem, or some other "system."

Carbon provides an example. All isotopes of an element have the same number of electrons, so they all interact with other atoms the same way. This means cells can use any isotope of carbon in reactions that require carbon atoms. For example, such reactions are part of the photosynthetic pathway. By putting plant cells in a medium enriched in ^{14}carbon, researchers identified the steps by which plants take up carbon and incorporate it into carbohydrates during photosynthesis. Researchers also are using tracers to identify how plants use naturally occurring nutrients and synthetic fertilizers. This knowledge may lead to improvements in crop production.

What about medical applications? Consider the human thyroid, the only gland of ours that takes up iodine. If a tiny amount of the radioisotope ^{123}iodine is injected into a patient's blood, the thyroid can be scanned with a photographic imaging device. Figure *b* shows examples of the resulting images.

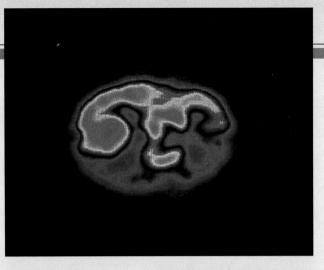

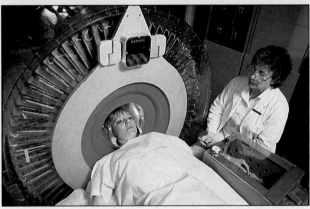

c A patient being moved into a PET scanner. The top photo shows a vivid image of a brain scan of a child who has a severe neurological disorder. Normally, different colors in a brain scan signify differences in metabolic activity in the brain. Notice how one half of this child's brain shows no activity.

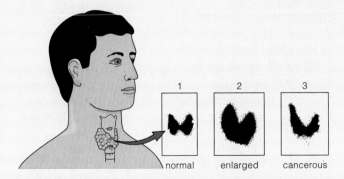

b Scans of human thyroid glands, taken after ^{123}iodine was injected into the bloodstream of three different patients. A normal thyroid takes up iodine (including radioisotopes of iodine) and uses it in hormone production. The scans show (1) a normal gland, (2) the enlarged gland of a patient with a thyroid disorder, and (3) a cancerous thyroid gland.

Saving Lives Radioisotopes are used in nuclear medicine to diagnose and treat diseases. For example, patients with irregular heartbeats are given artificial pacemakers, powered by energy from ^{238}plutonium. This dangerous radioisotope is sealed in a case to prevent its emissions from damaging body tissues.

As another example, PET (short for positron-emission tomography) yields images of metabolically active and inactive tissues. Radioisotopes are attached to glucose or some other biological molecule. Then they are injected into a patient, who is moved into a PET scanner. When cells in certain tissues absorb glucose, radioisotope emissions are used to produce a vivid image of variations or abnormalities in metabolic activity (Figure *c*).

Finally, in some cancer treatments, radioisotopes are used to destroy or impair the function of living cells. In radiation therapy, localized cancers are deliberately bombarded with energy from a ^{226}radium or ^{60}cobalt source.

2.2 THE NATURE OF CHEMICAL BONDS

We turn now to interactions between atoms. Take a moment to review Figure 2.4, which summarizes a few conventions used to describe these events.

What Is a Chemical Bond?

A **chemical bond** is a union between electron structures of atoms. In other words, *it is an energy relationship.* Most chemical bonds form when an atom gives up, gains, or shares one or more electrons with another atom. Some atoms enter into such relationships rather

---

Figure 2.4 Chemical bookkeeping.

We use symbols for elements when writing *formulas*, which identify the composition of compounds. For example, water has the formula H_2O. The subscript indicates that two hydrogen (H) atoms are present for every oxygen (O) atom. We use symbols and formulas in *chemical equations*—representations of reactions among atoms and molecules. An arrow in such an equation means "yields." Substances entering a reaction (reactants) are to the left of the arrow, and products are to the right, as shown by this equation for photosynthesis:

$$6CO_2 \; + \; 6H_2O \longrightarrow C_6H_{12}O_6 \; + \; 6O_2$$

6 carbons	12 hydrogens	6 carbons	12 oxygens
12 oxygens	6 oxygens	12 hydrogens	
		6 oxygens	

Notice there are as many atoms of each element to the right of the arrow as there are to the left, even though they are combined in different forms. Atoms taking part in chemical reactions may be rearranged, but they are never destroyed. By the *law of conservation of mass*, the total mass of all materials entering a reaction equals the total mass of all the products. Keep in mind, the equations that you use to represent cellular reactions must be balanced this way, because no atoms are lost.

Reactants and products of reactions can be expressed in moles. A "mole" is a certain number of atoms or molecules of any substance, just as "a dozen" can refer to any twelve cats, roses, and so forth. Its weight, in grams, equals the total atomic weight of the atoms that compose the substance.

For example, the atomic weight of carbon is 12, so one mole of carbon weighs 12 grams. A mole of oxygen (atomic weight 16) weighs 16 grams. Can you show why a mole of water (H_2O) weighs 18 grams, and why a mole of glucose ($C_6H_{12}O_6$) weighs 180 grams?

---

easily, but others do not. Whether one atom will bond with another depends on the *number* and *arrangement* of its electrons.

Electrons and Energy Levels

Picture three actresses arriving at the Academy Awards ceremony wearing the same bright red designer dress. Each seeks recognition but dreads being caught next to the others. Two might maneuver themselves *near* the center of attention while avoiding each other. But by unspoken agreement, all three never, ever are in the same place at the same time.

Electrons behave roughly the same way. They are attracted to an atom's protons but repelled by other electrons. They spend as much time as possible near the nucleus and far away from each other by moving in different orbitals. Think of **orbitals** as regions of space around an atom's nucleus in which electrons are likely to be at any instant. Each orbital has enough room for two electrons, at most.

In all atoms, the orbital closest to the nucleus is shaped like a ball (Figure 2.5*a*). Each hydrogen atom has a lone electron in that orbital; each helium atom has two. Any electron in the orbital closest to the nucleus is said to be at the *lowest energy level.*

Atoms larger than helium have two electrons in the first orbital. They also have other electrons that occupy different orbitals. On the average, those other electrons are farther away from the nucleus, and they are said to be at *higher energy levels.*

There is a simple although not quite accurate way to think about this. Imagine that all the different electron orbitals are arranged within a series of *shells* around the nucleus. As shown in Figure 2.5*b*, one of the orbitals, with space for two electrons, fits inside a shell closest to the nucleus. Four additional orbitals fit inside the second shell around the nucleus. Collectively, the four have space for a total of eight more electrons. Still more orbitals and electrons can occupy successive shells, as suggested by Table 2.2.

Take a look at the hydrogen and sodium atoms in Figure 2.6. Hydrogen, the simplest atom, has one electron in its first and only shell. Sodium, with eleven electrons, has a lone electron in its outermost shell. In other words, both kinds of atoms have a "vacancy" in an orbital in their outermost shell. Atoms with electron vacancies tend to react with other atoms. Hydrogen, oxygen, carbon, and nitrogen—*the most abundant components of organisms*—are like this.

An atom tends to react with other atoms when its outermost shell is only partly filled with electrons.

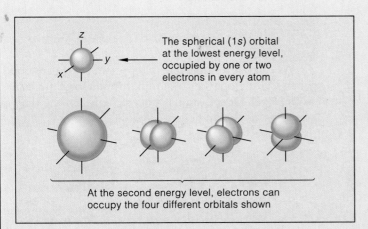

The spherical (1s) orbital at the lowest energy level, occupied by one or two electrons in every atom

At the second energy level, electrons can occupy the four different orbitals shown

Figure 2.5 Arrangement of electrons in atoms. (**a**) One or at most two electrons occupy a ball-shaped volume of space (an orbital) close to the nucleus. Electrons occupying this orbital are at the lowest energy level. At the next (higher) energy level, there can be as many as eight more electrons (four more orbitals, two electrons each). Orbital shapes get tricky, but for our purposes we can ignore them. Simply think of the total number of orbitals at a given energy level as being somewhere inside a "shell" around the nucleus. As suggested in (**b**), higher energy levels correspond to shells farther from the nucleus.

a Shapes of electron orbitals that occur at the first and second energy levels, which correspond to the first and second "shells" in (**b**).

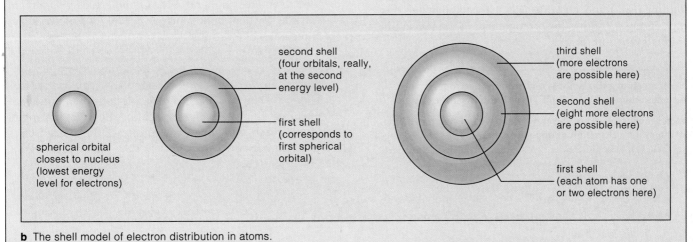

second shell (four orbitals, really, at the second energy level)

first shell (corresponds to first spherical orbital)

spherical orbital closest to nucleus (lowest energy level for electrons)

third shell (more electrons are possible here)

second shell (eight more electrons are possible here)

first shell (each atom has one or two electrons here)

b The shell model of electron distribution in atoms.

Table 2.2 Electron Distribution for a Few Elements

Element	Chemical Symbol	Atomic Number	Electron Distribution		
			First Shell	Second Shell	Third Shell
Hydrogen	H	1	1	—	—
Helium	He	2	2	—	—
Carbon	C	6	2	4	—
Nitrogen	N	7	2	5	—
Oxygen	O	8	2	6	—
Neon	Ne	10	2	8	—
Sodium	Na	11	2	8	1
Magnesium	Mg	12	2	8	2
Phosphorus	P	15	2	8	5
Sulfur	S	16	2	8	6
Chlorine	Cl	17	2	8	7

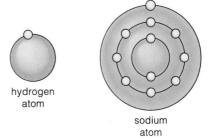

hydrogen atom

sodium atom

Figure 2.6 Distribution of electrons (*yellow dots*) in hydrogen and sodium atoms. Each atom has a lone electron (and room for more) in its outermost shell. Atoms having such partly filled shells tend to enter into reactions with other atoms.

IMPORTANT BONDS IN BIOLOGICAL MOLECULES

Ionic Bonding

Electrons can be knocked out of atoms, pulled away from them, or added to them. When an atom loses or gains one or more electrons, it becomes positively or negatively charged. In this state, it is an **ion**.

When do atoms lose or gain electrons? They do so when another atom of the right kind is nearby to accept or donate those electrons. Because one atom loses and one gains, *both* become ionized. Depending on the surroundings, two ions can go their separate ways or stay together through the mutual attraction of their opposite charges.

An association of two oppositely charged ions is known as an **ionic bond**. Table salt, or NaCl, has ions of sodium (Na^+) and chloride (Cl^-) linked together this way. Figure 2.7 shows the arrangement of the two kinds of ions in this substance.

When an atom gains or loses one or more electrons, it becomes an ion with an overall positive or negative charge.

In an ionic bond, a positive and a negative ion stay together by the mutual attraction of opposite charges.

Covalent Bonding

In a **covalent bond**, one atom cannot pull electrons completely away from another atom, and the two end up sharing electrons.

We can use a single line to represent a covalent bond when writing out the structural formula for a molecule, as in H—H. Or we can use dots to represent the shared electrons, as shown here:

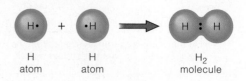

| H atom | H atom | H_2 molecule |

In a double covalent bond, two atoms share two pairs of electrons. This happens in the O_2 molecule, or O=O. In a triple covalent bond, such as N≡N, two atoms share three pairs of electrons.

Covalent bonds are nonpolar or polar. In a *nonpolar* covalent bond, both atoms exert the same pull on shared electrons. The term "nonpolar" implies no difference at the two ends (the two poles) of the bond. An example is the H—H molecule. The two hydrogen atoms, each with one proton, attract the shared electrons equally.

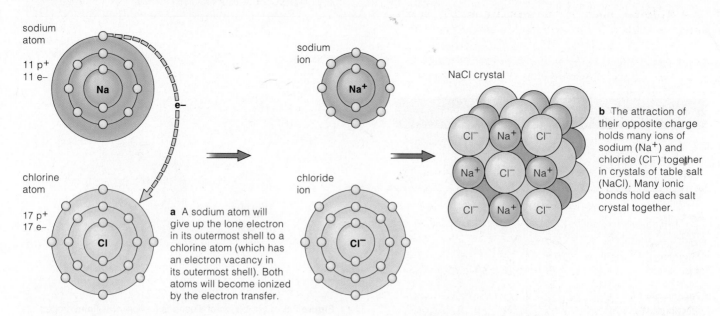

Figure 2.7 Ionic bonding in sodium chloride (NaCl).

sodium atom

11 p+
11 e–

Na

e–

chlorine atom

17 p+
17 e–

Cl

a A sodium atom will give up the lone electron in its outermost shell to a chlorine atom (which has an electron vacancy in its outermost shell). Both atoms will become ionized by the electron transfer.

sodium ion

Na⁺

chloride ion

Cl⁻

NaCl crystal

b The attraction of their opposite charge holds many ions of sodium (Na^+) and chloride (Cl^-) together in crystals of table salt (NaCl). Many ionic bonds hold each salt crystal together.

In a *polar* covalent bond, atoms of different elements (which have different numbers of protons) do not exert the same pull on shared electrons. The more attractive atom ends up with a slight negative charge (that atom is "electronegative"). It is balanced by the other atom, which ends up with a slight positive charge.

In other words, taken together, two atoms that are interacting in a polar covalent bond have no *net* charge, but the charge is distributed unevenly between the two ends of that bond.

There are two polar covalent bonds in a water molecule (H—O—H). In this case, the electrons are less attracted to the hydrogens than to the oxygen, which has more protons. Thus, even though a water molecule carries no *net* charge, it can weakly attract other atoms in the vicinity (because of its polarity).

In short, patterns of electron sharing in covalent bonds hold atoms together in specific arrangements in molecules. Additionally, some of these patterns are responsible for weak attractions and repulsions *between* molecules. As you will see in later chapters, all of these bonding forces are absolutely crucial to the structure and functioning of biological molecules.

In a covalent bond, atoms share electrons.

If electrons are shared equally, the bond is nonpolar. If they are not shared equally, the bond is polar (slightly positive at one end and slightly negative at the other).

Hydrogen Bonding

In a **hydrogen bond**, an atom of a molecule weakly interacts with a neighboring hydrogen atom that is already taking part in a polar covalent bond. (The hydrogen, with its slight positive charge, is attracted to the other atom's slight negative charge.) As Figure 2.8 shows, hydrogen bonds can form between two different molecules. Such bonds also can form between two different regions of the same molecule where it twists back on itself.

Hydrogen bonds are common in large biological molecules. For example, many occur between the two strands of a DNA molecule. Individually, the hydrogen bonds are easily broken, but collectively they help stabilize DNA's structure. In later chapters, we will look at the energy it takes to break these and other bonds. Here, we will turn next to the manner in which hydrogen bonds give water some of its life-sustaining properties.

In a hydrogen bond, an atom or molecule interacts weakly with a neighboring hydrogen atom that is already taking part in a polar covalent bond.

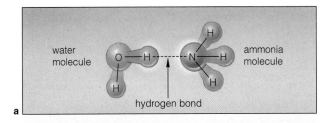

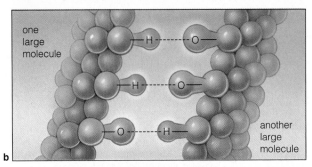

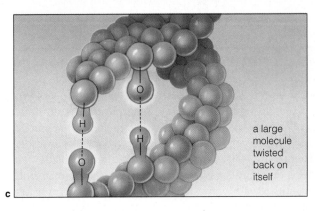

Figure 2.8 Examples of hydrogen bonds. Their collective action gives water and other substances some of their properties.

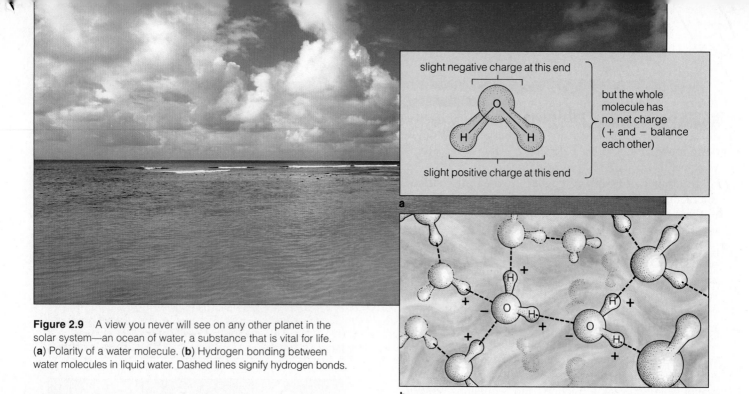

Figure 2.9 A view you never will see on any other planet in the solar system—an ocean of water, a substance that is vital for life. (**a**) Polarity of a water molecule. (**b**) Hydrogen bonding between water molecules in liquid water. Dashed lines signify hydrogen bonds.

2.4 PROPERTIES OF WATER

Life originated in water. Many organisms still live in it, and the ones that don't carry water around with them, in cells and tissue spaces. Many metabolic reactions require water as a reactant. The very shape and internal structure of cells depends on it. We will consider its effects on organisms in many chapters, so you may find it useful to study the following points concerning its major properties.

1. *The polarity of the water molecule influences the behavior of other substances.* A water molecule carries no *net* charge, but remember the charges it does carry are unevenly distributed. As a result of its electron arrangements and bond angles, the water molecule's oxygen "end" is a bit negative and its other end is a bit positive (Figure 2.9*a*). The polarity allows water molecules to form hydrogen bonds with one another and with other polar substances, such as sugars. All polar molecules are attracted to water; they are **hydrophilic** (water loving).

By contrast, water's polarity repels oil and other nonpolar substances, which are **hydrophobic** (water dreading). Shake a bottle containing water and salad oil, then put it on a counter. In time, hydrogen bonds reunite the water molecules. (They replace bonds that were broken when you shook the bottle.) As they do, they push oil molecules aside and force them to cluster in droplets or in a film at the water's surface. As you will see, such hydrophobic interactions help organize the rather oily, sheetlike layers of cell membranes.

2. *Water has temperature-stabilizing effects.* The first forms of life on earth originated in *liquid* water—not water in a frozen or gaseous form. And most of the earth's water tends to stay liquid because of hydrogen bonds between water molecules.

Consider that molecules of water or any other substance are in constant motion, and that energy inputs make them move faster. **Temperature** is a measure of this molecular motion. Compared to most other fluids, water requires a greater input of heat energy before its temperature increases measurably. Why? Hydrogen bonds in water absorb much of the incoming energy, so the motion of individual molecules does not increase as fast. This property helps stabilize temperatures in cells, which are mostly water. It helps cells resist temperature changes that could disrupt vital activities, such as enzyme action.

When water is liquid, its hydrogen bonds are constantly breaking and forming again. During **evaporation**, an energy input converts liquid water to the gaseous state. The energy input increases the molecular motion so much that hydrogen bonds stay broken, and molecules at the water's surface escape into the air. As molecules break free and depart in large numbers, they carry away energy and lower the water's surface temperature. That is why you may cool off when you work up a sweat on hot, dry days. Under such conditions, sweat—which is 99 percent water—evaporates from your skin.

Below 0°C, hydrogen bonds resist breaking and they lock water molecules in the bonding pattern of ice (Figure 2.10). During winter freezes, ice sheets form on ponds, lakes, and streams. They hold in water's heat and help protect aquatic organisms against freezing.

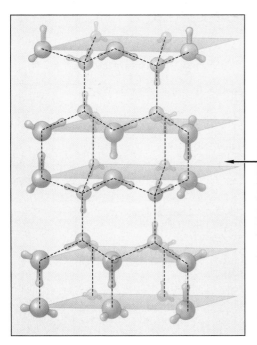

Figure 2.10 Hydrogen bonding between water molecules in ice. Below 0°C, each molecule becomes locked by four hydrogen bonds into a crystal lattice. In this bonding pattern, molecules are spaced farther apart than in liquid water at room temperature. When water is liquid, constant molecular motion usually prevents the maximum number of hydrogen bonds from forming.

3. *Water has cohesive properties.* Swimming on a hot summer night can be refreshing, if you don't mind the night-flying bugs that hit the water's surface and float about on it. Because of hydrogen bonds, the water shows cohesion. This means it resists rupturing when placed under tension—that is, stretched—as by weighty bugs (Figure 2.11). At the surface, hydrogen bonds pull the uppermost water molecules inward. A high surface tension results from the ongoing, collective bonding. In most land plants, cohesion helps move water through pipelines that extend from roots to leaves. When water evaporates from leaves, hydrogen bonds "pull" more water molecules up into leaf cells.

4. *Water has outstanding solvent properties.* Water is a great solvent, in that ions and polar molecules readily dissolve in it. Dissolved substances are called **solutes**. But what does "dissolved" mean? By way of example, pour some table salt into a glass of water. In time the salt crystals disappear, for they separate into Na^+ and Cl^- ions. Each Na^+ attracts the negative end of water molecules. Each Cl^- attracts the positive end of other water molecules. Many water molecules cluster as "spheres of hydration" around ions and keep them dispersed in fluid (Figure 2.12). A substance is "dissolved" when spheres of hydration form around its individual ions or molecules. This happens to solutes in cells, body fluids, maple tree sap, and so on.

Figure 2.11 Water's cohesion, as demonstrated by a water strider "walking" on water molecules held together by hydrogen bonds.

Water's internal cohesion, temperature-stabilizing effects, and capacity to dissolve many substances influence the structure and functioning of organisms.

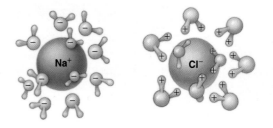

Figure 2.12 Spheres of hydration around charged ions.

2.5 WATER, DISSOLVED IONS, AND PH VALUES

The pH Scale

A variety of ions are dissolved in the fluids inside and outside of cells, and these profoundly influence the structure and activities of biological molecules. Among the most influential kinds are **hydrogen ions** (H^+), which are the same thing as free (unbound) protons.

At any given time in liquid water, a few molecules are breaking apart into hydrogen ions and **hydroxide ions** (OH^-). This ionization of water is the basis of the **pH scale**, shown in Figure 2.13. The pH scale is used to measure the concentration of H^+ in blood, tree sap, or any other aqueous solution.

At 25°C, pure water always has just as many H^+ as OH^- ions. Whether for water or any other fluid, this condition represents *neutrality* on the pH scale, and it is assigned a value of 7. It is the midpoint of the pH scale, which ranges from 0 (highest H^+ concentration) to 14 (lowest concentration). *The greater the H^+ concentration, the lower the pH value.*

Starting at neutrality, each change by one unit of the pH scale corresponds to a *tenfold* increase or decrease in H^+ concentration. To get a sense of this logarithmic difference between units, dissolve some baking soda (pH 9) on your tongue or taste some egg white (pH 8.0). Next, sip pure water (pH 7), then vinegar (3) or lemon juice (2.3).

The fluid in most of your cells hovers around pH 7. The pH of your blood and most tissue fluids ranges between 7.35 and 7.45. The pH of an aquatic habitat may be higher or lower than it is inside organisms dwelling there. Certain airborne industrial wastes are acidic (Figure 2.14), and they affect the pH of rain. We will consider the effects of acid rain in Chapter 50.

The pH value of any solution represents its concentration of hydrogen ions.

Acids, Bases, and Salts

Acids are any substances that release H^+ ions when they dissolve in water, and **bases** are substances that combine with those ions. *Acidic* solutions such as lemon juice have more H^+ than OH^- ions, and their pH is below 7. *Basic* (or alkaline) solutions, such as egg whites, have fewer H^+ than OH^- ions, and their pH is above 7.

Think of what happens when you sniff, chew, then swallow fried chicken and so send it on its way to the gastric fluid in your saclike stomach. As an outcome of

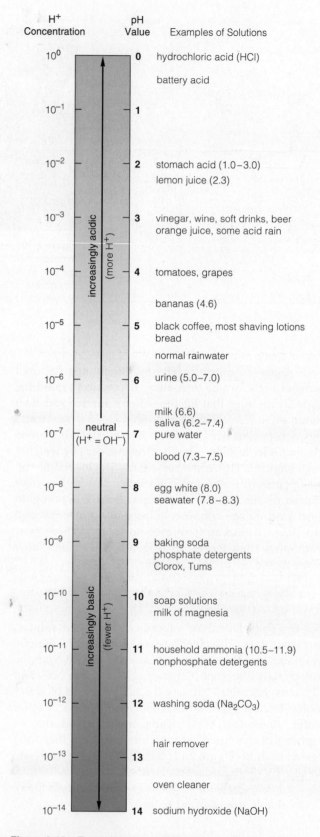

H^+ Concentration	pH Value	Examples of Solutions
10^0	0	hydrochloric acid (HCl)
		battery acid
10^{-1}	1	
10^{-2}	2	stomach acid (1.0–3.0)
		lemon juice (2.3)
10^{-3}	3	vinegar, wine, soft drinks, beer orange juice, some acid rain
10^{-4}	4	tomatoes, grapes
		bananas (4.6)
10^{-5}	5	black coffee, most shaving lotions bread
		normal rainwater
10^{-6}	6	urine (5.0–7.0)
		milk (6.6)
		saliva (6.2–7.4)
10^{-7}	7	pure water
		blood (7.3–7.5)
10^{-8}	8	egg white (8.0) seawater (7.8–8.3)
10^{-9}	9	baking soda phosphate detergents Clorox, Tums
10^{-10}	10	soap solutions milk of magnesia
10^{-11}	11	household ammonia (10.5–11.9) nonphosphate detergents
10^{-12}	12	washing soda (Na_2CO_3)
		hair remover
10^{-13}	13	
		oven cleaner
10^{-14}	14	sodium hydroxide (NaOH)

increasingly acidic (more H^+)

neutral ($H^+ = OH^-$)

increasingly basic (fewer H^+)

Figure 2.13 The pH scale, in which a liter of fluid is assigned a number according to the number of hydrogen ions in it. The scale ranges from 0 (most acidic) to 14 (most basic). A change of only 1 on the scale means a tenfold change in the H^+ concentration.

Figure 2.14 Sulfur dioxide emissions from a coal-burning power plant. Special camera filters revealed these otherwise invisible emissions. Together with other airborne pollutants, sulfur dioxides dissolve in atmospheric water to form acidic solutions. They are a major component of acid rain.

your feeding activities, cells in the stomach's lining are stimulated to secrete hydrochloric acid (HCl), which separates into H^+ and Cl^-. These ions make the gastric fluid more acidic, and a good thing, too. Increased acidity switches on enzymes that can digest the chicken proteins. It also helps kill bacteria that may be lurking in or on the chicken.

If you eat too much of the fried chicken, you may end up with an "acid stomach." And you might reach for an antacid tablet. Milk of magnesia is one kind of antacid. When dissolved, it releases magnesium ions and hydroxide ions. The hydroxide ions may then *combine with* some of the excess hydrogen ions in your gastric fluid and help settle things down.

Acids commonly combine with bases. The results are ionic compounds called **salts**. Salts often dissolve and form again, depending on the pH. Consider how sodium chloride forms, then dissolves:

$$HCl \quad + \quad NaOH \quad \longrightarrow \quad NaCl \quad + \quad H_2O$$

hydrochloric acid · sodium hydroxide (a base) · sodium chloride (a salt)

$$Na^+ \quad Cl^-$$

Many other salts also dissolve into ions. Such ions serve key functions in cells. For example, many help maintain the body's acid-base balance, described next.

Buffers and the pH of Body Fluids

Chemical reactions in cells are sensitive to even slight shifts in pH. Yet hydrogen ions are continually being added to the cellular environment and withdrawn from

it. Control mechanisms counter the potentially disruptive shifts and help maintain cellular pH.

In multicelled organisms, controls also maintain the pH of blood and tissue fluids, which together constitute the body's internal fluid environment. Later in the book, you will see how lungs and kidneys help maintain the acid-base balance of this internal environment, at levels suitable for cellular life. For now, simply keep in mind that many of the control mechanisms involve buffer molecules.

A **buffer** is any molecule that can combine with hydrogen ions, release them, or both, and so help stabilize pH. Some weak acids and bases work as a buffer system. Consider how bicarbonate (HCO_3^-) helps restore pH when blood becomes too acidic. It combines with excess H^+ to form carbonic acid:

$$HCO_3^- \quad + \quad H^+ \quad \longrightarrow \quad H_2CO_3$$

bicarbonate · carbonic acid

Conversely, bicarbonate releases H^+ when blood is not acidic enough. Its buffering action is crucial. For instance, some lung diseases interfere with carbon dioxide elimination, so the blood level of carbonic acid (hence of H^+) increases. This abnormal condition, a form of *acidosis*, makes breathing difficult and weakens the body.

In multicelled organisms, buffering mechanisms help counter slight shifts in the pH of the internal fluid environment that sustains the body's living cells.

2.6 CHEMICAL INTERACTIONS AND THE WORLD OF CELLS

Electrons and protons. Atoms, ions, and molecules. With this chapter, we have started thinking about interactions among these bits of matter. These interactions will occupy our attention in many other parts of the book, for they are the foundation for the organization and activities characteristic of life.

For now, merely reflect on the manner in which water influences the functions of a biological molecule —a protein, for instance. The surface of a protein that is dissolved in some cellular fluid may be positively or negatively charged, overall. The charged regions and polar groups attract water molecules. They also attract ions—which in turn attract more water molecules. In this way, an electrically charged "cushion" of ions and water forms around the protein. Figure 2.15 is a simple way of showing this cushioning effect.

Through such interactions with water and ions, the protein remains dispersed in the cellular fluid rather than randomly settling against some cell structure. Why is it important to prevent settling? Many chemical reactions proceed on specific molecular regions of proteins. *Cells must have access to those surfaces, for they are stages for life-sustaining tasks.*

Many more examples could be given of interactions between the polar water molecule and other substances characteristic of the cellular world. For now, the point to keep in mind is this:

Interactions among atoms, ions, and molecules are the start of the organization and behavior of substances that make up cells and the cellular environment.

Figure 2.15 Simple diagram of a dissolved protein, surrounded by an electrically charged cushion of ions and water molecules.

SUMMARY

1. Protons, neutrons, and electrons are building blocks of atoms. All atoms of an element have the same number of protons and electrons. Interaction between atoms depends on the number and arrangement of their electrons, which carry a negative charge.

2. All atoms of an element have the same number of protons and electrons. The ones that differ slightly in their number of neutrons are isotopes. Radioisotopes decay spontaneously into atoms of different types.

3. An orbital is a region of space around an atom's nucleus that can be occupied by one or at most two electrons. There are many different orbitals. We can think of them as being arranged inside a series of imaginary shells around the nucleus. Electrons in the shell closest to the nucleus are at the lowest energy level. Electrons in shells farther away are at higher energy levels.

4. Atoms of hydrogen, carbon, nitrogen, and oxygen are the main elements of biological molecules. They have one or more unfilled orbitals in their outermost shell and tend to form bonds with other elements.

5. Atoms have no net charge. If an atom gains or loses one or more electrons, it becomes an ion, with an overall positive or negative charge.

6. In ionic bonds, a positive ion and a negative ion stay together by mutual attraction of their opposite charges. In covalent bonds, atoms share one or more electrons. In hydrogen bonds, an atom or molecule weakly interacts with a neighboring hydrogen atom that is already taking part in a polar covalent bond.

7. The pH scale is a measure of the concentration of hydrogen ions (H^+) that are dissolved in some fluid. Acids are any substances that release H^+, and bases are any substances that combine with them. At pH7, the H^+ and OH^- concentrations in a solution are equal.

8. Buffers and other mechanisms maintain pH values of blood, tissue fluids, and the fluid inside cells.

9. A water molecule shows polarity. Due to its electron arrangements, one end of the molecule carries a partial negative charge, and the other end carries a partial positive charge.
   a. Other polar molecules are attracted to water; they are hydrophilic.
   b. Nonpolar molecules are repelled by water; they are hydrophobic.

10. Water takes part in many reactions and helps give cells shape and internal organization. Because of hydrogen bonds between its molecules, water resists temperature changes. It shows internal cohesion and a notable capacity to dissolve other substances.

Review Questions

1. Define element, atom, molecule, and compound. What are the six main elements (and their symbols) in most organisms? *18–19*

2. Define an atom, an ion, and an isotope. *19, 24*

3. Explain the differences among covalent, ionic, and hydrogen bonds. *24–25*

4. What is the difference between a hydrophilic and a hydrophobic interaction? Is a film of oil on water an outcome of bonding between the molecules making up the oil? *26*

5. What type of bond is associated with the temperature-stabilizing, cohesive, and solvent properties of water? Is that bond also important in hydrophobic interactions? *26–27*

6. Define an acid, a base, and a salt. On a pH scale from 0 to 14, what is the acid range? Why are buffers important in cells? *28–29*

Self-Quiz *(Answers in Appendix IV)*

1. Atoms are constructed of protons, neutrons, and ____ .

2. An _____ has a net charge of zero; an _____ has gained or lost one or more electrons, and so has become negatively or positively charged.
   a. ion; ion
   b. ion; atom
   c. atom; atom
   d. atom; ion

3. Interactions between atoms as they give up, acquire, or share _____ help determine the organization and activities of living things.

4. _____ are atoms of the same element that vary only in the number of neutrons they possess.

5. The main chemical elements found in biological molecules are:
   a. hydrogen, sulfur, nitrogen, oxygen
   b. phosphorus, hydrogen, carbon, oxygen
   c. carbon, oxygen, hydrogen, nitrogen
   d. carbon, oxygen, nitrogen, sulfur

6. Orbitals within shells around the nucleus of an atom can each hold no more than _____ electrons.
   a. one
   b. two
   c. three
   d. four

7. Electrons are shared unequally in a(n) _____ bond.
   a. nonpolar covalent
   b. ionic
   c. hydrogen
   d. polar covalent

8. Polar substances are _____; nonpolar substances are _____ .
   a. hydrophilic; also hydrophilic
   b. hydrophilic; hydrophobic
   c. hydrophobic; also hydrophobic
   d. hydrophobic; hydrophilic

9. Which characterizes the internal pH of most cells?
   a. high concentration of H^+
   b. nearly equal concentration of H^+ and OH^-
   c. high concentration of OH^-
   d. both b and c are correct

10. A(n) _____ can combine with hydrogen ions or release them in response to changes in cellular pH.
    a. acid
    b. salt
    c. base
    d. buffer

11. Match these chemistry concepts appropriately:
    ___ water molecule's polarity
    ___ common bonds in biological molecules
    ___ cellular pH
    ___ hydrogen bonds between water molecules
    ___ salt

    a. close to neutral
    b. temperature-stabilizing and cohesive properties
    c. permits ions and polar molecules to dissolve more easily
    d. produced by reaction between acid and base
    e. ionic, covalent, and hydrogen

Selected Key Terms

acid *28*	hydroxide ion *28*
atom *19*	ion *24*
atomic number *19*	ionic bond *24*
base *28*	isotope *19*
buffer *29*	mass number *19*
chemical bond *22*	molecule *19*
compound *19*	neutron *19*
covalent bond *24*	orbital *22*
electron *19*	pH scale *28*
element *18*	proton *19*
evaporation *26*	radioisotope *19*
hydrogen bond *25*	salt *29*
hydrogen ion *28*	solute *27*
hydrophilic substance *26*	temperature *26*
hydrophobic substance *26*	tracer *21*

Readings

Lehninger, A., D. Nelson, and M. Cox. 1993. *Principles of Biochemistry.* Second edition. New York: Worth. Many excellent illustrations in this new edition of a highly respected textbook.

Science. 26 June 1992. "A New Blueprint for Water's Architecture." 256:1,764.

3 CARBON COMPOUNDS IN CELLS

Mom, Dad, and Clogged Arteries

Butter! Bacon and eggs! Ice cream! Cheesecake! Possibly you think of such foods as enticing, off-limits, or both. After all, who in America doesn't know about animal fats and the dreaded cholesterol?

Soon after you feast on these fatty foods, cholesterol from them enters your bloodstream. Cholesterol has useful roles. It is a structural component of animal cell membranes, and without membranes, there would be no cells. Cells also remodel cholesterol into various molecules, including the vitamin D that is necessary for good bones and teeth. Normally, however, your liver synthesizes enough cholesterol for your cells.

Also circulating in the blood are Apos (short for apolipoproteins). These protein molecules combine

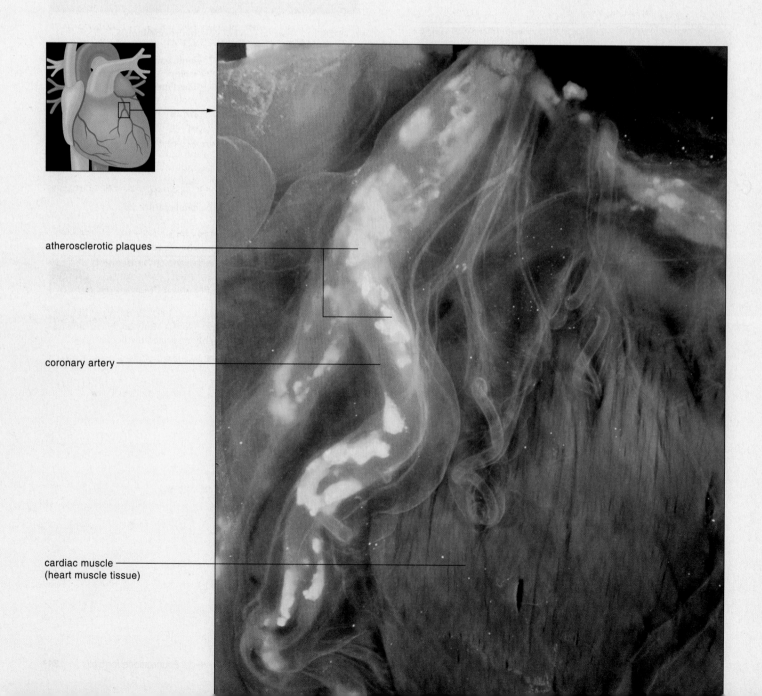

atherosclerotic plaques

coronary artery

cardiac muscle
(heart muscle tissue)

with cholesterol and other substances, forming particles called lipoproteins. Some particles are HDLs (high-density lipoproteins). They collect cholesterol, then transport it to the liver for metabolism. Other particles are VDLs and LDLs (very-low and low-density lipoproteins). If all goes well, these end up in cells that store or use cholesterol.

When too many LDLs form, the excess infiltrates the walls of blood vessels called arteries. There, calcium accumulates and a fibrous net forms. The resulting abnormal masses are atherosclerotic plaques, and they may interfere with blood flow (Figure 3.1). Plaques can narrow arteries, a condition called *atherosclerosis*. When they clog a small-diameter artery that delivers blood to the heart, symptoms can range from mild chest pains to a heart attack.

Do high-cholesterol diets put you at risk? That largely depends on what you inherited from your parents. For instance, each parent provided you with DNA instructions (a gene) for building a certain Apo molecule. However, there are three slightly different versions of this gene in the human population. They specify three different forms of the molecule—Apo 2, 3, or 4—which behave in different ways. If both genes specify Apo E2, your blood cholesterol level will remain so low that your arteries won't get clogged no matter what you eat. If one specifies Apo E2 and the other, Apo E3, the level will be higher. If they both specify Apo E4 , you definitely should forgo the butter and bacon.

This example takes us into the world of large biological molecules—the complex carbohydrates, proteins, lipids, and nucleic acids. These molecules are the foundation for the structure and function of every cell, every organism. Your own body uses them as building materials, as "worker" molecules such as enzymes and transporters, and as storehouses of energy that drive all of your activities—from chewing on an apple to thinking about the hazards of too much cholesterol.

KEY CONCEPTS

1. In organic compounds, carbon atoms covalently bonded to each other serve as a backbone to which hydrogen, oxygen, nitrogen, and other atoms are attached.

2. Cells assemble simple sugars, fatty acids, amino acids, and nucleotides. All of the large biological molecules—the complex carbohydrates, lipids, proteins, and nucleic acids—can be assembled from these four families of small organic compounds.

3. The simple sugars, including glucose, are carbohydrates. So are organic compounds composed of two or more sugar units, of the same or different kind, that are covalently bonded together. The most complex carbohydrates are the polysaccharides, many of which consist of hundreds or thousands of sugar units.

4. Lipids are greasy or oily compounds that dissolve in one another but show little tendency to dissolve in water. They include neutral fats, phospholipids, waxes, and sterols.

5. Cells use carbohydrates and lipids as sources of energy and as building blocks.

6. Proteins have truly diverse roles. Many are structural materials. Many are enzymes that enhance the rate of specific metabolic reactions. Other kinds transport cell substances, contribute to cell movements, trigger changes in cell activities, and defend the body against disease.

7. The nucleic acids called DNA and RNA are the basis of inheritance and cell reproduction.

Figure 3.1 Potentially life-threatening plaques (*bright yellow*) inside one of the arteries that deliver blood to the heart. Such plaques are a legacy of abnormally high levels of an otherwise vital organic compound—cholesterol.

3.1 PROPERTIES OF ORGANIC COMPOUNDS

An **organic compound** consists of carbon and one or more additional elements, covalently bonded to one another. The term is a holdover from a time when chemists thought "organic" substances were the ones they obtained from animals and vegetables, as opposed to the "inorganic" substances obtained from minerals. The term persists even though researchers now synthesize organic compounds in laboratories. And it persists even though there are reasons to believe that organic compounds were present on the earth *before* organisms were.

Carbon-to-Carbon Bonds and the Stability of Organic Compounds

By far, you and all other organisms are composed of oxygen, hydrogen, and carbon. Much of the oxygen and hydrogen is in the form of water. Remove the water, and carbon makes up more than half of what's left.

Carbon's importance in life arises from its versatile bonding behavior. *Each carbon atom can share pairs of elec-trons with as many as four other carbon atoms.* Each covalent bond formed this way is quite stable. Such bonds link carbon atoms in chains and rings, and these serve as a backbone to which hydrogen, oxygen, and other elements become attached (Figure 3.2).

Carbon-to-Carbon Bonds and the Shape of Organic Compounds

The flattened structural formulas in Figure 3.2 might lead you to believe that the molecules they represent are flat, like paper dolls. They actually show stunning diversity in their three-dimensional shapes, which begin with bonding arrangements in carbon backbones.

A carbon atom taking part in four single covalent bonds has this arrangement:

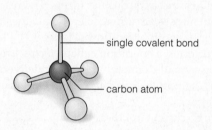

single covalent bond

carbon atom

Figure 3.2 Carbon compounds. There is a Tinkertoy quality to carbon compounds, for a single atom can be the start of diverse molecules assembled from "straight-stick" covalent bonds. Start with the simplest hydrocarbon, methane (CH_4). Stripping one hydrogen from methane leaves a methyl group, which occurs in fats, oils, and waxes:

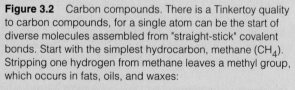

methane → methyl group

Imagine stripping a hydrogen atom from each of two methane molecules, then bonding the two molecules together. If the resulting structure were to lose a hydrogen atom, you would end up with an ethyl group:

ethyl group

You could go on building a continuous chain:

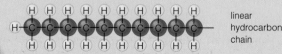

linear hydrocarbon chain

And you could add branches to the chain:

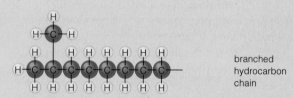

branched hydrocarbon chain

You might have the chain coil back on itself into a ring, which you could diagram in various ways:

or

carbon rings

You might even join many carbon rings together and so produce larger molecules:

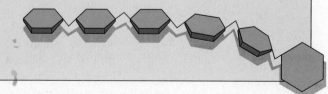

Often, the carbon atoms can rotate freely such bonds:

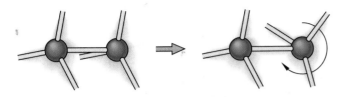

By contrast, a *double* covalent bond uniting a carbon atom with another atom restricts rotation. Where such bonds exist in a carbon backbone, the two atoms rigidly hold their position in space:

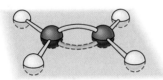

Now imagine a long carbon backbone, with some atoms locked in rigid positions and others free to rotate at different angles in space. The resulting orientations of atoms attached to the backbone may enhance or discourage interactions with neighboring atoms or molecules in the vicinity. As you will see shortly, such interactions help give rise to the three-dimensional shapes and functions of biological molecules.

Flexible and rigid bonding arrangements in carbon backbones are a starting point for the three-dimensional shapes and functions of organic compounds.

Hydrocarbons and Functional Groups

In **hydrocarbons**, only hydrogen atoms are attached to the carbon backbone. Hydrocarbons don't break apart easily, and they form stable portions of most biological molecules. These molecules also have other kinds of atoms attached to the backbone, either alone or in clusters. Atoms covalently bonded to a carbon backbone are **functional groups**. Many influence the electron arrangements of neighboring atoms and so affect the structure and behavior of the whole molecule. Table 3.1 lists some of the major functional groups.

Consider sugars and other organic compounds that have one or more hydroxyl groups (—OH) attached to carbon atoms. These compounds, classified as **alcohols**, readily dissolve in water, because water molecules can form hydrogen bonds with —OH groups. Or consider proteins. Their backbone forms by repeated reactions

Table 3.1	Functional Groups Common to Biological Molecules	
Group	Structural Formula	Common Locations
Methyl	H, H, C, H (—C— with H top and bottom, H right)	Fats, oils, waxes
Hydroxyl	—OH	Sugars, other alcohols
Aldehyde	—C⟨H, O	Sugars
Ketone	—C=O	Sugars
Carboxyl	—C⟨O, OH	Sugars, fats, amino acids
Amino	—N⟨H, H or —N(—H)(—H)H	Amino acids, proteins
Phosphate	—O—P(O⁻)(=O)—O⁻ Symbolized as —Ⓟ	DNA, RNA, ATP
Sulfhydryl	—S—H	Proteins

between many amino groups and carboxyl groups. And the particular bonding patterns associated with this backbone contribute to the three-dimensional structure of proteins. Amino groups also can combine with H^+ and so act as buffers against decreases in pH.

The structure and behavior of an organic compound are influenced by the functional groups covalently bonded to its carbon backbone.

3.2 HOW CELLS USE ORGANIC COMPOUNDS

Five Classes of Reactions

Enzymes, a special class of proteins, speed up specific metabolic reactions. We will study these remarkable proteins in later chapters. For now, it is enough to know that they mediate five categories of reactions by which most biological molecules are assembled, rearranged, and broken apart:

---

1. **Functional-group transfer**. One molecule gives up a functional group, which another molecule accepts.

2. **Electron transfer**. One or more electrons stripped from one molecule are donated to another molecule.

3. **Rearrangement**. A juggling of internal bonds converts one type of organic compound into another.

4. **Condensation**. Through covalent bonding, two molecules combine to form a larger molecule.

5. **Cleavage**. A molecule splits into two smaller ones.

To get a sense of what goes on, consider just two examples of these reactions. In many condensation reactions, enzymes remove a hydroxyl group from one molecule and an H atom from another, then a covalent bond forms between the two molecules at the exposed sites (Figure 3.3a). The discarded atoms (now H^+ and OH^- ions) combine to form a water molecule. Cells assemble starch and other polymers by repeated condensation reactions. A "polymer" is a large molecule composed of three to millions of subunits, which may or may not be identical. The individual subunits are often called "monomers," as in the sugar monomers of starch.

One type of cleavage reaction, **hydrolysis**, is like condensation in reverse (Figure 3.3b). Enzyme action breaks covalent bonds at functional groups and splits a molecule into two or more parts. At the same time, —H and —OH derived from a water molecule are attached to the exposed sites. Cells commonly hydrolyze starch and other polymers, then use the released subunits as building blocks or energy sources.

The Molecules of Life

Under the physical conditions that now exist on earth, only living cells can synthesize the organic compounds known as carbohydrates, lipids, proteins, and nucleic acids. Together, *these are the molecules characteristic of life.* Different kinds function as energy packets, energy stores, structural materials, metabolic workers, and libraries of hereditary information.

Some biological molecules are rather small organic compounds. The simple sugars, fatty acids, amino acids, and nucleotides are like this. As you will see, cells use these four families of small organic compounds as subunits for the synthesis of larger carbohydrates, lipids, proteins, and nucleic acids.

The large molecules of life are composed of one or more units called simple sugars, fatty acids, amino acids, and nucleotides.

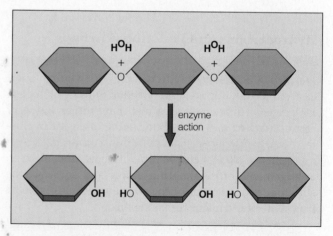

a By two condensation reactions, these three molecules covalently bond into a larger molecule; two water molecules form as by-products.

b Hydrolysis, a water-requiring cleavage reaction. Covalent bonds in a molecule are broken, and H^+ and OH^- derived from a water molecule bond to the molecular fragments.

Figure 3.3 Examples from two categories of enzyme-mediated reactions: condensation and hydrolysis.

3.3 THE SMALL CARBOHYDRATES

A **carbohydrate** is a simple sugar or a molecule composed of two or more sugar units. Carbohydrates are the most abundant biological molecules. All cells use them as structural materials, transportable packets of energy, and stored forms of energy.

We recognize three classes of carbohydrates. These are **monosaccharides** (simple sugars), **oligosaccharides**, and **polysaccharides** (complex carbohydrates).

Monosaccharides—The Simple Sugars

"Saccharide" comes from a Greek word meaning sugar. A *mono*saccharide, or one sugar unit, is the simplest carbohydrate of all. It has at least two —OH groups and an aldehyde or ketone group. Most simple sugars are sweet tasting, and they dissolve readily in water.

The most common monosaccharides have a backbone of five or six carbon atoms that tends to form a ring structure when dissolved in body fluids or cells. Ribose and deoxyribose (sugar components of RNA and DNA, respectively) have five carbon atoms. Glucose has six (Figure 3.4a). You will encounter glucose repeatedly in this book. It is the main energy source for most organisms. It is a precursor (parent molecule) of many compounds and a building block for larger carbohydrates.

You also will encounter three other compounds derived from sugar monomers. Glycerol (a sugar) is a component of fats. Vitamin C (a sugar acid) has roles in nutrition. Glucose-6-phosphate (a sugar phosphate) is a premier entrant into major reaction pathways, including aerobic respiration.

Oligosaccharides

An *oligo*saccharide is a short chain of two or more covalently bonded sugar units. Among those with two sugars—the *di*saccharides—are lactose, sucrose, and maltose. You probably know that lactose (a glucose and a galactose unit) is present in milk.

Sucrose, the most plentiful sugar in nature, consists of a glucose and a fructose unit (Figure 3.4). Leafy plants continually convert carbohydrates to sucrose, which is easily transported through the fluid-filled pipelines that service all living cells in leaves, stems, and roots. Crystallizing sucrose extracts from sugar cane and other plants gives us table sugar.

When seeds of barley and other plants first start to sprout, the forthcoming seedlings produce an abundance of hydrolytic enzymes. These enzymes break down starch, stored in tissues inside the seed coat, into

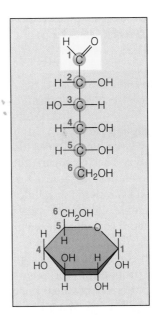

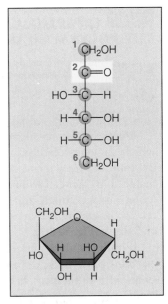

a Glucose   **b** Fructose

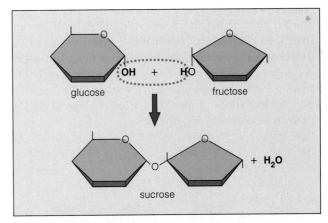

c Formation of sucrose

Figure 3.4 Straight-chain and ring forms of glucose (**a**) and fructose (**b**). For reference purposes, the carbon atoms of sugars are often numbered in sequence, starting at the end of the molecule closest to the aldehyde or ketone group.

(**c**) Condensation of two monosaccharides (glucose and fructose) into a disaccharide (sucrose).

maltose (two glucose units). Seedlings take up maltose as a source of energy for rapid growth. In malt breweries, the hydrolytic enzymes in large batches of carefully tended sprouting seeds hasten the conversion of starch to many sugar molecules that can be fermented.

Proteins and other large molecules often have oligosaccharides attached to them as side chains. These chains are composed of three or more sugar monomers, but they are all short. Some have roles in cell membrane function and in immunity, which are topics of later chapters.

3.4 COMPLEX CARBOHYDRATES: THE POLYSACCHARIDES

A *poly*saccharide is a straight or branched chain of sugar units—hundreds or thousands of the same or different kinds. The most common of these—starch, cellulose, and glycogen—consist only of glucose.

Plant cells store sugars as large starch molecules, which enzymes readily hydrolyze back to sugar units. By contrast, they incorporate cellulose as a structural material in their cell walls. Fibers composed of cellulose molecules are tough and insoluble. Like steel rods in reinforced concrete, they withstand considerable weight and stress (Figure 3.5).

If both cellulose and starch consist of glucose, why do they have such different properties? The answer starts with differences in the pattern of covalent bonding between adjacent glucose units. Long chains of glucose form in both substances. In cellulose, however, the chains stretch out, side by side, and hydrogen-bond to one another at —OH groups (Figure 3.6). The many hydrogen bonds help stabilize the chains in tight bundles that resist breakdown. In starch, a flexible bonding arrangement allows the chains to twist into a coil, with many —OH groups facing outward. The linkages become oriented in positions that are accessible to hydrolytic enzymes (Figure 3.7).

Figure 3.5 A few uses of polysaccharides. Stems gain strength from cellulose, a structural material in plant cell walls. Cells in flowers store starch, which is readily hydrolyzed into sugars that enrich bird-attracting nectar. Birds can convert nectar sugars into glycogen, the animal's equivalent of starch.

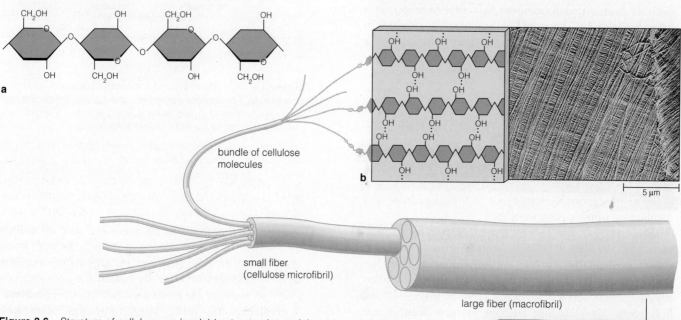

Figure 3.6 Structure of cellulose, an insoluble structural material composed of glucose units. Very few organisms have enzymes that can digest it. (**a**) Bonding pattern between the adjacent glucose units. (**b**) In large fibers of cellulose, chains of glucose units are stretched out, side by side. The chains hydrogen-bond to one another at —OH groups. The bonds stabilize the chains in tight bundles that are arranged into increasingly larger fibers. The micrograph shows large cellulose fibers from the cell wall of *Cladophora*, a green alga.

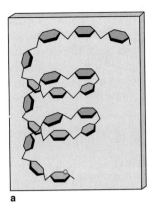

a

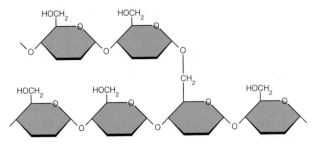

b

Figure 3.7 (**a**) Structure of amylose, a readily soluble form of starch composed of glucose units. (**b**) Bonding pattern between adjacent glucose units. These particular linkages cause chains of glucose to coil. Coiling orients the linkages in such a way that they are accessible to enzymes of hydrolysis.

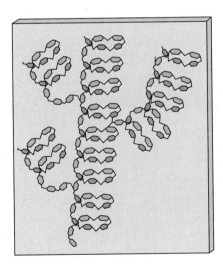

Figure 3.8 Branching structure of glycogen. Like cellulose and starch, glycogen also is composed of glucose units.

a

b

Figure 3.9 (**a**) Scanning electron micrograph of a tick. Its body covering is a protective, chitin-reinforced cuticle. (**b**) Big laughing mushroom (*Gymnophilus*), the cell walls of which are structurally reinforced with chitin.

Like starch in plants, glycogen is a sugar storage form in animals—notably so in liver and muscle tissues. When blood sugar levels fall, liver cells break down glycogen and release glucose units to the blood. During exercise, muscle cells tap into their glycogen stores for quick access to energy. Figure 3.8 shows a few of glycogen's numerous branchings.

Cells of many animals and most fungi secrete chitin. This polysaccharide has nitrogen atoms attached to the backbone. Chitin is the main structural material in external skeletons and other hard body parts of many animals, including crabs and insects (Figure 3.9*a*). It also is the main structural material in the cell walls of many fungal species (Figure 3.9*b*).

3.5 LIPIDS

The greasy or oily compounds called **lipids** dissolve readily enough in one another but show very little tendency to dissolve in water. In nearly all organisms, certain lipids function as the main reservoirs of stored energy. Other lipids function as structural materials in cellular components (such as membranes) and products (such as surface coatings). Here, we consider neutral fats, phospholipids, and waxes, all of which have fatty acid components. We also consider sterols, each with a backbone of four carbon rings.

Fatty Acids

The hydrocarbons called **fatty acids** have a backbone of up to thirty-six carbon atoms, a carboxyl (—COOH) group at one end, and hydrogen atoms occupying most or all of the remaining bonding sites. When combined with other molecules, fatty acids typically stretch out, like flexible tails. Tails with one or more double bonds in their backbone are said to be *unsaturated*. Those with single bonds only are said to be *saturated*. Figure 3.10 shows examples.

When many saturated fatty acid tails are present in a substance, weak attractions cause them to snuggle in parallel, and this gives the substance a rather solid consistency. By contrast, the double and triple bonds in unsaturated fatty acids put rigid kinks in the tails. The packing arrays are less stable, and this imparts fluidity to substances.

Neutral Fats (Triglycerides)

Butter, lard, and oils are examples of **triglycerides**, or neutral fats—the body's most abundant lipids and its richest source of energy. These lipids have fatty acid tails attached to a backbone of glycerol (Figure 3.11). Gram for gram, triglycerides yield more than twice as much energy as carbohydrates. Energy is released when bonds are broken—and triglycerides have far more covalent bonds than carbohydrates do.

In vertebrates, cells of adipose tissue store great quantities of triglycerides as fat droplets. In penguins, seals, and some other animals, a very thick layer of triglycerides also functions as insulation against near-freezing environmental temperatures (Figure 3.12).

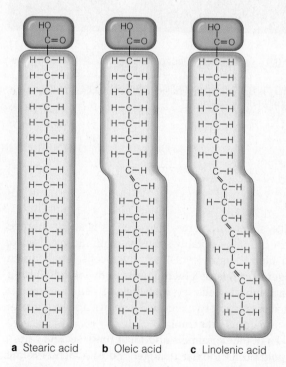

a Stearic acid **b** Oleic acid **c** Linolenic acid

Figure 3.10 Structural formulas for three fatty acids. (**a**) Stearic acid's carbon backbone is fully saturated with hydrogen atoms. (**b**) Oleic acid, with its double bond in the carbon backbone, is unsaturated. (**c**) Linolenic acid, with three double bonds, is a "polyunsaturated" fatty acid.

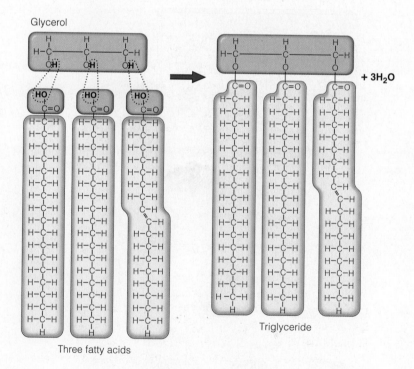

Three fatty acids

Triglyceride

Figure 3.11 Condensation of fatty acids into a triglyceride.

Figure 3.12 Penguins taking the plunge. They can swim in icy waters for long periods, thanks to a thick, insulative layer of triglycerides under their skin.

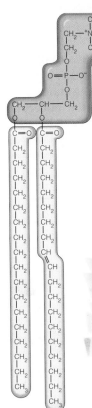

Figure 3.13 Structural formula of a typical phospholipid of animal cell membranes. The hydrophilic head is shaded orange. Are the hydrophobic tails (*yellow parts*) saturated or unsaturated?

a   b

Figure 3.14 (**a**) Honeycomb, constructed from a firm, water-repellant, waxy secretion called beeswax. (**b**) Demonstration of the water-repelling attribute of the cherry cuticle.

Phospholipids

A **phospholipid** has a backbone of glycerol, two fatty acid tails, and a hydrophilic "head" that includes a phosphate group (Figure 3.13). It is the main component of the two lipid layers of cell membranes. The phospholipid heads of one layer are dissolved in cellular fluids, those of the other layer are dissolved in the surroundings, and all fatty acid tails are sandwiched between the two.

Waxes

The lipids called **waxes** have long-chain fatty acids, tightly packed and linked to long-chain alcohols or to carbon rings. Waxes have a firm consistency and repel water. In many animals, cells secrete waxes for coatings that keep skin or hair protected, lubricated, and pliable. In waterfowl and other birds, wax secretions help keep feathers dry. Honeycomb, shown in Figure 3.14a, consists of beeswax (after *weax*, an archaic English word meaning "the honeycomb material"). In many plants, waxes and another lipid (cutin) form a cuticle, a surface covering on aboveground parts that helps restrict water loss and hinder some parasites. The waxy cherry cuticle is an example (Figure 3.14b).

Sterols and Their Derivatives

Sterols are among the lipids that have no fatty acid tails. Sterols differ in the number, position, and type of their functional groups, but they all have a rigid backbone of four fused-together carbon rings:

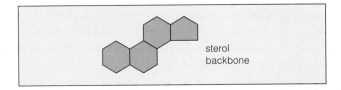

sterol backbone

Sterols are structural components of eukaryotic cell membranes. Cholesterol, described at the start of this chapter, is the major kind in animal tissues. It is the precursor of steroid hormones such as testosterone and estrogen, which influence gamete formation and sexual traits in humans. Some athletes and bodybuilders use hormonelike steroids to increase muscle mass, but these have bad side effects (page 649). Bile salts, which assist fat digestion in the small intestine, also are derived from cholesterol. So is vitamin D, which has roles in calcium and phosphate metabolism.

3.6 PROTEINS

Of all the large biological molecules, **proteins** are by far the most diverse. The ones called enzymes make reactions proceed much faster than they otherwise would. The structural types are the stuff of bone and cartilage, webs and feathers, and a dizzying variety of other biological parts and products. Transport proteins move cargo across cell membranes or through body fluids. Nutritious proteins abound in milk, eggs, and assorted seeds. Regulatory types, including protein hormones, function as signals for change in cellular activities. Many proteins even function in the body's defense against invasion or predation. Amazingly, cells build all of these diverse molecules from their pools of only twenty or so kinds of amino acids.

An **amino acid** is a small organic compound having an amino group, an acid group, a hydrogen atom, and one or more atoms called its R group. All of these parts are covalently bonded to the same carbon atom:

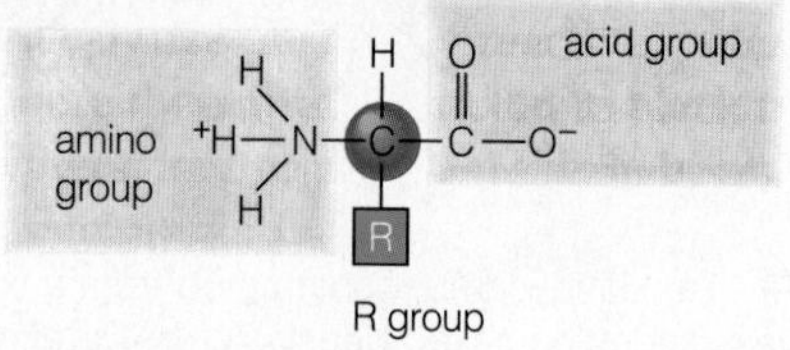

Figure 3.15 shows some amino acids that we will consider later in the book.

Primary Structure of Proteins

When cells synthesize proteins, amino acids become linked, one after the other, by *peptide* bonds. As Figure 3.16 shows, this type of covalent bond forms between the amino group of one amino acid and the acid (carboxyl) group of another. Three or more amino acids joined this way form a **polypeptide chain**. The backbone of such chains incorporates nitrogen atoms in this regular pattern: —N—C—C—N—C—C—.

Amino acid units are not chosen at random for a given chain. They are specifically selected, one at a time, from the twenty kinds available. *And the resulting sequence of amino acids is unique for each particular kind of protein.*

Figure 3.17 shows the sequence for two chains that make up insulin molecules in cattle. (Insulin, a protein hormone, prods cells to take up glucose.) Although you and other vertebrates also produce insulin molecules, the sequences in your versions are not quite the same as the ones in cattle. The overall sequence of amino acids is unique for each kind of protein and represents its *primary* structure.

A protein consists of one or more chains of amino acids.

The particular amino acids that follow one another in sequence in such chains are unique for each kind of protein and represent its primary structure.

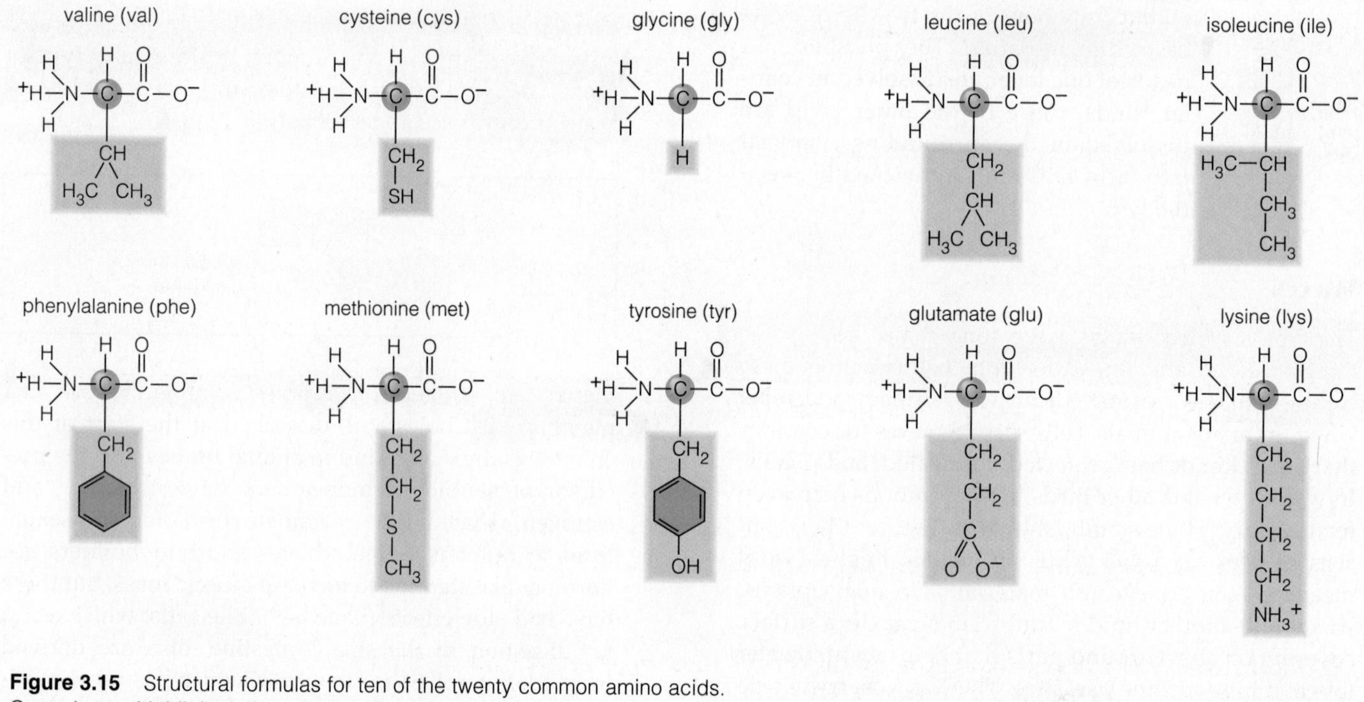

Figure 3.15 Structural formulas for ten of the twenty common amino acids. Green boxes highlight their particular R groups.

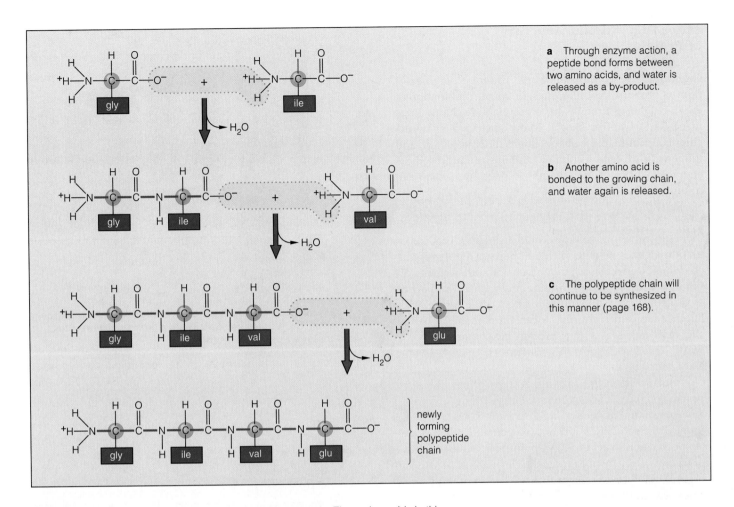

a Through enzyme action, a peptide bond forms between two amino acids, and water is released as a by-product.

b Another amino acid is bonded to the growing chain, and water again is released.

c The polypeptide chain will continue to be synthesized in this manner (page 168).

newly forming polypeptide chain

Figure 3.16 Peptide bond formation during protein synthesis. The amino acids in this example are the first four in the sequence for one of the two polypeptide chains that make up the protein insulin in cattle (Figure 3.17).

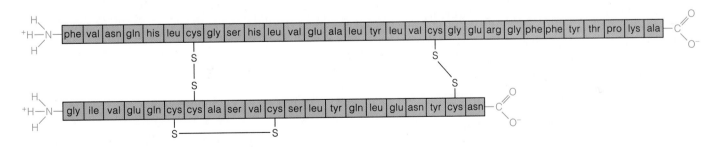

Figure 3.17 Linear sequence of amino acids in cattle insulin, as deduced by Frederick Sanger in 1953. This protein is composed of two polypeptide chains. Disulfide bridges (—S—S—), formed by condensation at two sulfhydryl groups, link the two chains together.

Three-Dimensional Structure of Proteins

Cells synthesize thousands of different proteins—fibrous proteins, globular ones, or some combination of the two. Fibrous proteins have polypeptide chains organized as strands or sheets. Many such protein molecules contribute to the shape, internal organization, and movement of cells. Globular proteins, including most enzymes, have their chains folded into compact, rounded shapes.

Each protein's shape and function arise from its primary structure—that is, from chemical information inherent in its unique amino acid sequence. That information dictates whether different parts of a polypeptide chain will fold, coil, or stretch out with other chains, in parallel array. A folded, coiled, or stretched-out part of a protein might allow the protein to interact with another molecule or a cell. A certain part might be the site that makes a certain protein an enzyme. In such ways, structure dictates the protein's chemical behavior and functions.

Primary structure influences a protein's shape in two major ways. *First*, it gives rise to patterns of hydrogen bonding between different amino acids in the sequence. Such bonds form at oxygen and other atoms of specific amino acids. *Second*, it puts R groups in positions that allow them to interact. Through these interactions, a polypeptide chain is forced to bend and twist into its three-dimensional shape.

Think of a polypeptide chain as a set of dominoes held together by links that can swivel a bit. Each domino corresponds to a peptide group in the chain, as shown here:

a peptide group

Such groups are rigidly oriented in the same plane. They cannot rotate around covalent bonds, owing to the way that electrons are being shared. But atoms on either side of these planes can rotate somewhat. And some can form bonds with neighboring atoms.

In many cases, hydrogen bonds form between every third amino acid, so that the "dominoes" become helically coiled (Figure 3.18*a*). The polypeptide chains of hemoglobin, an oxygen-carrying protein, have such coils. In other cases, the chain is extended, and hydrogen bonds hold two or more chains side by side in a sheetlike structure (Figure 3.18*b*). The proteins of silk are like this.

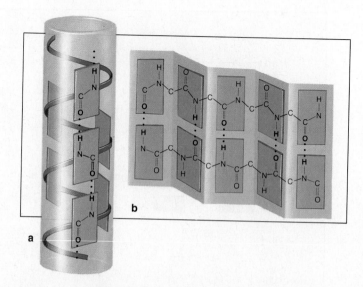

Figure 3.18 Hydrogen bonds (dotted lines) in a polypeptide chain. Such bonds can give rise to a coiled chain (**a**) or to a sheetlike array of chains (**b**).

Thus, proteins have *secondary* structure—a coiled or extended pattern, brought about by hydrogen bonds at regular intervals along polypeptide chains.

Most coiled chains continue to fold into distinctive shapes when one R group interacts with another R group some distance away, with the backbone of the chain itself, or with substances present in the cell. The folding that arises through interactions among the R groups of a polypeptide chain represents the *tertiary* structure of protein molecules. Figure 3.19*a* shows an example of this.

Finally, some proteins have *quaternary* structure, meaning that they incorporate two or more polypeptide chains. Hemoglobin is a good example of a globular protein. Keratin, the structural material of hair and fur, is an example of a fibrous one (Figure 3.20). Collagen, the most common animal protein, also is fibrous. Skin, bone, tendons, cartilage, blood vessels, heart valves, corneas—these and other structural components of the animal body depend on the strength inherent in collagen.

The amino acid sequence in polypeptide chains gives rise to the three-dimensional structure of proteins. That structure influences a protein's functions.

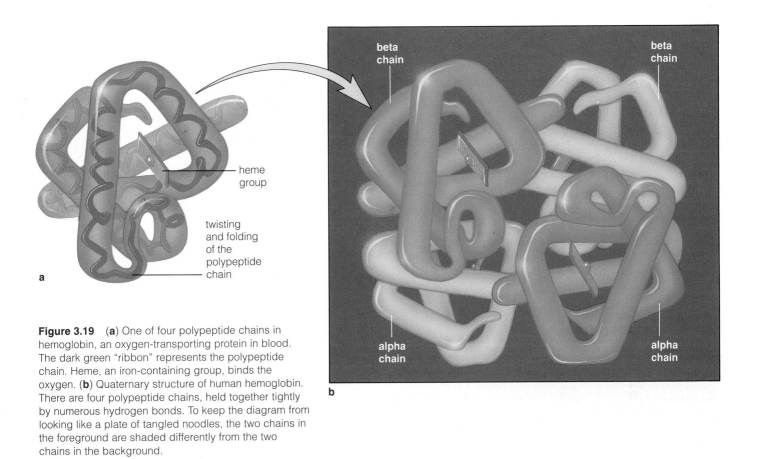

Figure 3.19 (**a**) One of four polypeptide chains in hemoglobin, an oxygen-transporting protein in blood. The dark green "ribbon" represents the polypeptide chain. Heme, an iron-containing group, binds the oxygen. (**b**) Quaternary structure of human hemoglobin. There are four polypeptide chains, held together tightly by numerous hydrogen bonds. To keep the diagram from looking like a plate of tangled noodles, the two chains in the foreground are shaded differently from the two chains in the background.

In Figure 3.19a: heme group; twisting and folding of the polypeptide chain; a

In Figure 3.19b: beta chain; beta chain; alpha chain; alpha chain; b

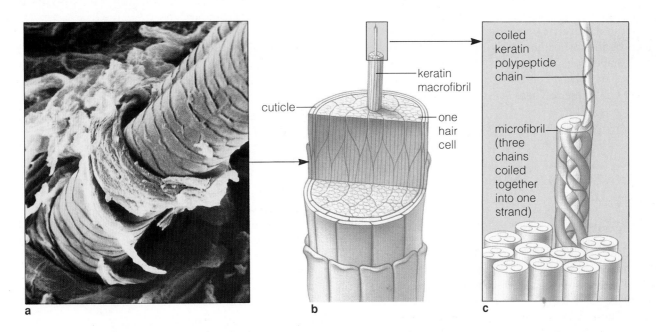

Labels: cuticle; keratin macrofibril; one hair cell; coiled keratin polypeptide chain; microfibril (three chains coiled together into one strand); a; b; c

Figure 3.20 Structure of hair. Polypeptide chains of the protein keratin are synthesized inside hair cells, which are derived from cells of the skin. The chains become organized into fine fibers (microfibrils), which become bundled together into larger, cablelike fibers (macrofibrils). These practically fill the cells, which eventually die. Dead, flattened cells form a tubelike cuticle around the developing hair shaft.

Lipoproteins and Glycoproteins

Many proteins combine with other types of molecules. Think back on the **lipoproteins**, described at the start of this chapter. They form when apolipoproteins that circulate freely in the bloodstream encounter and combine with cholesterol, triglycerides, and phospholipids that have been absorbed from the gut. Lipoproteins differ in their combinations of protein and lipid components.

By contrast, most **glycoproteins** form when newly synthesized polypeptide chains are undergoing modification into mature proteins. These are proteins to which oligosaccharides are covalently bonded. Some of the attached oligosaccharides are linear chains; others are branched. Nearly all proteins on the outer surface of animal cells are glycoproteins. So are most protein secretions from cells and most of the proteins in blood.

Protein Denaturation

If a protein, nucleic acid, or any other molecule loses its three-dimensional shape following the disruption of weak bonds, we call this event **denaturation**. For example, being individually weak, sensitive to heat, and sensitive to pH, hydrogen bonds that help hold a protein in its normal, three-dimensional shape can be disrupted. Then, its polypeptide chains unwind or change shape, and the protein can no longer function.

Consider the protein albumin, concentrated in the "egg white" of uncooked chicken eggs. When you cook an egg, the heat doesn't disrupt the strong covalent bonds of albumin's primary structure. But it destroys the weaker bonds contributing to albumin's three-dimensional shape. For some proteins, denaturation can be reversed when normal conditions are restored—but albumin isn't one of them. There is no way to uncook a cooked egg.

3.7 NUCLEOTIDES AND NUCLEIC ACIDS

Nucleotides with Key Roles in Metabolism

The small organic compounds called **nucleotides** have three parts: a five-carbon sugar, a phosphate group, and a nitrogen-containing base. The sugar is either ribose or deoxyribose. The base has either a single- or double-carbon ring structure (Figure 3.21).

One nucleotide is absolutely central to metabolism; we call it **ATP** (adenosine triphosphate). ATP can deliver energy from one reaction site to virtually any other reaction site in cells. Other kinds of nucleotides

function as **coenzymes**, or enzyme helpers. They accept hydrogen atoms and electrons that are being stripped from molecules and transfer them elsewhere. The coenzymes NAD^+ (nicotinamide adenine dinucleotide) and FAD (flavin adenine dinucleotide) are examples. Still other nucleotides function as **chemical messengers** within and between cells. You will encounter one of these (cyclic adenosine monophosphate, or cAMP) later in the book.

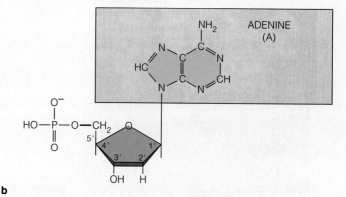

Figure 3.21 Examples of a nucleotide having (**a**) a single ring structure and (**b**) a double ring structure. (**c**) Pattern of covalent bonding between the nucleotides *within* a strand of RNA or DNA.

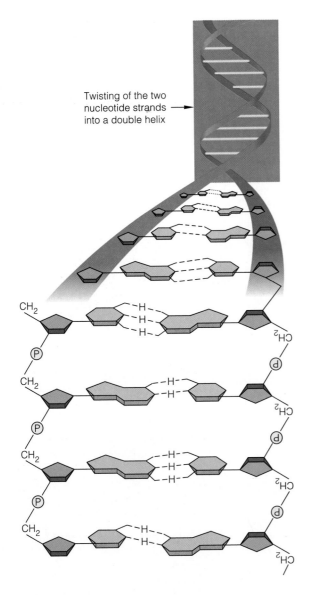

Twisting of the two nucleotide strands into a double helix

Figure 3.22 Pattern of hydrogen bonding *between* the two nucleotide strands of a DNA molecule. Hydrogen bonds connect bases of one strand with the bases of the other strand.

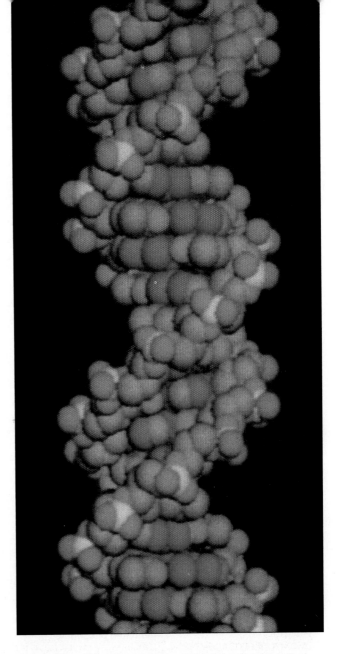

Figure 3.23 Computer-generated model showing the helical coiling of DNA, the double-stranded molecule that is central to maintaining and reproducing the cell.

Arrangement of Nucleotides in Nucleic Acids: DNA and RNA

In **nucleic acids**, four different kinds of nucleotides are strung together, forming large single- or double-stranded molecules. Each strand's backbone consists of joined-together sugars and phosphate groups of adjacent nucleotides. The nucleotide bases stick out to the side (Figure 3.21*b*). As is true of the amino acid sequence of proteins, the sequence of particular nucleotides is unique to each kind of nucleic acid.

You have probably heard of **RNA** (ribonucleic acid) as well as **DNA** (deoxyribonucleic acid). RNA is a sin-

gle nucleotide strand, most often. DNA is usually a double-stranded molecule that twists helically, like a spiral staircase, with hydrogen bonds holding the two strands together. Figures 3.22 and 3.23 show this helical configuration.

You will read more about RNA and DNA in later chapters. For now, it is enough to know this:

Genetic instructions are encoded in the sequence of bases in DNA.

RNA molecules function in the processes by which genetic instructions are used to build proteins.

Table 3.2 Summary of the Main Carbon Compounds in Living Things

Category	Main Subcategories	Some Examples and Their Functions	
CARBOHYDRATES *contain an aldehyde or a ketone group, and one or more hydroxyl groups*	**Monosaccharides** (simple sugars)	Glucose	Energy source
	Oligosaccharides	Sucrose (a disaccharide)	Form of sugar transported in plants
	Polysaccharides (complex carbohydrates)	Starch Cellulose	Energy storage Structural roles
LIPIDS *are largely hydrocarbon; generally do not dissolve in water but dissolve in nonpolar substances*	**Lipids with fatty acids:**		
	Glycerides: one, two, or three fatty acid tails attached to glycerol backbone	Fats (e.g., butter) Oils (e.g., corn oil)	Energy storage
	Phospholipids: phosphate group, another polar group, and (often) two fatty acids attached to glycerol backbone	Phosphatidylcholine	Key component of cell membranes
	Waxes: long-chain fatty acid tails attached to alcohol	Waxes in cutin	Water retention by plants
	Lipids with no fatty acids:		
	Sterols: four carbon rings; the number, position, and type of functional groups vary	Cholesterol	Component of animal cell membranes; can be converted into a variety of steroid molecules; precursor of vitamin D
PROTEINS *are polypeptides (up to several thousand amino acids, covalently linked)*	**Fibrous proteins:**		
	Individual polypeptide chains, often linked into tough, water-insoluble molecules	Keratin Collagen	Structural element of hair, nails Structural element of bones and cartilage
	Globular proteins:		
	One or more polypeptide chains folded and linked into globular shapes; many roles in cell activities	Enzymes Hemoglobin Insulin Antibodies	Increase in rates of reactions Oxygen transport Control of glucose metabolism Tissue defense
NUCLEIC ACIDS (AND NUCLEOTIDES) *are chains of units (or individual units) that each consist of a five-carbon sugar, phosphate, and a nitrogen-containing base*	**Adenosine phosphates**	ATP	Energy carrier
	Nucleotide coenzymes	NAD^+, $NADP^+$	Transport of protons (H^+) and electrons from one reaction site to another
	Nucleic acids:		
	Chains of thousands to millions of nucleotides	DNA, RNAs	Storage, transmission, translation of genetic information

SUMMARY

1. Organic compounds consist of carbon and one or more additional elements. Each carbon atom can form up to four covalent bonds with other atoms. Often they are bonded together in linear or ring structures that serve as the backbone of organic compounds.

2. A hydrocarbon has only hydrogen atoms bonded to the carbon backbone. Most organic compounds in cells are hydrocarbon derivatives, in that they also have functional groups covalently bonded to the backbone. Such groups are reactive arrangements of atoms that impart specific chemical properties.

3. Cells assemble, rearrange, and break apart most organic compounds by four kinds of enzyme-mediated reactions: functional-group transfers, electron transfers, internal rearrangements, condensation reactions, and cleavages (including hydrolysis).

4. Cells have pools of simple sugars, fatty acids, amino acids, and nucleotides, all of which are small organic compounds that include no more than twenty or so carbon atoms. They are building blocks for the large biological molecules—the polysaccharides, lipids, proteins, and nucleic acids.

5. The structure and function of proteins and nucleic acids arise from chemical information contained in the precise sequences of their subunits.

6. Table 3.2 summarizes the main categories of biological molecules described in this chapter. We will have occasion to return to the nature and roles of these molecules in diverse life processes.

Review Questions

1. Pantothenic acid is a type of vitamin. Identify the functional groups attached to its backbone: 35

2. Identify the carbohydrate, fatty acid, amino acid, and polypeptide in the following list: 42, 37, 40
   a. $^+NH_3$—CHR—COO^- c. (glycine)$_{20}$
   b. $C_6H_{12}O_6$ d. $CH_3(CH_2)_{16}COOH$

3. Explain how the amino acid sequence of one or more polypeptides gives rise to the three-dimensional structure and function of proteins. 44–45

4. Distinguish between the following:
   a. monosaccharide, polysaccharide 37–39
   b. peptide bond, polypeptide 42
   c. glycerol, fatty acid 40
   d. nucleotide, nucleic acid 46–47

5. A clerk in a health-food store tells you that certain "natural" vitamin C tablets extracted from rose hips are better for you than synthetic vitamin C tablets. Given your understanding of the structure of organic compounds, what would be your response?

Self-Quiz (Answers in Appendix IV)

1. The backbone of organic compounds is formed by the chemical bonding of _Carbon_ atoms into chains and rings.

2. A carbon atom can form up to _____ bonds with other atoms.
   a. four c. eight
   b. six d. sixteen

3. Four categories of large biological molecules are the _____ , _____ , _____ , and _____ .

4. All of the following *except* _____ are small organic compounds that serve as building blocks for large biological molecules or as energy sources.
   a. fatty acids d. nucleotides
   b. simple sugars e. amino acids
   c. triglycerides

5. Which of the following would *not* be included in the family of carbohydrates?
   a. glucose molecules c. waxes
   b. simple sugars d. polysaccharides

6. _____ molecules enhance the rate of specific reactions.
   a. DNA c. Steroid
   b. Amino acid d. Enzyme

7. Chemical information that is the basis of inheritance and cell reproduction resides in the structure of _____ .
   a. polysaccharides c. proteins
   b. DNA and RNA d. simple sugars

8. Match each molecule with the correct description.
   _c_ long sequence of amino acids a. carbohydrate
   _e_ the main energy carrier b. phospholipid
   _b_ glycerol, fatty acids, phosphate c. protein
   _d_ two strands of nucleotides d. DNA
   _a_ one or more sugar monomers e. ATP

Selected Key Terms

alcohol 35
amino acid 42
ATP 46
carbohydrate 37
chemical messenger 46
cleavage 36
coenzyme 46
condensation 36
denaturation 46
DNA 47
electron transfer 36
enzyme 36
fatty acid 40
functional group 35
functional-group transfer 36
glycoprotein 46
hydrocarbon 35

hydrolysis 36
lipid 40
lipoprotein 46
monosaccharide 37
nucleic acid 47
nucleotide 46
oligosaccharide 37
organic compound 34
phospholipid 41
polypeptide chain 42
polysaccharide 37
protein 42
rearrangement 36
RNA 47
sterol 41
triglyceride 40
wax 41

Readings

Goodsell, D. September–October 1992. "A Look Inside the Living Cell." *American Scientist.* 80:457–465. Current models of biological molecules.

Scientific American. October 1985. "The Molecules of Life." Entire issue devoted to biological molecules.

4 CELL STRUCTURE AND FUNCTION

Animalcules and Cells Fill'd With Juices

Early in the seventeenth century, Galileo Galilei arranged two glass lenses in a cylinder. With this instrument he happened to look at an insect, and afterward he described the stunning geometric patterns of its tiny eyes. Thus Galileo, who was not a biologist, was the first to record a biological observation made through a microscope. The study of the cellular basis of life was about to begin. First in Italy, then in France and England, biologists set out to explore a world whose existence had not even been suspected.

At midcentury Robert Hooke, Curator of Instruments for the Royal Society of England, was at the forefront of these studies. When Hooke first turned one of his microscopes to thinly sliced cork from a mature tree, he observed tiny compartments (Figure 4.1*a*). He gave them the Latin name *cellulae* (meaning small rooms); hence the origin of the biological term "cell." They were actually the walls of dead cells, which is what cork is made of, but Hooke did not think of them as being dead because he did not know cells could be alive. In other plant tissues, he discovered cells "fill'd with juices" but could not imagine what they represented.

Given the simplicity of their instruments, it is amazing that the pioneers in microscopy saw as much as they did. Antony van Leeuwenhoek, a Dutch shopkeeper, had great skill in constructing lenses and possibly the keenest vision (Figure 4.1*b*). By the late 1600s he was turning up wonders everywhere, including "many very small animalcules, the motions of which were very pleasing to behold," in scrapings of tartar from his teeth. He observed diverse protistans, sperm, even a bacterium—an organism so small it would not be seen again for another two centuries!

In the 1820s, improvements in lens design brought cells into sharper focus. Robert Brown, a botanist, noticed the constant presence of an opaque spot in egg cells, pollen cells, then cells of the growing tissues of orchid plants. He called the spot a "nucleus." In 1838 Matthias Schleiden, another botanist, suggested that the nucleus and cell development are closely related. On the basis of his own studies, Schleiden decided further that each plant cell leads a double life—one independent, pertaining to its own development, the other as an integral part of the plant.

By 1839, after years of studying the structure and growth of animal tissues, the zoologist Theodor Schwann had this to say: Animals as well as plants consist of cells and cell products—and even though the cells are part of a whole organism, they have, to some extent, an individual life of their own.

Yet a question remained: Where do cells come from? A decade later Rudolf Virchow, a physiologist, completed studies of cell growth and reproduction—that is, their division into two cells. Every cell, he concluded, comes from an already existing cell.

And so, by the middle of the nineteenth century, microscopic analysis yielded these insights: *the cell is the smallest unit of life—and the very continuity of life arises directly from the growth and division of single cells.* The insights still hold true.

This chapter provides an overview of our current understanding of cell structure and function. It introduces some of the modern microscopes that transport us ever more deeply into the spectacular worlds of juice-fill'd cells and animalcules. The chapter provides background information for answering a question that you might have asked of Virchow—*If every cell comes from a cell, then where did the very first cell come from?*—but that's another story.

Figure 4.1 Early glimpses into the world of cells. (**a**) Robert Hooke's compound microscope and his drawing of cell walls from cork tissue. (**b**) Antony van Leeuwenhoek, microscope in hand, and (**c**) one of his sketches of sperm cells. (**d**) Cartoon evidence of the startling impact of microscopic observations on nineteenth-century London.

a

b

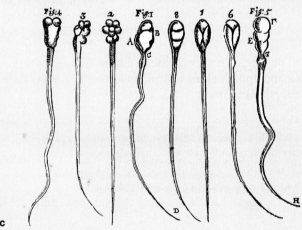

c

d

1. Cells are the smallest units that still retain the characteristics of life, including complex organization, metabolic activity, and reproductive behavior.

2. All cells have an outermost, plasma membrane that keeps their interior distinct from the surroundings. They have a region of DNA. They have cytoplasm, a region that is structurally and functionally organized for energy conversions, protein synthesis, cell movements, and other activities necessary for survival.

3. Eukaryotic cells contain a nucleus and other organelles (compartments bounded by cell membranes). Organelles separate different metabolic reactions and allow them to proceed in orderly fashion. Prokaryotic cells (bacteria) do not have comparable organelles.

4.1 THE CELL THEORY

Within your body and at its moist surfaces, trillions of cells live together in interdependency. In scummy pondwater, a single-celled amoeba thrives on its own. For humans, amoebas, and all other organisms, the **cell** is the smallest biological entity that still retains the characteristics of life. A cell either can survive on its own or has the potential to do so. It is structurally organized in specific ways, and it shows metabolic behavior. It senses and responds to specific changes in the surrounding environment. And its inherited DNA instructions give it the potential to reproduce.

Under the physical and chemical conditions that have prevailed on earth for about the past 2 billion years, living cells have not been able to arise spontaneously from nonliving matter. Only cells that already exist have been able to divide and give rise to new cells. The early microscopists sensed this when they observed cells growing and dividing in plant and animal tissues. They came up with the following generalizations, which constitute the **cell theory**:

All organisms are composed of one or more cells.

The cell is the basic unit of life.

New cells arise only from cells that already exist.

4.2 THE NATURE OF CELLS

Basic Aspects of Cell Structure and Function

Cells differ greatly in size, shape, and activities, as you might gather by comparing a bacterium with one of your liver cells. Yet they are alike in three respects. All cells start out life with a plasma membrane, a region of DNA, and a region of cytoplasm:

1. **Plasma membrane.** This outermost membrane maintains the cell as a distinct entity, apart from the environment, and allows metabolic events to proceed in organized, controlled ways. The plasma membrane does not *isolate* the cell interior; substances and signals continually move across it.

2. **DNA-containing region.** DNA occupies part of the cell interior, along with molecules that can copy or read its hereditary instructions.

3. **Cytoplasm.** The cytoplasm is everything enclosed by the plasma membrane, *except* for the region of DNA. It is a semifluid substance in which particles, filaments, and often membranous parts are organized.

This chapter introduces two fundamentally different kinds of cells. **Eukaryotic cells** contain distinctive arrays of organelles (sacs and other compartments formed by internal membranes). One organelle houses the DNA. It is the **nucleus**, and it is the key defining feature of eukaryotic cells:

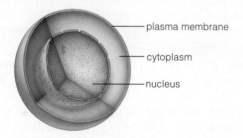

plasma membrane

cytoplasm

nucleus

By contrast, **prokaryotic cells** have no nucleus; no membranes intervene between the region of DNA and the surrounding cytoplasm. Bacteria are the only prokaryotic cells. Outside the realm of bacteria, all other organisms—from amoebas to peach trees and puffball mushrooms to zebras—are eukaryotes.

Structure and Functions of Cell Membranes

Two thin sheets of lipid molecules serve as the structural framework for cell membranes. Figure 4.2 shows this "lipid bilayer" arrangement. The bilayer of a plasma membrane is a continuous boundary that bars the free passage of water-soluble substances into and out of the cell. Within many cell types, membrane bilayers also subdivide the cytoplasm into compartments in which specific substances accumulate.

Proteins, positioned in the lipid bilayer or at its surfaces, carry out most membrane functions. Some of the proteins are passive channels for water-soluble substances. Others carry electrons or pump substances across the bilayer. The receptor types latch onto hormones and other substances that trigger alterations in cell activities. In multicelled organisms, certain proteins even help cells recognize each other and stick together in a tissue.

The next chapter provides a closer look at the structure and functions of cell membranes. For now, keep these points in mind:

A lipid bilayer imparts structure to cell membranes and serves as a barrier to water-soluble substances.

Proteins embedded in the bilayer or positioned at its surfaces carry out most membrane functions.

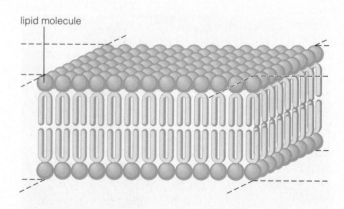

lipid molecule

Figure 4.2 The arrangement of lipid molecules in a layer two molecules in thickness. The double layer provides the framework of all biological membranes.

Surface-to-Volume Constraints on the Size and Shape of Cells

Can any cell be seen with the unaided human eye? There are a few, including the "yolks" of bird eggs, cells in the red part of a watermelon, and the fish eggs we call caviar. Generally, however, cells are too small to be observed without microscopes (see Figure 4.3 and the *Focus* essay on the next page). To give you a sense of cell sizes, one of your red blood cells is only about 8 *millionths* of a meter in diameter. You could fit a string of about 2,000 of them across your thumbnail!

Why are most cells so small? A physical relationship, the **surface-to-volume ratio**, constrains increases in cell size. By this relationship, an object's volume increases with the cube of the diameter, but its surface area increases with the square (Figure 4.4). Simply put, *when a cell expands in diameter, its volume increases more rapidly than its surface area does.*

Suppose we figure out a way to make a round cell grow four times wider. Its volume increases sixty-four times (4^3) and its surface area increases sixteen times (4^2). Unlike fat cells and chicken eggs (which are chockful of fat, food, and so on), our expanded cell is chockful of cytoplasmic machinery. Unfortunately, each unit of plasma membrane must serve four times as much cytoplasm as before! Past a certain point, the inward flow of nutrients and the outward flow of wastes will not be fast enough, and the cell will die.

A very large, round cell also would have trouble moving nutrients and wastes *through* the cytoplasm. By contrast, the random, tiny motions of molecules easily distribute substances through small or skinny cells. If a cell isn't small, it probably is long and thin, or has outfoldings and infoldings that increase its surface relative to its volume. *The smaller or more stretched out or frilly-surfaced the cell, the more efficiently materials can cross its surface and become distributed through the interior.*

We also see the influence of surface-to-volume constraints in the body plans of multicelled organisms. For example, in some algae, cells are attached end to end, forming delicate strands. The arrangement allows each cell to interact directly with the environment. Other algae and a few protistans are sheetlike, with all cells at or near the body surface.

Complex plants and animals also provide evidence of surface-to-volume constraints. For example, they have transport systems that move materials to and from millions, billions, or trillions of cells packed together in tissues. That is the point of having an incessantly pumping heart and an elaborate network of blood vessels inside your own body. This efficient circulatory system quickly delivers materials from the environment to all tissues and sweeps away wastes. Its "highways" cut through the volume of tissue and so shrink the distance to and from individual cells.

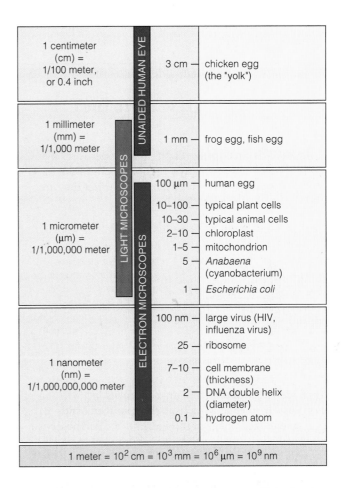

1 centimeter (cm) = 1/100 meter, or 0.4 inch	3 cm —	chicken egg (the "yolk")
1 millimeter (mm) = 1/1,000 meter	1 mm —	frog egg, fish egg
1 micrometer (µm) = 1/1,000,000 meter	100 µm —	human egg
	10–100 —	typical plant cells
	10–30 —	typical animal cells
	2–10 —	chloroplast
	1–5 —	mitochondrion
	5 —	*Anabaena* (cyanobacterium)
	1 —	*Escherichia coli*
1 nanometer (nm) = 1/1,000,000,000 meter	100 nm —	large virus (HIV, influenza virus)
	25 —	ribosome
	7–10 —	cell membrane (thickness)
	2 —	DNA double helix (diameter)
	0.1 —	hydrogen atom

Vertical bands: UNAIDED HUMAN EYE, LIGHT MICROSCOPES, ELECTRON MICROSCOPES

$$1 \text{ meter} = 10^2 \text{ cm} = 10^3 \text{ mm} = 10^6 \text{ µm} = 10^9 \text{ nm}$$

Figure 4.3 Units of measure used in microscopy. Biologists use the micrometer when describing whole cells or large cell structures. They use the nanometer when describing smaller cell structures and large organic molecules.

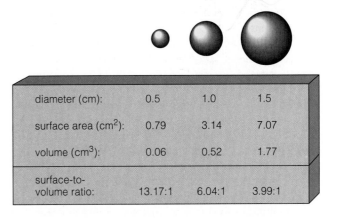

diameter (cm):	0.5	1.0	1.5
surface area (cm^2):	0.79	3.14	7.07
volume (cm^3):	0.06	0.52	1.77
surface-to-volume ratio:	13.17:1	6.04:1	3.99:1

Figure 4.4 Relationship between surface area and volume when a sphere is enlarged. As the diameter increases, the volume increases more rapidly than the surface area does.

Microscopes—Gateways to the Cell

Light Microscopes Light microscopes work by bending (refracting) light rays. Light rays pass straight through the center of a curved lens. The farther they are from the center, the more they bend:

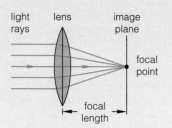

light rays / lens / image plane / focal point / focal length

The angle at which a ray of light enters a glass lens and the molecular structure of the glass dictate the extent to which the ray will bend

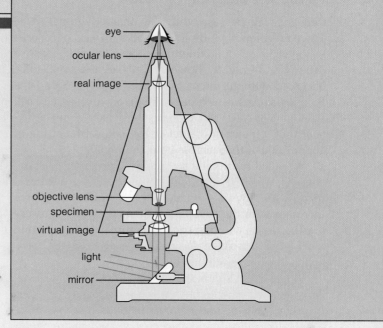

eye / ocular lens / real image / objective lens / specimen / virtual image / light / mirror

a Light microscope

When viewing a specimen with a *compound light microscope*, light emanating from the specimen bends to form an enlarged image of it (Figure *a*). You can observe a living cell if it is small or thin enough for light to pass through. Its internal structures will be visible only if they differ in color and contrast from the surroundings. Because most cell structures are nearly colorless and optically uniform in density, you may wish to stain the specimen (expose it to dyes that react with some cell structures but not others). Bear in mind, staining usually alters the structures and kills the cells. Also, dead cells begin to break down at once, so they must be preserved (fixed) before staining. Most micrographs show dead, fixed, or stained cells.

A **micrograph** is simply a photograph of an image formed with a microscope. Fine cell structures are the first to fall apart when cells die, so some of the structures that show up in micrographs may be artifacts—they are not really present in cells.

If living cells are mostly transparent, you can observe them with a *phase-contrast microscope*. This microscope converts small differences in the way different structures bend light to larger variations in brightness.

Suppose you keep magnifying a cell with a light microscope until its image diameter is 2,000 times larger. Cell structures are no longer clear. Why? Think of what happens when you hold a magnifying glass close to a newspaper photograph. You see only black dots. You cannot see a detail as small as or smaller than a dot; the dot covers it up. In microscopy, something like dot size intervenes to limit resolution—the property that determines whether small objects close together can be discerned as separate entities. That limiting factor is the physical size of wavelengths of visible light.

Light travels as different **wavelengths**, which correspond to different colors. Imagine a train of waves moving across an ocean. Each "wavelength" is the distance from one wave's peak to the peak of the wave behind it. For a given wavelength of light, that distance remains constant. It is about 750 nanometers for red wavelengths and 400 nanometers for violet; all other colors fall in between. Suppose a cell structure is less than one-half of a wavelength long. Light rays passing by it will overlap so much that the object won't be visible. The best light microscopes resolve detail only to about 200 nanometers.

Transmission Electron Microscopes Electrons are particles that also behave like waves. Electron microscopes employ electrons with wavelengths of about 0.005 nanometer—about 100,000 times shorter than those of visible light! Ordinary glass lenses would scatter rather than focus such accelerated streams of electrons. But the electric charge of each electron responds to the force of a magnetic field. In *transmission electron microscopes*, a magnetic field acts as a lens; it diverts electrons along defined paths and channels them to a focal point (Figure *b*).

Electrons must travel in a vacuum (molecules in the air would randomly scatter them). Cells can't live in a vacuum, so they don't stay alive in a transmission electron microscope. To observe fine details, you must slice cells extremely thin. Then, electrons will be scattered in patterns that correspond to the density of different structures. Dense cell parts will be the darkest areas in the final image formed. Most cell structures are somewhat transparent to

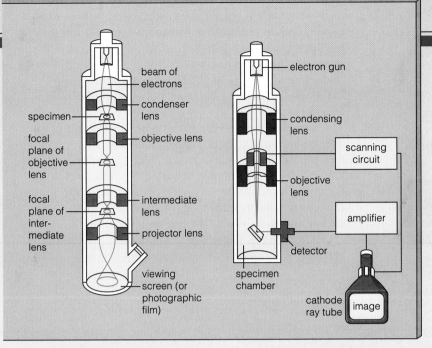

b Transmission electron microscope   **c** Scanning electron microscope

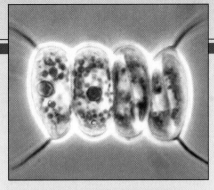

d Light micrograph (phase-contrast)

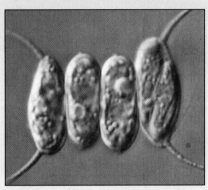

e Light micrograph (Nomarski process)

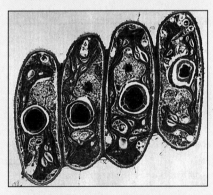

f Transmission electron micrograph, thin section

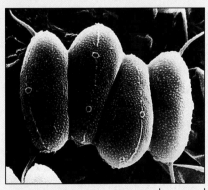

g Scanning electron micrograph

electrons, so they must first be stained with heavy metal "dyes."

Compared to standard electron microscopes, *high-voltage* electron microscopes excite electrons until they are ten times more energetic. The energy boost allows electrons to penetrate intact cells several micrometers thick. Like an x-ray plate, the resulting image shows some of the internal organization of cells.

Scanning Electron Microscopes   A narrow beam of electrons is used in *scanning electron microscopes* (Figure *c*). The beam moves back and forth across a specimen's surface, which has been coated with a thin metal layer. The metal responds by emitting some of its own electrons. Equipment similar to a television camera detects the emission patterns, and an image is formed. Scanning electron microscopy does not approach the high resolution of transmission instruments. However, its images have fantastic depth.

We conclude this essay with a comparison of how different types of microscopes reveal different aspects of the same organism—*Scenedesmus*, a green alga. All four specimens in Figures *d* through *g* are at the same magnification. The usefulness of light micrographs has been enhanced by phase-contrast and Nomarski processes, which create optical contrasts without staining the cells.

As for other micrographs in the book, the short bar below the micrograph in Figure *g* provides a reference for size. A micrometer (µm) is 1/1,000,000 of a meter.

4.3 PROKARYOTIC CELLS— THE BACTERIA

We turn now to specific cell types, starting with bacteria. All bacterial cells are prokaryotic; their DNA is not enclosed in a nucleus. *Prokaryotic* means "before the nucleus." The word implies that bacteria existed on earth before the nucleus evolved in the forerunners of all other cells.

Bacteria are the smallest cells. Usually they are not much more than a micrometer wide, and even the rod-shaped ones aren't much more than a few micrometers long. Many have one or more long, threadlike motile structures that extend from the cell surface, as shown in Figure 4.5a. These structures are called **bacterial flagella** (singular, flagellum). They permit rapid movements through fluid environments.

In structural terms, bacteria are the simplest cells to think about. Most types have a continuous, semirigid or rigid **cell wall** around the plasma membrane. The wall supports the cell and imparts shape to it (Figure 4.5b). Often, polysaccharides cover the wall and help its owner attach to rocks, teeth, and other surfaces. In many disease-causing bacteria, polysaccharides form a jellylike capsule that deters counterattacks.

The plasma membrane controls the movement of substances into and out of the cytoplasm. It has receptors for molecules that activate built-in machinery for key metabolic reactions, including the breakdown of energy-rich molecules. Photosynthetic types have special clusters of membrane proteins that harness light energy and convert it to the chemical energy of ATP.

Bacterial cells have a small volume of cytoplasm. Suspended within the cytoplasm is an irregularly

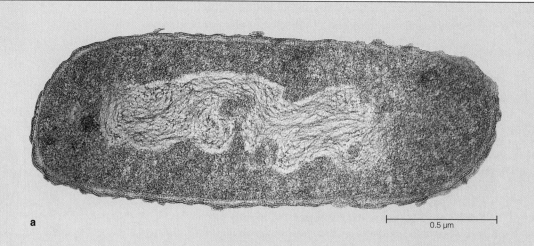

a

0.5 µm

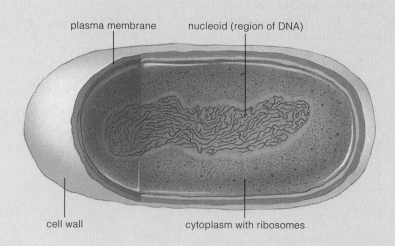

plasma membrane   nucleoid (region of DNA)

cell wall   cytoplasm with ribosomes

Figure 4.5 Prokaryotic body plans. (**a**) Micrograph and sketch of a common bacterium, *Escherichia coli*. (**b**) This *E. coli* cell has been shocked into releasing its circular molecule of DNA.

Your own gut is home to a large population of a normally harmless strain of *E. coli*. In 1993, a dangerous strain contaminated meat that was sold to some fast-food restaurants. The same strain also contaminated hard apple cider sold at a few roadside stands. Cooking the meat thoroughly (or boiling the cider) would have killed the bacterial cells. Where this was not done, people who ate the meat or drank the cider became quite sick, and some died.

shaped region of DNA. No cell membranes surround this region, which sometimes is called a nucleoid. It contains a single, circular DNA molecule. In the best-studied bacterium, *Escherichia coli*, the circle of DNA is about 1,500 µm long.

The cytoplasm often is densely stained, owing to the presence of large numbers of ribosomes. A **ribosome** consists of two subunits, each composed of RNA and protein molecules. Each is about 20 to 30 nanometers in diameter. *In all living cells, not just bacteria, proteins are synthesized at ribosomes located in the cytoplasm.* At the ribosomal surface, enzymes speed the assembly of polypeptide chains. Each new protein consists of one or more of those chains (page 44).

Figure 4.6 shows only a few representative bacteria. As a group, bacteria are the most metabolically diverse of all organisms. They have managed to exploit energy and raw materials in just about every kind of environment. Besides this, ancient members of their kingdom gave rise to all the protistans, plants, fungi, and animals ever to appear on earth.

The evolution, structure, and functioning of bacteria are topics that will be addressed later in the book. In the pages that follow, our focus will be on nucleated cells, the eukaryotes.

Bacteria alone are prokaryotic cells; their DNA is not confined in a membranous envelope. Their cytoplasm contains many ribosomes and is enclosed in a plasma membrane, which in most species is enclosed within a cell wall.

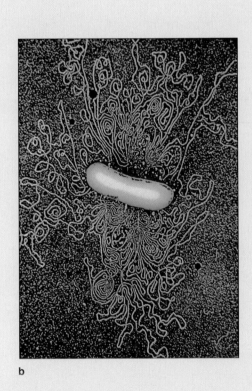

b

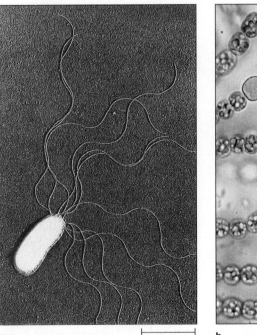

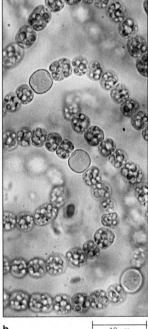

a |—— 1 µm ——| b |—— 10 µm ——|

Figure 4.6 Interesting variations on the bacterial body plan. (**a**) Like the *Pseudomonas marginalis* cell shown here, many species have surface appendages such as bacterial flagella, which propel the cell through fluid environments. (**b**) Cells of assorted species are shaped like rods, corkscrews, or balls. The ball-shaped cells of *Nostoc*, a photosynthetic bacterium, stick together inside a thick, gelatin-like sheath. Chapter 22 gives other splendid examples.

4.4 EUKARYOTIC CELLS

Functions of Organelles

By definition, an **organelle** is an internal, membrane-bounded sac or some other kind of compartment that has a specific metabolic function within a cell. As Robert Brown perceived when he tracked that opaque spot through generations of orchid cells, the nucleus is a most conspicuous organelle. The cells that have it are *eukaryotic* (which means "true nucleus").

No human-built apparatus matches the eukaryotic cell for the sheer number of chemical activities that can proceed simultaneously in so small a space. For example, many metabolic reactions are incompatible with others. Think about a plant cell putting together a starch molecule by some reactions and breaking it down by others. The cell would gain nothing if the synthesis and breakdown reactions proceeded at the same time on the same starch molecule! Such reactions can proceed smoothly at the same time, largely because organelle membranes keep them physically separated.

Organelle membranes also permit compatible, interconnected reactions to proceed at different times. Thus starch molecules are produced and stored by reactions in one organelle—then later released for use in other reactions in the same plant cell.

Organelles physically separate chemical reactions, many of which are incompatible, inside the cell.

Organelles separate different reactions in time, as when molecules are produced in one organelle, then used later in other reaction sequences.

Organelles Characteristic of Plants

Figure 4.7 can start you thinking about the location of organelles in a typical plant cell. Keep in mind that calling this cell "typical" is like calling a cactus or a crocus a "typical" plant. As is true of animal cells, variations on the basic plan are mind-boggling. With this qualification in mind, also take a look at Figure 4.8. The sketches accompanying the micrograph outline the functions of organelles and other structures that you are likely to find in the plant kingdom.

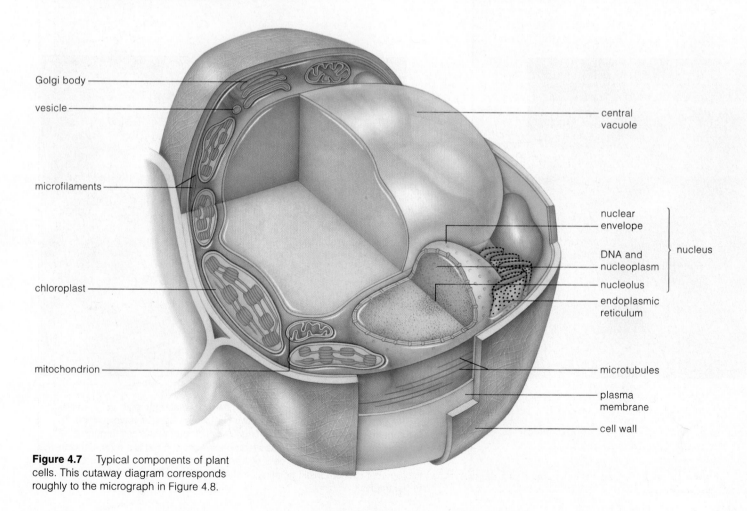

Golgi body

vesicle

microfilaments

chloroplast

mitochondrion

central vacuole

nuclear envelope

DNA and nucleoplasm

nucleolus

endoplasmic reticulum

nucleus

microtubules

plasma membrane

cell wall

Figure 4.7 Typical components of plant cells. This cutaway diagram corresponds roughly to the micrograph in Figure 4.8.

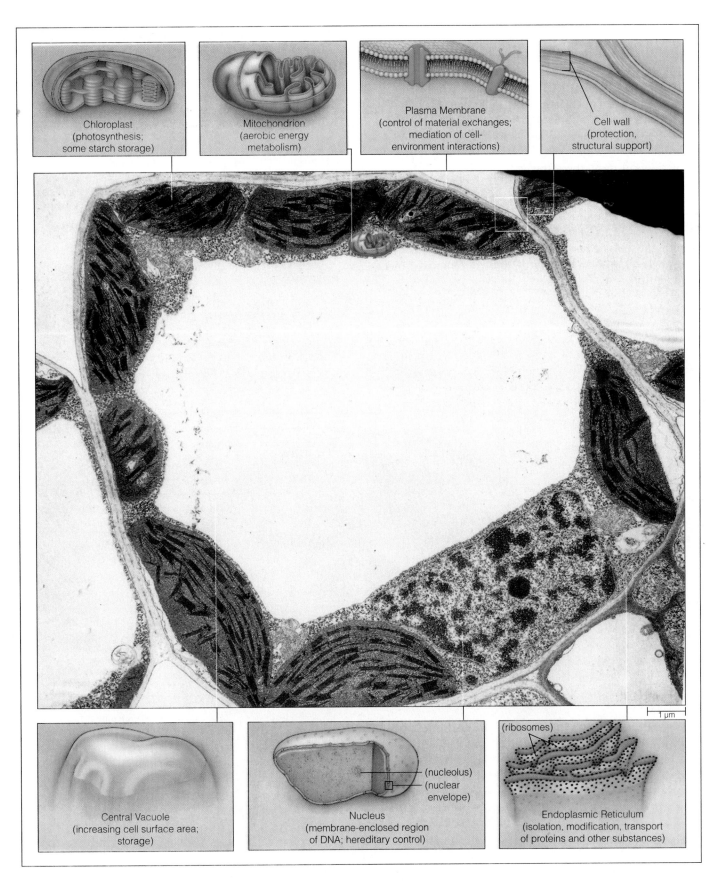

Chloroplast
(photosynthesis;
some starch storage)

Mitochondrion
(aerobic energy
metabolism)

Plasma Membrane
(control of material exchanges;
mediation of cell-
environment interactions)

Cell wall
(protection,
structural support)

1 µm

Central Vacuole
(increasing cell surface area;
storage)

Nucleus
(membrane-enclosed region
of DNA; hereditary control)

(nucleolus)
(nuclear
envelope)

(ribosomes)

Endoplasmic Reticulum
(isolation, modification, transport
of proteins and other substances)

Figure 4.8   Transmission electron micrograph of a plant cell, cross-section, from a blade of Timothy grass.

Organelles Characteristic of Animals

Now take a look at the organelles of a typical animal cell, as shown in Figures 4.9 and 4.10. Right away, you can see that it is similar to the plant cell illustrated earlier, in that it has a nucleus, mitochondria, and the other components listed in Table 4.1. Such similarities point to basic functions that are necessary for survival and reproduction, regardless of the cell type. We will return to this concept throughout the book.

Comparisons of Figures 4.7 through 4.10 also give you an initial idea of the ways in which plant and animal cells are structurally different. For example, you won't ever see an animal cell surrounded by a cell wall. (You might see assorted fungal and protistan cells with one, however.) What other structural differences can you identify?

Table 4.1	Features Typical of Most Eukaryotic Cells
Nucleus	Physical isolation and organization of DNA
Ribosomes	Synthesis of polypeptide chains
Endoplasmic reticulum (ER)	Initial modification of new polypeptide chains; lipid synthesis
Golgi bodies	Further modification of polypeptide chains into mature proteins; sorting, shipping of proteins and lipids for secretion or use in cell
Diverse vesicles	Transport or storage of substances; digestion inside cell; other functions
Mitochondria	Efficient ATP formation
Cytoskeleton	Cell movement, shape, internal organization

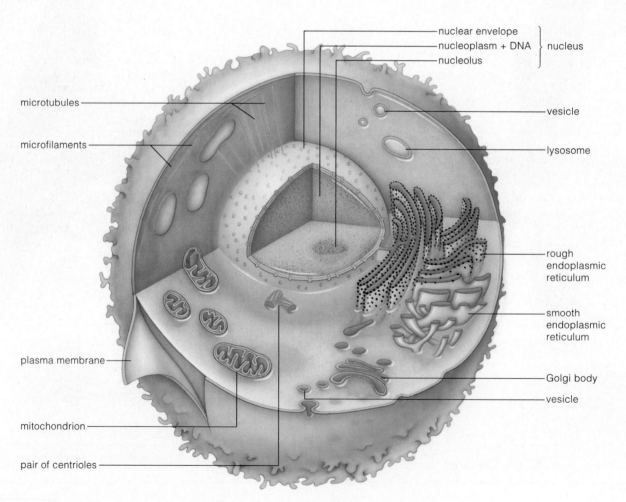

Figure 4.9 Typical components of animal cells. This cutaway diagram corresponds roughly to the micrograph in Figure 4.10.

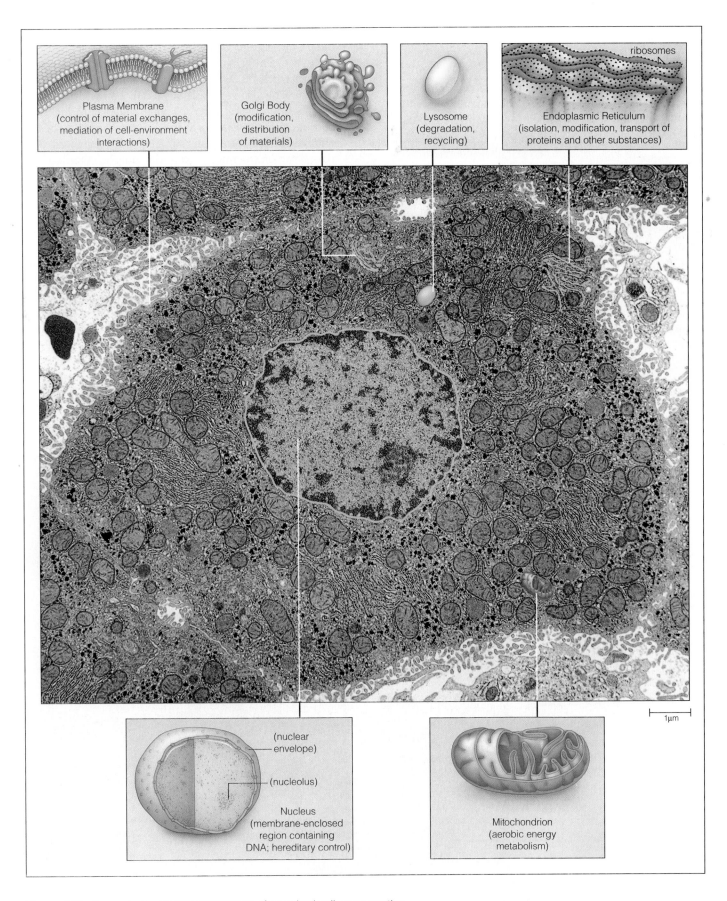

Plasma Membrane (control of material exchanges, mediation of cell-environment interactions)

Golgi Body (modification, distribution of materials)

Lysosome (degradation, recycling)

ribosomes

Endoplasmic Reticulum (isolation, modification, transport of proteins and other substances)

1μm

(nuclear envelope)

(nucleolus)

Nucleus (membrane-enclosed region containing DNA; hereditary control)

Mitochondrion (aerobic energy metabolism)

Figure 4.10 Transmission electron micrograph of an animal cell, cross-section. This cell is from a rat liver.

4.5 THE NUCLEUS

There would be no cells whatsoever without complex carbohydrates, lipids, proteins, and nucleic acids. It takes a special class of proteins—enzymes—to build and use those molecules. Thus, *cell structure and function begin with proteins—and instructions for building the proteins themselves are contained in DNA.*

Unlike bacteria, eukaryotic cells have their hereditary instructions distributed in several to many DNA molecules of various lengths. Your own body cells, for example, contain forty-six DNA molecules. Stretched out end to end, they would be about 1 meter long. The DNA in frog cells would be 10 meters, end to end. Compared to the single molecule in bacteria, that's a lot of DNA!

Eukaryotic DNA resides in the nucleus. This type of organelle has a characteristic structure, shown by the example in Figure 4.11, and it serves two functions. First, the nucleus sequesters all of the DNA molecules, out of the way of the complex metabolic machinery in the cytoplasm. This makes it easier to sort out hereditary instructions when the time comes for a cell to divide. The DNA molecules can be assorted into parcels—one parcel for each new cell that forms. Second, membranes that form the outermost portion of the nucleus help control the signals and substances that pass between the nuclear material and the cytoplasm.

Nucleolus

You may be wondering about the dense, globular mass of material within the nucleus shown in Figure 4.11. As eukaryotic cells grow, one or more of these appear in the nucleus. Each mass is a **nucleolus** (plural, nucleoli). Nucleoli are sites where the proteins and RNA subunits of ribosomes are assembled. These subunits are shipped out of the nucleus, into the cytoplasm. When proteins are about to be synthesized, they join together into intact, functional ribosomes.

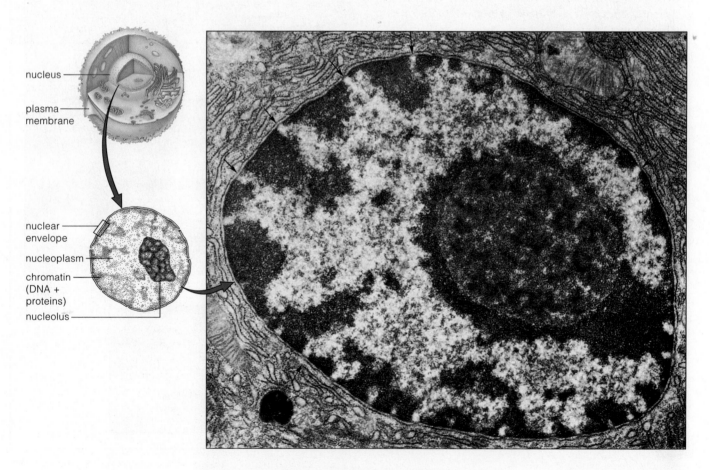

nucleus

plasma membrane

nuclear envelope

nucleoplasm

chromatin (DNA + proteins)

nucleolus

Figure 4.11 Transmission electron micrograph of the nucleus from a pancreatic cell, thin section. Arrows point to pores in its nuclear envelope.

Nuclear Envelope

Notice, in Figures 4.11 and 4.12, that the outermost part of the nucleus has *two* lipid bilayers, one wrapped around the other. They completely surround the nucleoplasm, which is the fluid portion of the nucleus. This double-membrane system is the **nuclear envelope**. As is true of all cell membranes, its lipid bilayers act as a barrier to water-soluble substances. In this case, clusters of membrane proteins span both bilayers, forming pore complexes (Figure 4.12). The pore complexes allow ions and small molecules to enter and leave the nucleus—but they selectively control the passage of larger molecules and particles. For example, they permit ribosomal subunits to pass into the cytoplasm, but they restrict enzymes of the nucleus and the cytoplasm to their respective "compartments" in the cell.

The innermost surface of the nuclear envelope has attachment sites for tiny filaments that help organize the DNA. Attached to the outermost surface are ribosomes, which take part in protein synthesis.

Chromosomes

Between cell divisions, eukaryotic DNA is threadlike, with many enzymes and other proteins attached to it like beads on a string. Except at extreme magnification, the beaded threads look grainy, as they do in Figure 4.11. Prior to division, however, DNA molecules are duplicated (so each new cell will get a set of hereditary instructions). Before the molecules are sorted into two sets, they fold and twist into condensed structures, proteins and all.

Early microscopists named the grainy material *chromatin* and the condensed structures *chromosomes*. Today, we define **chromatin** as the total collection of DNA molecules and associated proteins in the nucleus. We define a **chromosome** as an individual DNA molecule and its associated proteins, regardless of whether it is in threadlike or condensed form.

The point is, the "chromosome" does not always look the same during the life of a eukaryotic cell. We will consider different aspects of chromosomes in chapters to come. It will help to keep this point, and the following sketch, in mind:

| unduplicated, uncondensed chromosome (a DNA double helix + proteins) | duplicated, uncondensed chromosome (two DNA double helices + proteins) | duplicated, condensed chromosome |

Table 4.2 summarizes the components of the nucleus.

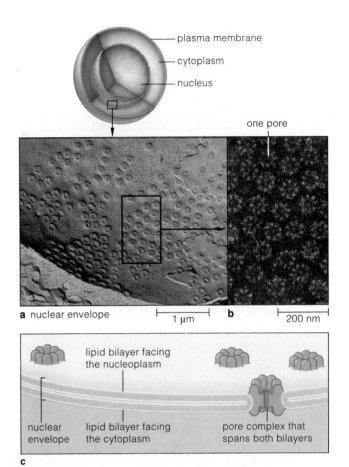

Figure 4.12 (**a**) Surface view of a nuclear envelope, which consists of two pore-studded lipid bilayers. This specimen was deliberately fractured for microscopy, by the method described on page 80. (**b**) Closer look at the pores, each a protein complex that permits the selective transport of larger molecules into and out of the nucleus. (**c**) Diagram of the nuclear envelope's structure.

Table 4.2	Components of the Nucleus
Nuclear envelope	Pore-riddled double membrane system that selectively controls passage of substances into and out of the nucleus
Nucleolus	Dense cluster of RNA and proteins used in the assembly of ribosomal subunits
Nucleoplasm	Fluid portion of interior of nucleus
Chromosome	One DNA molecule and the proteins that are associated with it
Chromatin	Total collection of all DNA molecules and their associated proteins in the nucleus

4.6 CYTOMEMBRANE SYSTEM

In eukaryotic cells, enzymes assemble polypeptide chains on the many thousands of ribosomes that are present in the cytoplasm. What happens to the new chains? Many become stockpiled in the cytoplasm. Others pass through a series of organelles called the **cytomembrane system**. These include the endoplasmic reticulum, Golgi bodies, and certain vesicles (Figure 4.13). In this system, lipids as well as proteins take on final form, then become packaged in vesicles.

Some vesicles deliver proteins and lipids to regions where new membrane must be built. Others accumulate and store proteins or lipids for specific uses. Still other vesicles are part of a "secretory pathway." They move to the plasma membrane, fuse with it, and release their contents to the surroundings. Figure 4.13 shows where the secretory pathway fits in the system.

Figure 4.13 (**a**) Destinations of the proteins and lipids that are modified or assembled in this system, then packaged and shipped out of it. (**b**) The cytomembrane system.

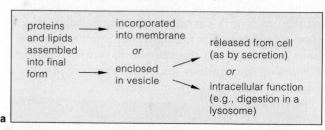

a

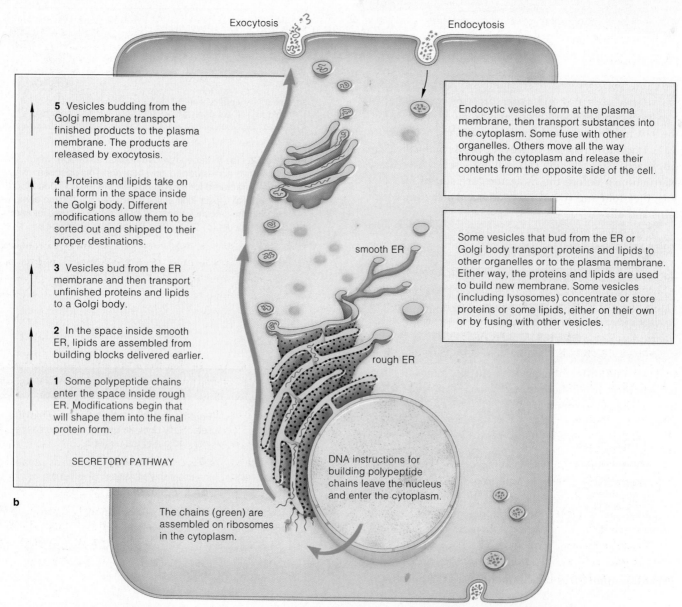

Exocytosis   Endocytosis

5 Vesicles budding from the Golgi membrane transport finished products to the plasma membrane. The products are released by exocytosis.

4 Proteins and lipids take on final form in the space inside the Golgi body. Different modifications allow them to be sorted out and shipped to their proper destinations.

3 Vesicles bud from the ER membrane and then transport unfinished proteins and lipids to a Golgi body.

2 In the space inside smooth ER, lipids are assembled from building blocks delivered earlier.

1 Some polypeptide chains enter the space inside rough ER. Modifications begin that will shape them into the final protein form.

SECRETORY PATHWAY

Endocytic vesicles form at the plasma membrane, then transport substances into the cytoplasm. Some fuse with other organelles. Others move all the way through the cytoplasm and release their contents from the opposite side of the cell.

Some vesicles that bud from the ER or Golgi body transport proteins and lipids to other organelles or to the plasma membrane. Either way, the proteins and lipids are used to build new membrane. Some vesicles (including lysosomes) concentrate or store proteins or some lipids, either on their own or by fusing with other vesicles.

smooth ER

rough ER

DNA instructions for building polypeptide chains leave the nucleus and enter the cytoplasm.

The chains (green) are assembled on ribosomes in the cytoplasm.

b

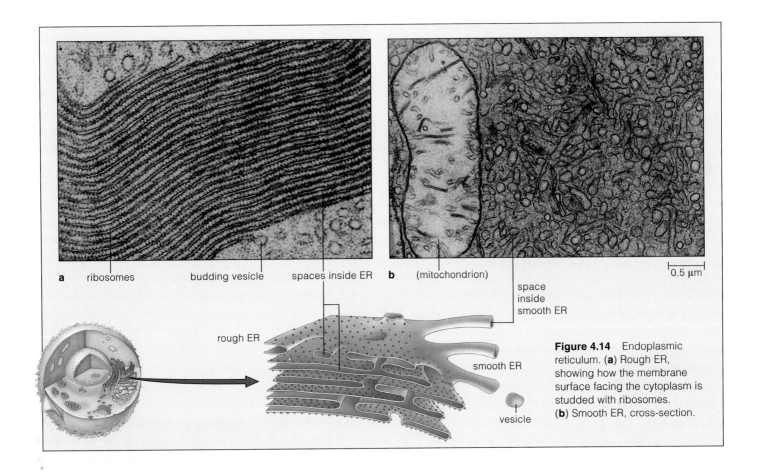

a ribosomes   budding vesicle   spaces inside ER   **b** (mitochondrion)   0.5 μm

rough ER

space inside smooth ER

smooth ER

vesicle

Figure 4.14 Endoplasmic reticulum. (**a**) Rough ER, showing how the membrane surface facing the cytoplasm is studded with ribosomes. (**b**) Smooth ER, cross-section.

Endoplasmic Reticulum

In animal cells, **endoplasmic reticulum**, or **ER**, is a membrane that begins at the nucleus and curves through the cytoplasm. It has rough and smooth regions, due largely to the presence or absence of ribosomes on the side facing the cytoplasm.

Rough ER often is arranged as stacked, flattened sacs and has many ribosomes attached (Figure 4.14*a*). Polypeptide chains are assembled on the ribosomes. Only the newly forming chains that incorporate a "signal" of about fifteen to twenty specific amino acids can enter spaces inside rough ER. Once inside, some acquire side chains (oligosaccharides, mostly).

Many kinds of cells specialize in secreting proteins, and rough ER is notably abundant in them. In your own body, for example, some ER-rich cells of the pancreas produce and secrete specific enzymes that end up in your small intestine, where they function in digestion.

Smooth ER is free of ribosomes and curves through the cytoplasm like connecting pipes (Figure 4.14*b*). It is the main site of lipid synthesis in many cells. Smooth ER is highly developed in seeds. The smooth ER of liver cells inactivates certain drugs and harmful by-products of metabolism. Sarcoplasmic reticulum, a type of smooth ER in skeletal muscle cells, stores and releases calcium ions for muscle contraction.

Peroxisomes

Among the vesicles that form in cells are **peroxisomes**. These sacs of enzymes have a specialized metabolic role: they break down fatty acids and amino acids. The reactions produce hydrogen peroxide, a potentially harmful substance. Before hydrogen peroxide can do harm, another enzyme converts it to water and oxygen or uses it to break down alcohol. If you drink alcohol, nearly half of it is degraded in peroxisomes of your liver and kidney cells.

Similar vesicles, called glyoxysomes, form in plant cells. Their enzymes help convert stored fats and oils to the sugars necessary for rapid, early growth. Glyoxysomes are abundant in lipid-rich seeds, such as those of peanut plants.

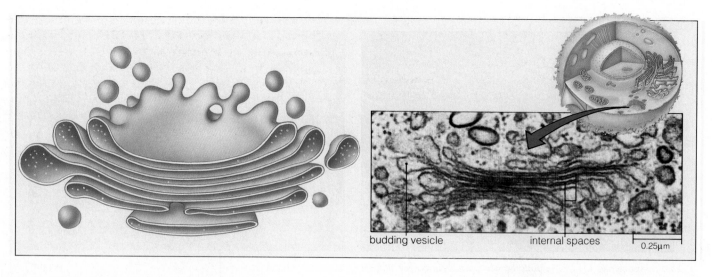

Figure 4.15 Sketch and micrograph of a Golgi body.

budding vesicle   internal spaces   0.25μm

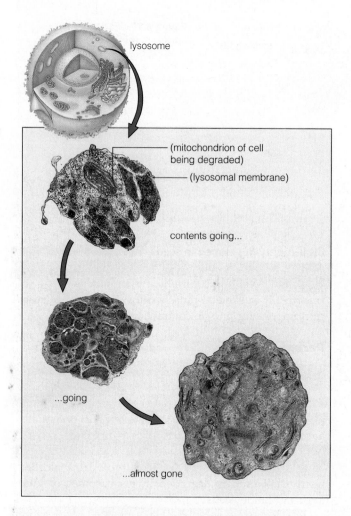

lysosome

(mitochondrion of cell being degraded)

(lysosomal membrane)

contents going...

...going

...almost gone

Figure 4.16 Digestion of "foreign" organelles inside one of the lysosomes of a phagocytic cell. Some of your own white blood cells are phagocytic (which means "cell-eater"). They engulf and digest bacteria and other disease-causing cells that enter your body.

Golgi Bodies

Outwardly, a **Golgi body** resembles a stack of pancakes (Figure 4.15). The "pancakes" actually are flattened sacs in which lipids and proteins are modified. Specific modifications allow them to be sorted out and packaged in vesicles for transport to specific locations. For example, an enzyme in one of the sacs may attach a phosphate group to a particular protein, and this serves as a mailing tag to the protein's proper destination. At the topmost Golgi membrane, vesicles form when a portion of the membrane bulges out, then breaks away.

Lysosomes

Among the vesicles that bud from the Golgi membranes of animal cells and certain fungal cells are **lysosomes**. Lysosomes are organelles of intracellular digestion. They contain different enzymes that can break down virtually all polysaccharides, proteins, and nucleic acids, and some lipids.

In many types of cells, lysosomes fuse with endocytic vesicles, which contain substances or foreign cells that docked against the plasma membrane. As described in the next chapter, such vesicles form when the plasma membrane dimples inward and forms a membrane sac around the trapped particles or cells. Figure 4.16 shows what can happen from that point on.

Often, lysosomes take part in the digestion of some or all of the cell's own parts. For example, when a tadpole turns into an adult frog, its tail disappears as lysosomal enzymes destroy tail cells in a controlled, genetically programmed way.

4.7 MITOCHONDRIA

Energy that ATP molecules carry from one reaction site to another drives nearly all cell activities. In **mitochondria** (singular, mitochondrion), energy stored in breakdown products of glucose and other organic compounds is used to form *many* ATP molecules. These organelles use oxygen to extract far more energy than can be done by any other means. When you breathe in, you are taking in oxygen primarily for mitochondria.

As Figure 4.17 shows, a mitochondrion has two membranes. The outer one faces the cytoplasm. The inner membrane usually folds inward repeatedly. Each fold is a crista (plural, cristae). This double-membrane system creates two compartments in the mitochondrion. As described in Chapter 8, enzymes and other proteins associated with the inner membrane serve as the machinery for ATP formation, and oxygen helps keep the machinery running.

All eukaryotic cells have one or more mitochondria. You might find only one in a single-celled yeast. You might find a thousand or more in energy-demanding cells, such as those in many muscles. Take another look at all the mitochondria in Figure 4.10, which is a micrograph of merely one thin slice from a liver cell. This alone tells you that your liver must be an exceptionally active organ.

In their size and biochemistry, mitochondria resemble bacteria. They have their own DNA and some ribosomes. They even divide on their own. Possibly they evolved from ancient bacteria that were engulfed by a predatory, amoebalike cell, yet managed to escape digestion (page 331). Perhaps they were able to reproduce inside the cell and its descendants. If they became permanent, protected residents, they may have lost many structures and functions required for independent life while becoming mitochondria.

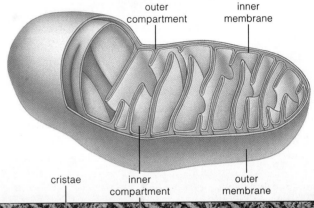

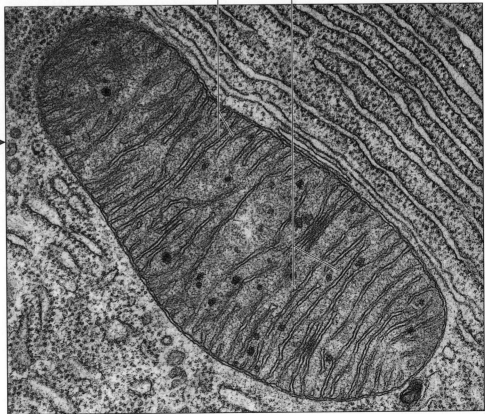

Figure 4.17 Transmission electron micrograph and sketch of a typical mitochondrion, an organelle that specializes in producing large quantities of ATP molecules.

Each mitochondrion has two membranes, each a lipid bilayer with associated proteins. The double-membrane system creates two compartments that function in ATP formation. Enzymes and other molecules embedded in the inner membrane carry out this task. In this micrograph, the dark spots between the membrane folds may be lipid deposits.

0.5 μm

SPECIALIZED PLANT ORGANELLES

Chloroplasts and Other Plastids

Many types of plant cells contain one or more plastids, which are organelles of photosynthesis or storage. Three kinds are common: chloroplasts, chromoplasts, and amyloplasts.

Only plant cells that specialize in photosynthesis have **chloroplasts**. Within these organelles, sunlight energy is absorbed, ATP forms, and organic compounds are synthesized from simple raw materials (water and carbon dioxide).

Chloroplasts often are oval or shaped like disks. Their outermost portion consists of two membrane layers, one wrapped around the other. These surround a semifluid interior, the stroma. *Another* membrane weaves elaborately through the stroma and forms an inner system of compartments that connect with one another (Figure 4.18). Often you will observe disk-shaped compartments stacked together. Each stack is called a granum (plural, grana).

The energy-acquiring and ATP-forming stage of photosynthesis proceeds at light-trapping pigments, enzymes, and other proteins of the inner membrane system. The synthesis stage proceeds in the stroma, where sugars, starch, and other organic compounds are put together. Here also, starch grains (clusters of new starch molecules) may be stored temporarily.

Chloroplast pigments impart green, yellow-green, or golden-brown coloration to plant parts. Chlorophyll molecules (green) are the most abundant of these, followed by the carotenoids (yellow, orange, and red).

In many ways, chloroplasts resemble certain photosynthetic bacteria. Like mitochondria, they may have evolved from ancient bacterial ancestors that became engulfed but not digested by predatory cells. We return to this idea in Chapter 21.

Unlike chlorophylls, the **chromoplasts** have an abundance of carotenoids but no chlorophylls. They often are the source of the yellow-to-red colors of many flowers, autumn leaves, ripening fruits, and carrots and other roots. Their pigments also may visually attract animals that pollinate plants and disperse seeds.

The **amyloplasts** have no pigments whatsoever. They often store starch grains and are abundant in cells of stems, potato tubers, and many seeds.

Central Vacuole

Mature, living plant cells often have a fluid-filled **central vacuole** (Figure 4.7). It usually occupies 50 to 90 percent of the cell interior, leaving room for only a narrow zone of cytoplasm. A central vacuole stores amino acids, sugars, ions, and toxic wastes. It also increases cell size and surface area. During growth, fluid pressure builds up in the vacuole and forces the still-pliable cell wall to enlarge. The cell itself enlarges under this force, and its increased surface area enhances the rate at which substances can move into and out of it.

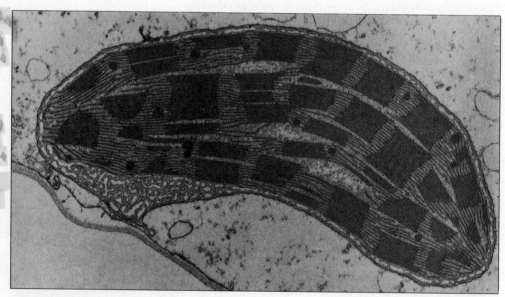

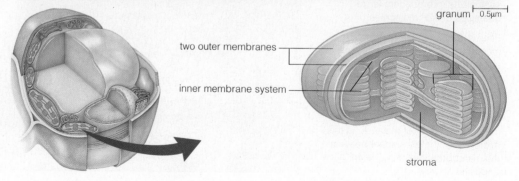

Figure 4.18 Transmission electron micrograph and generalized structure of a chloroplast, an organelle that specializes in trapping sunlight energy, converting it to the chemical energy of ATP—then using the ATP to form sugars and other compounds.

granum · 0.5μm

two outer membranes

inner membrane system

stroma

4.9 THE CYTOSKELETON

Extending from the plasma membrane and nucleus of eukaryotic cells is an interconnected system of bundled fibers, slender threads, and lattices. This system, the **cytoskeleton**, gives eukaryotic cells their internal organization, overall shape, and capacity to move. Some parts of the cytoskeleton reinforce the plasma membrane and hold its proteins in place; other parts do the same for the nuclear envelope. Still other parts are scaffolds for specific cytoplasmic regions. They may even stabilize the machinery—enzyme systems—of protein synthesis and other metabolic activities.

In all eukaryotic cells, the cytoskeleton consists mainly of **microtubules** and **microfilaments**. The cytoskeletons of animal cells also include various **intermediate filaments**. All three types of components are assembled from protein subunits (Figure 4.19). Microtubules consist of tubulin subunits, arranged in parallel rows. The subunits of microfilaments differ, depending on cell type, but they all contain actin and myosin subunits. Keratin subunits derived from intermediate filaments occur in hair cells.

Although many parts of the cytoskeleton are permanent, other parts appear only at certain times in a cell's life. Before a cell divides, for instance, new microtubules are assembled to form a "spindle" structure that moves chromosomes about, and then they disassemble when the task is done.

Figure 4.19 (**a**, **b**) Examples of cytoskeletons, which give eukaryotic cells their internal organization, shape, and capacity for movements. (**c**) Their main components are microtubules and microfilaments. Cytoskeletons of animal cells also contain intermediate filaments.

The cytoskeleton in (**a**) is from a cell of the African blood lily (*Haemanthus*). The cell took up molecules previously labeled with fluorescent dyes. One type of labeled molecule bound only to tubulins, the protein subunits of microtubules. These glowing, green-stained microtubules extend from the cell's nucleus (with purple-stained chromosomes) to its plasma membrane. The cytoskeleton in (**b**) is from a fibroblast, a type of cell that gives rise to certain animal tissues. In this composite of three images, green identifies microtubules, and blue and red identify two different kinds of microfilaments.

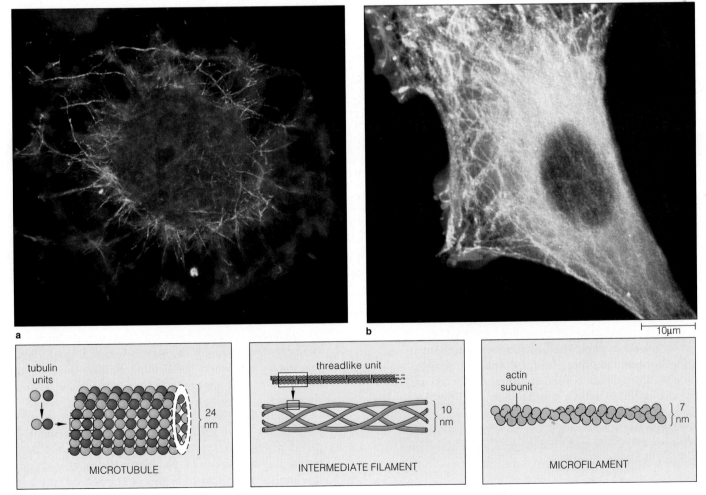

a

b

10μm

tubulin units

24 nm

MICROTUBULE

threadlike unit

10 nm

INTERMEDIATE FILAMENT

actin subunit

7 nm

MICROFILAMENT

c

4.10 THE STRUCTURAL BASIS OF CELL MOVEMENTS

If a cell lives, it moves. This is true even of cells embedded in the tissues of multicelled organisms. The simplest movements are slow, limited rearrangements of internal structures. With more intricate movements, a cell might propel itself through a fluid environment, rapidly shunt organelles or chromosomes to different locations, bulge forward, or produce skinny, temporary surface lobes called **pseudopods** ("false feet").

Microfilaments, microtubules, or both produce most movements. And they do so by three mechanisms:

---

1. *Through the controlled assembly and disassembly of their subunits, microtubules and microfilaments grow or shrink in length, and structures attached to them are thereby pushed or dragged through the cytoplasm.* For example, this is how spindle microtubules separate chromosomes before a cell divides.

2. *Microfilaments or microtubules actively slide past one another.* As an example, muscle cells contain permanent arrays of actin and myosin microfilaments, arranged in parallel, that bring about **contraction**. Briefly, tiny molecular heads projecting from the myosin microfilaments repeatedly bind and release their actin neighbors. When bound to actin, each head bends in a short, ATP-driven power stroke that slides the actin past it in a given direction. The cumulative, directional sliding causes the arrays of microfilaments to shorten (contract).

A similar kind of sliding mechanism produces the **amoeboid motion** of certain free-living cells and cancer cells. In this case, contractions in one part of a cell squeeze some cytoplasm into other parts, causing the cell body to bulge forward or pseudopods to form. Still another kind of sliding mechanism underlies the beating of flagella and cilia, as described shortly.

3. *Microtubules or microfilaments shunt organelles from one location to another.* For example, between dawn and dusk, the sun's overhead position changes. In response to the changing angle of incoming rays of light, chloroplasts in photosynthetic cells seem to flow to new light-intercepting positions! Actually, myosin filaments attached to the chloroplasts are "walking" over bundles of actin microfilaments, carrying the chloroplasts with them. This mechanism is the basis of **cytoplasmic streaming**. The term refers to an active flowing of cytoplasmic components that is especially pronounced in plant cells.

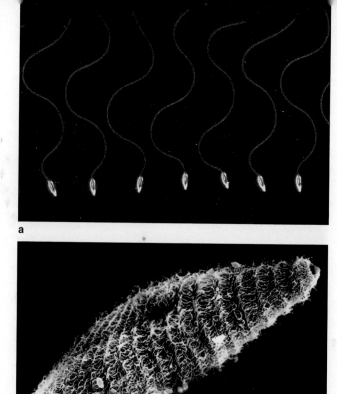

a

b

20 µm

Figure 4.20 (**a**) Composite micrographs of the beating pattern of a sperm flagellum. The waves start at the base of the flagellum and move toward the tip. (**b**) Scanning electron micrograph of *Paramecium*, a ciliated protistan.

The Internal Structure of Flagella and Cilia

Later chapters provide closer looks at the mechanisms just described. For now, let's limit our attention to the structure and operation of the **flagellum** (plural, flagella) and **cilium** (plural, cilia). Both structures project from the cell surface and beat in distinctive patterns. Flagella are whiplike tails that propel many free-living cells, such as sperm, through fluid environments (Figure 4.20*a*). Cilia are shorter and usually more profuse than flagella (Figure 4.20*b*). In multicelled organisms, some ciliated epithelial cells stir their surroundings. For example, thousands of these cells line airways to your respiratory tract, and the beating of their cilia directs mucus-trapped bacteria and other particles away from the lungs.

In both flagella and cilia, nine pairs of microtubules ring two central microtubules. This arrangement is called a "9 + 2 array." A system of spokes and links holds the arrangement together. When the sliding mechanism described in Figure 4.21 operates, the system forces the flagellum or cilium to bend.

Figure 4.21 Internal organization of flagella and cilia. Both kinds of motile structures have a system of microtubules and a connective system of spokes and linking elements that serve as the basis of beating movements. Nine pairs (doublets) of the microtubules are arranged as an outer ring around two central microtubules. This is called a 9 + 2 array.

In an unbent flagellum (or cilium), all microtubule doublets extend the same distance into the tip. When the flagellum bends, the doublets on the side that is bending the most are displaced farthest from the tip, as shown here:

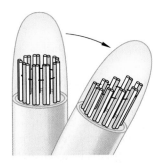

When a flagellum bends, it is not contracting. Rather, a sliding mechanism is operating between microtubule doublets of the outer ring. A series of short arms, composed of the protein dynein, extend from each doublet toward the next doublet in the ring. Inputs of ATP energy cause the dynein arms to attach to the doublet in front of it, tilt in a short power stroke that pulls the attached doublet along with it, then release its hold. Through repeated power strokes, doublets in front of the arms are forced to slide in a prescribed direction. As one doublet slides down toward the base of the flagellum, its cross-bridging arms force the next doublet in line to slide down farther, and so on in sequence.

Because the system of connecting spokes and links extends the length of the flagellum, it prevents the doublets from sliding completely out of their 9 + 2 array. This restriction forces the entire flagellum to bend to accommodate the internal displacement of sliding doublets. Thus, energized dynein arms cause doublet sliding, and restrictions imposed by the spokes and links convert doublet sliding into bending movements. That, in essence, is the model proposed by cell biologists K. Summers and R. Gibbons.

Microtubule Organizing Centers

Microtubules of a flagellum or cilium arise from **centrioles**, which remain at the base of the completed structure as a basal body (Figure 4.21). Centrioles often are present in some kinds of **MTOCs** (short for microtubule organizing centers). These are sites of dense material that generate large numbers of microtubules. (For example, centrosomes, a type of MTOC near the cell nucleus, give rise to the spindle microtubules.)

In short, this is the point to remember about the cytoskeleton of eukaryotic cells:

Microtubules and other cytoskeletal elements are the basis of a cell's shape, internal structure, and capacity for movements.

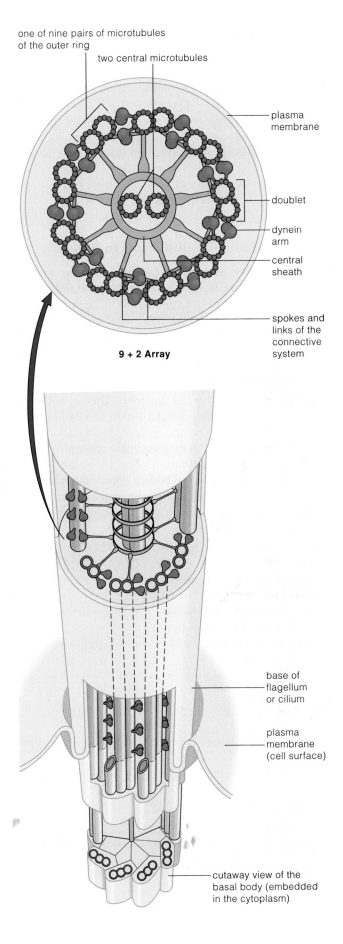

one of nine pairs of microtubules of the outer ring

two central microtubules

plasma membrane

doublet

dynein arm

central sheath

spokes and links of the connective system

9 + 2 Array

base of flagellum or cilium

plasma membrane (cell surface)

cutaway view of the basal body (embedded in the cytoplasm)

4.11 CELL SURFACE SPECIALIZATIONS

Single-celled eukaryotes interact directly with the environment. Like bacteria, many are supported or protected by a cell wall around the plasma membrane. Cell walls are porous, so water and solutes can move to and from the plasma membrane. In multicelled species, walls or other surface features allow adjacent cells to interact with one another and with their physical surroundings, as a few examples will illustrate.

Cell Walls and Cell Junctions in Plants

Cells of leafy plants stick together, wall to wall. In growing plant parts, bundles of cellulose strands form a *primary* cell wall, of the sort shown in Figure 4.22. Primary walls are pliable, so the surface area of cells can enlarge under incoming water pressure.

Later, in many cell types, more layers are deposited inside the first wall. They form a rigid, *secondary* wall that helps maintain the cell's shape. Deposits of pectin cement the adjacent walls together. (They also thicken jams and jellies.) Other deposits, such as waxes, protect and reduce water loss from cells at the plant's surface. Numerous channels cross adjacent walls and connect the cytoplasm of neighboring cells. These are plasmodesmata (singular, plasmodesma). The number of channels affects how fast substances are transported through the plant.

Intercellular Material in Animals

Cell secretions and other materials often intervene between cells of animal tissues. Think of the cartilage at the knobby ends of your leg bones. Cartilage consists of cells scattered in a "ground substance" (of modified polysaccharides) and protein fibers (of collagen or elastin). Through this material, nutrients and other substances diffuse from cell to cell. In mature bone and other tissues, intercellular material accounts for much of your body weight.

Cell Junctions in Animals

Three cell-to-cell junctions are common in animal tissues. *Tight* junctions link cells of epithelial tissues, which line the body's outer surface, inner cavities, and organs. The junctions form seals that keep molecules from freely crossing the tissue, as when they keep acids from leaking out of the stomach. *Adhering* junctions are like spot welds. They keep cells together in tissues of the skin, heart, and other organs that are subject to stretching. *Gap* junctions link the cytoplasm of adjacent cells; they are open channels for the rapid flow of signals and substances.

We will return to these and other cell-to-cell interactions in later chapters. For now, the point to remember is this: *In multicelled species, coordinated cell activities depend on specialized forms of linkages and communication between cells.*

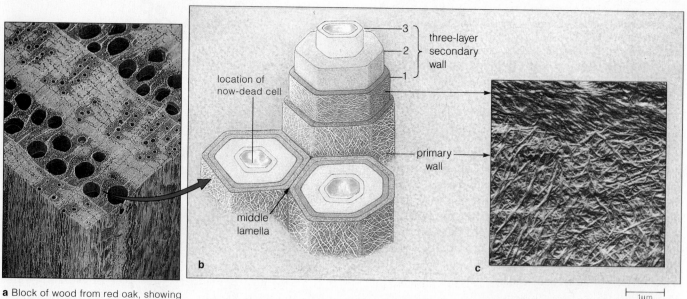

a Block of wood from red oak, showing water-conducting tubes

location of now-dead cell

3
2 } three-layer secondary wall
1

primary wall

middle lamella

b

c

1μm

Figure 4.22 Primary and secondary walls of plant cells. Many cell types have only a primary wall. Other types form a secondary wall after they have stopped growing and their primary wall is completed. The cells in the sketch died; only their walls remain. Walls of many such cells, arranged end to end and side by side, conduct water and strengthen plant parts.

Table 4.3 Summary of Typical Components of Prokaryotic and Eukaryotic Cells

Cell Component	Function	Prokaryotic Moneran	Eukaryotic Protistan	Eukaryotic Fungus	Eukaryotic Plant	Eukaryotic Animal
Cell wall	Protection, structural support	✓*	✓*	✓	✓	none
Plasma membrane	Control of substances moving into and out of cell	✓	✓	✓	✓	✓
Nucleus	Physical separation and organization of DNA	none	✓	✓	✓	✓
DNA	Encoding of hereditary information	✓	✓	✓	✓	✓
RNA	Transcription, translation of DNA messages into specific proteins	✓	✓	✓	✓	✓
Nucleolus	Assembly of ribosomal subunits	none	✓	✓	✓	✓
Ribosome	Protein synthesis	✓	✓	✓	✓	✓
Endoplasmic reticulum (ER)	Initial modification of many newly forming proteins; lipid synthesis	none	✓	✓	✓	✓
Golgi body	Final modification of proteins, lipids; sorting and packaging them for use inside cell or for export	none	✓	✓	✓	✓
Lysosome	Intracellular digestion	none	✓	✓*	✓*	✓
Mitochondrion	ATP formation	**	✓	✓	✓	✓
Photosynthetic pigment	Light–energy conversion	✓*	✓*	none	✓	none
Chloroplast	Photosynthesis, some starch storage	none	✓*	none	✓	none
Central vacuole	Increasing cell surface area, storage	none	none	✓*	✓	none
Cytoskeleton	Cell shape, internal organization, basis of cell motion	none	✓*	✓*	✓*	✓
9 + 2 flagellum, cilium	Movement	none	✓*	✓*	✓*	✓

*Known to occur in at least some groups.
**Aerobic reactions do occur in many groups, but mitochondria are not involved.

SUMMARY

1. The cell theory has three key points: All living things are made of one or more cells. Each cell is a basic living unit (it lives independently or has the inherent capacity to do so). A new cell arises only from cells that already exist.

2. At the minimum, each newly formed cell has a plasma membrane, a region of cytoplasm, and a region of DNA.

3. Cell membranes are composed of lipids and proteins. The lipids become arranged in two layers, and these serve as a barrier to water-soluble substances. The pro-teins carry out most membrane functions, such as transport of large molecules into and out of the cell.

4. In eukaryotic cells, internal membranes divide the cytoplasm into functional compartments, which are called organelles. The nucleus is a prime example. Prokaryotic cells (bacteria) do not have comparable organelles.

5. Organelle membranes separate different chemical reactions in the space of the cytoplasm and so allow them to proceed in orderly fashion.

6. Table 4.3 summarizes the organelles and other structures of both prokaryotic and eukaryotic cells.

1. Label the organelles indicated in this diagram of a typical plant cell. *58*

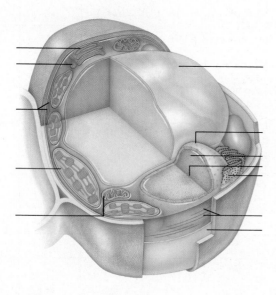

2. Label the organelles indicated in this diagram of a typical animal cell. *60*

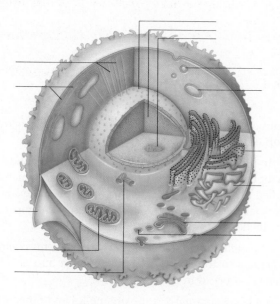

3. State the three key points of the cell theory. *51*

4. Why is it likely that you will never encounter a predatory two-ton living cell on the sidewalk? *53*

5. Suppose you want to observe the three-dimensional surface of an insect's eye. Would you benefit most from a compound light microscope, transmission electron microscope, or scanning electron microscope? *54–55*

6. Describe the three features that all cells have in common. Then, after reviewing pp. 56–57 and Table 4.1, write a paragraph describing the key differences between prokaryotic and eukaryotic cells. *52, 56–58, 60*

7. Which organelles are part of the cytomembrane system? Sketch their arrangement in an animal cell, from the nuclear envelope to the plasma membrane. *64*

8. Is the following statement true or false? Plant cells contain chloroplasts, but not mitochondria. *67–68*

9. What are the functions of the central vacuole in mature, living plant cells? *68*

10. What is a cytoskeleton? How does it aid in cell functioning? *69*

11. Are all components of the cytoskeleton permanent? *69*

12. What gives rise to the 9 + 2 microtubular array of cilia and flagella? *71*

13. Cell walls are features of which organisms: bacteria, protistans, plants, fungi, or animals? Are cell walls solid or porous? *72–73*

14. In plants, is a secondary wall deposited inside or outside the surface of the primary cell wall? Do all plant cells have secondary walls? *72*

15. What are some functions of the intercellular matrix in animal tissues? *72*

16. In multicelled organisms, coordinated interactions depend on linkages and communications between adjacent cells. What types of junctions occur between adjacent animal cells? Plant cells? *72*

1. The plasma membrane _____ .
   a. surrounds cytoplasm
   b. separates nucleus from cytoplasm
   c. acts as a nucleus in prokaryotic cells
   d. both a and b

2. Unlike eukaryotic cells, prokaryotic cells _____ .
   a. do not have a plasma membrane
   b. have RNA, not DNA
   c. do not have a nucleus
   d. all of the above

3. The _____ is responsible for cell shape, internal structural organization, and cell movement.
   a. flagellum   c. cytoskeleton
   b. cilium   d. both a and b

4. Cell membranes consist mainly of a _____ .
   a. carbohydrate bilayer and proteins
   b. protein bilayer and phospholipids
   c. phosopholipid bilayer and proteins
   d. none of the above

5. Organelles _____ .
   a. are membrane-bound compartments
   b. are typical of eukaryotic cells, not prokaryotic cells
   c. separate chemical reactions in time and space
   d. all of the above are functions of organelles

6. Plant cells but not animal cells have _____ .
   a. mitochondria   c. ribosomes
   b. a plasma membrane   d. a cell wall

7. Eukaryotic DNA is contained within the _____ .
   a. central vacuole   d. Golgi body
   b. nucleus   e. b and d are correct
   c. lysosome

8. The cytomembrane system does *not* include:
   a. ER   d. plastids
   b. transport vesicles   e. all of the above are
   c. Golgi bodies   parts of the system

9. Match each cell component with its function.
   _h_ mitochondrion   a. synthesis of polypeptide chains
   _g_ chloroplast   b. movement
   _a_ ribosome   c. digestion in cell
   _b_ flagellum   d. initial modification of new
   _d_ rough ER   polypeptide chains
   _e_ Golgi body   e. modify, sort, ship proteins and
   _i_ nucleolus   lipids
   _f_ cytoskeleton   f. cell shape, organization,
   _c_ lysosome   movement
                          g. photosynthesis
                          h. ATP formation
                          i. ribosome subunit assembly

Selected Key Terms

amoeboid motion 70
amyloplast 68
bacterial flagella 56
cell 51
cell theory 51
cell wall 56
central vacuole 68
centriole 71
chloroplast 68
chromatin 63
chromoplast 68
chromosome 63

cilium 70
contraction 70
cytomembrane system 64
cytoplasm 52
cytoplasmic streaming 70
cytoskeleton 69
ER (endoplasmic reticulum) 65
eukaryotic cell 52
flagellum 70
Golgi body 66
intermediate filament 69
lysosome 66
microfilament 69
micrograph 54
microtubule 69
mitochondrion (mitochondria)
MTOC (microtubule organizing center) 71
nuclear envelope 63
nucleolus 62
nucleus 52
organelle 58
peroxisome 65
plasma membrane 52
prokaryotic cell 52
pseudopod 70
ribosome 57
surface-to-volume ratio 53
wavelength 54

Readings

Bloom, W., and D. Fawcett. 1986. *A Textbook of Histology*. Eleventh edition. Philadelphia: Saunders.

Burgess, J., M. Marten, and R. Taylor. 1987. *Under the Microscope—A Hidden World Revealed*. Cambridge: Cambridge University Press. Paperback.

deDuve, C. 1985. *A Guided Tour of the Living Cell*. New York: Freeman. Beautifully illustrated introduction to the cell; two short volumes.

Weibe, H. 1978. "The Significance of Plant Vacuoles." *Bioscience* 28: 327–331.

Wolfe, S. 1993. *Molecular and Cellular Biology*. Belmont, California: Wadsworth. Outstanding reference text.

5 A CLOSER LOOK AT CELL MEMBRANES

It Isn't Easy Being Single

As small as it may be, a cell is a *living* thing engaged in the risky business of survival. Consider how something as ordinary as water can challenge its very existence. Water bathes cells inside and out, donates its molecules to metabolic reactions, and dissolves vital ions. If all goes well, each cell holds on to enough water and dissolved ions—not too little, not too much—to survive. But who is to say that life consistently goes well?

Think of a goose barnacle attached to the submerged side of a log that is drifting offshore, at the mercy of ocean currents. At feeding time, the barnacle opens its hinged shell and extends featherlike appendages, which trap bacteria and other bits of food suspended in the water (Figure 5.1*a*). The fluid bathing each living cell in the barnacle's body is salty, rather like the salt composition of seawater. And seawater usually is in balance with the salty fluid inside barnacle cells.

But suppose the log drifts into dilute waters from a melting glacier and stays there (Figure 5.1*b*). For reasons explored in this chapter, salts inevitably move out of the barnacle's body—and out of its cells. The salt-water balance is gradually destroyed, and the cells die. And so, in time, does the barnacle.

The same thing can happen to burrowing worms and other soft-bodied organisms that live between the high and low tide marks along a rocky shore. During an unpredictably heavy storm, when the seawater becomes dilute with runoff from the land, the balance between the surrounding water and the body fluids that bathe the organisms' cells is similarly upset. At such times, the resulting death toll in the intertidal zone can be catastrophic.

With these examples we begin to see the cell for what it is: a tiny, organized bit of life in a world that is, by

Figure 5.1 (**a**) Goose barnacles, which live attached to logs and other floating objects in the seas. As is true of every organism, drastic changes in salt concentration can threaten the cells of these marine animals. Such changes result when the barnacles accidentally end up in glacial meltwaters. (**b**) The dark-blue seawater in this photograph is quite salty. The lighter blue water flowing out from the glacier in the background is much, much lower in salts.

a

comparison, unorganized and sometimes harsh. How exquisitely adapted a living cell must be to its environment! *The cell must be built in such a way that it can bring in certain substances, release or keep out other substances, and conduct its internal activities with great precision.*

For this bit of life, precision begins at the plasma membrane—a flimsy bilayer of lipids, dotted with diverse proteins, that surrounds the cytoplasm. Across this membrane, every cell exchanges substances with its surroundings in highly selective ways. For eukaryotic cells, precision continues at the internal membranes of those specialized compartments called organelles. Photosynthesis, respiration, and many other metabolic reactions depend absolutely on the selective movement of substances across organelle membranes. Understand the structure and function of a cell's membranes, and you will gain insight into survival at life's most fundamental level.

b

5.1 MEMBRANE STRUCTURE AND FUNCTION

The Lipid Bilayer

A cell's organization as well as its activities depends on membranes. Fluid bathes both surfaces of each cell membrane. How, then, does the membrane remain structurally intact? Is it like a solid wall separating two fluid regions? Not at all. Push a fine needle into and out of a cell, and cytoplasm will not ooze out. Instead, the cell surface will appear to flow over the puncture and seal it. This observation suggests that the plasma membrane shows fluid behavior!

How does a "fluid" cell membrane remain distinct from fluid surroundings? For the answer, start with phospholipids—the most abundant component of cell membranes. A **phospholipid**, recall, consists of a hydrophilic (water-loving) head and two fatty acid tails that are mainly hydrophobic (water-dreading):

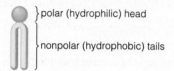

polar (hydrophilic) head

nonpolar (hydrophobic) tails

Like other lipids, phospholipid molecules cluster together spontaneously when immersed in water. Hydrophobic interactions may even force them into two layers, with all the fatty acid tails sandwiched between the hydrophilic heads. This arrangement, a **lipid bilayer**, is the structural basis of cell membranes:

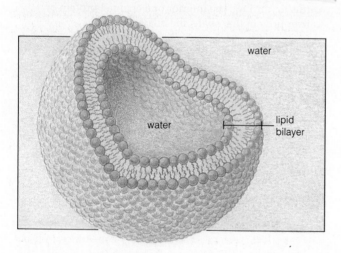

water

water

lipid bilayer

Because lipid bilayers minimize the number of hydrophobic groups exposed to water, the fatty acid tails do not have to spend a lot of energy fighting the water molecules, so to speak. In fact, that plasma membrane you "punctured" earlier exhibited sealing behavior precisely because a puncture is energetically

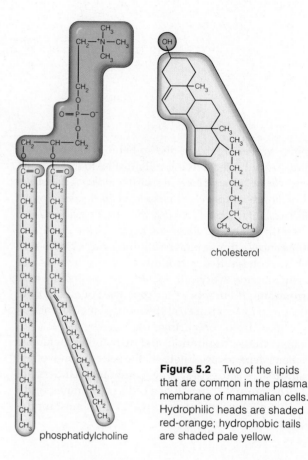

cholesterol

phosphatidylcholine

Figure 5.2 Two of the lipids that are common in the plasma membrane of mammalian cells. Hydrophilic heads are shaded red-orange; hydrophobic tails are shaded pale yellow.

unfavorable—it leaves too many hydrophobic groups exposed to the surrounding water.

Ordinarily, few cells are ever punctured with fine needles. But the self-sealing behavior of membrane lipids is useful for more than damage control. It has roles in the formation of those vesicles described in the preceding chapter. For example, as vesicles bud off ER or Golgi membranes, hydrophobic interactions with cytoplasmic water push lipid molecules together and close off the ruptured site. The same thing happens during exocytosis and endocytosis, as you will see later in the chapter.

Lipid bilayers of cell membranes incorporate a variety of phospholipids, glycolipids, and sterols. The phospholipids differ in their hydrophilic heads and in the length and saturation of their fatty acid tails. (Unsaturated fatty acids have one or more double bonds in the carbon backbone; fully saturated ones have none.) Figure 5.2 shows the structure of phosphatidylcholine, one of the most abundant phospholipids. Glycolipids of the bilayer are structurally similar to this, and their head region also incorporates one or more sugar monomers. Cholesterol is an abundant sterol of animal cell membranes; phytosterols are common to plant cell membranes.

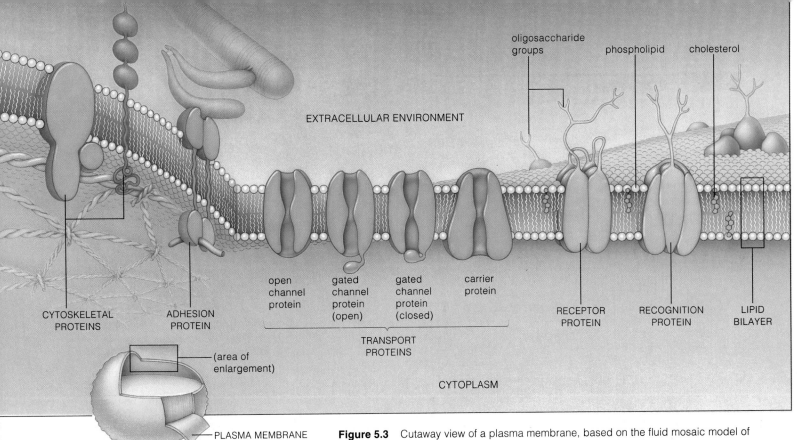

Figure 5.3 Cutaway view of a plasma membrane, based on the fluid mosaic model of membrane structure.

The Fluid Mosaic Model of Membrane Structure

Figure 5.3 shows the **fluid mosaic model** of membrane structure. By this model, a membrane bilayer shows "fluid" behavior as a result of the constant motion of lipid molecules and their interactions with one another. Membrane lipids spin about their long axis, they move sideways, and their tails flex back and forth—all of which helps keep neighboring lipids from packing into a solid layer. Lipids with short tails and unsaturated (kinked) tails further disrupt the packing. Being a composite of lipids and proteins, the membrane is said to have a "mosaic" quality. The proteins carry out most membrane functions, such as the transport of substances into and out of the cell.

The mosaic is not symmetrical. The two sides of a given membrane are not the same; they differ in terms of the number, kind, and arrangement of their lipids and proteins. In later chapters, you will see how the asymmetry relates to differences in the tasks carried out on the two sides of the membrane.

Bear in mind, the fluid mosaic model is only a starting point for discussing membranes—which differ in fluidity as well as in composition and molecular arrangements. Why bother to think about this fluidity? Cell survival depends on it, as a simple example will make clear. When outside temperatures fall,

membranes tend to stiffen and disrupt the functions of membrane proteins. Bacterial and yeast cells rapidly synthesize unsaturated fatty acids at such times. Infusion of these kinky lipids into a membrane helps keep it from stiffening up.

The first fluid mosaic model of membrane structure was put together in 1972 by S. J. Singer and G. Nicolson. Evidence favoring this model comes from many sources, including freeze-fracture microscopy, as described in the *Focus* essay on the next page. The features of this model can be summarized this way:

1. Cell membranes are composed mostly of lipids (especially phospholipids) and proteins.

2. The lipid molecules have their hydrophilic heads at the two outer faces of a bilayer and their fatty acid tails sandwiched in between.

3. A lipid bilayer imparts structure to cell membranes and serves as a hydrophobic barrier between two solutions. (A plasma membrane separates the fluids inside and outside the cell. Internal cell membranes separate different fluids in the cytoplasm.)

4. Proteins embedded in the bilayer or positioned at its surfaces carry out most membrane functions.

5.2 FUNCTIONS OF MEMBRANE PROTEINS

Spanning the bilayer of all plasma membranes and organelle membranes are proteins with transport functions. Plasma membranes also incorporate proteins that function in signal reception, adhesion, and cell-to-cell recognition.

Transport proteins allow water-soluble substances to move through their interior, which opens on both sides of the bilayer. They are either *channel* proteins or *carrier* proteins. As you will see, most types weakly bind ions or molecules on one side of the membrane and release them on the other.

Some channel proteins remain open at all times. Others have molecular gates that open and close in controlled ways. (For example, gated channels help control the directional flow of ions across the plasma membrane of neurons. This is the basis of "messages" that travel through nervous systems.) The interior of a carrier protein remains closed off to solutes until it undergoes changes in shape. As you will see shortly, some types of carrier proteins passively transport their weakly bound cargo. Other types require an energy input. Once energized, they actively pump their cargo across the membrane.

Receptor proteins have binding sites for hormones and other extracellular substances that can trigger alterations in cell activities. For example, enzymes that crank up machinery for cell growth and division are switched on when the hormone somatotropin binds to cell receptors. Different cell types have different combinations of receptors. Among vertebrates, some kinds of receptors are present at the surface of most cells. Others are restricted to only a few cell types.

Recognition proteins are like molecular fingerprints at the cell surface; they identify a cell as being of a specific type. For example, your body cells bear "self" recognition proteins. Infection-fighting white blood cells recognize these and leave the cells alone. They also recognize certain proteins on bacterial cells and other invaders as being "nonself"—and destroy them.

Adhesion proteins help cells stay connected to one another in a given tissue. They are glycoproteins (with oligosaccharides attached). While the tissues are forming, glycoproteins connect neighboring cells. Some of these attachment sites become the cell-to-cell junctions mentioned earlier, on page 72.

Different classes of membrane proteins function in the transport of substances across the lipid bilayer and in signal reception, cell-to-cell recognition, and cell-to-cell adhesion.

Focus on Science

Discovering Details About Membrane Structure

Suppose you'd like to explore for yourself some of the methods that researchers used to deduce the structure of the plasma membrane. You could start by attempting to identify its molecular components. Your first challenge is to secure a large sample of plasma membrane that isn't contaminated with organelle membranes. Using red blood cells will simplify the task. These are abundant, easy to collect—and structurally simple. While red blood cells are maturing, they discard their nucleus. They have only a few components, such as hemoglobin and ribosomes, but these are enough to keep them functioning for their four-month life span.

By placing a sample of red blood cells in a test tube filled with distilled water, you can separate the plasma membranes from the rest of the cell components. Red blood cells have more solutes and fewer water molecules than distilled water. The difference in solute concentrations between the two regions causes water to move across a plasma membrane, into the cells. These particular cells have no mechanisms for actively expelling the excess water, so they swell up. In time they burst and their contents spill out.

Now you have a mixture of membranes, hemoglobin, and other cell parts in the water. How do you separate the bits of membrane? You can place a tube containing a special solution of cell parts in a **centrifuge**, a device that spins test tubes at high speed. Each component of a cell has its own molecular composition that gives it a characteristic density. As the centrifuge spins at the appropriate speed, components of the greatest density move toward the bottom of the tube. Other components take up layered positions above them, according to their relative densities.

With proper centrifugation, one layer in the solution will be shreds of cell membrane only. You can carefully draw off this layer and examine it with a microscope to verify that your membrane sample is not contaminated. Standard chemical analysis tells you that the plasma membranes consist of lipids and proteins.

In the past, there were two competing models of membrane structure. According to the "protein coat" model, the membrane was composed of a bilayer of lipids, coated on both surfaces with a layer of proteins. According to the "fluid mosaic" model, proteins were largely embedded within the bilayer.

Suppose you decide to test the protein coat model. First you calculate how much protein actually would be required to coat the inner and outer surface of a known number of red blood cells. Then you separate and measure a membrane sample into its lipid and protein fractions. In

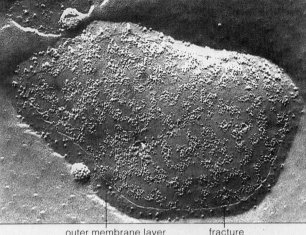

1. With freeze-fracture techniques, specimens being prepared for electron microscopy are rapidly frozen, then fractured by a sharp blow from the edge of a fine blade.

outer membrane layer exposed by etching

fracture edge

2. Fractured membranes commonly split down the middle of the lipid bilayer. Typically, one inner surface is studded with particles and depressions, and the other is a complementary pattern of depressions and particles. The particles are membrane proteins.

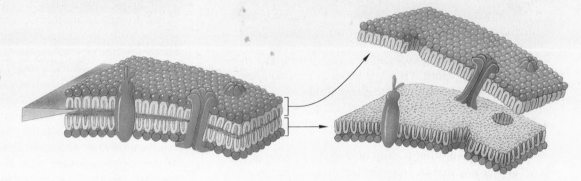

exposed by etching

3. Specimens are often freeze-etched. More ice is evaporated from the fracture face to expose the outer membrane surface.

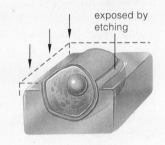

this way, you can compare the *observed* ratio of proteins to lipids against the ratio *predicted* on the basis of the model. Such tests were done with membrane samples. They reveal that there is far too little protein to cover both surfaces of a lipid bilayer. They provide evidence against the protein coat model, at least for red blood cells.

Suppose you now want to do a direct test of the fluid mosaic model. You could employ the *freeze-fracturing* and *freeze-etching* methods of preparing cells for electron microscopy. You first immerse a sample of cells in liquid nitrogen, an extremely cold fluid. The cells freeze instantly. You can now strike them with a very small chisel. As shown in Figure *a*, a properly directed blow will fracture the cell in such a way that one layer of the lipid bilayer separates from the other. These preparations can then be inspected under extremely high magnifications.

If the *protein coat* model were correct, what do you suppose examination of one such layer would reveal?

What you actually see is not a perfectly smooth, pure lipid layer. Freeze-fractured cell membranes reveal many bumps and other irregularities in lipid layers. The bumps are proteins, incorporated directly in the bilayer—just as the fluid mosaic model suggests.

(deposition of carbon and metal in thin layer on specimen surface)

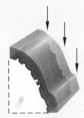

4. With metal shadowing techniques, the fractured surface is coated with a layer of carbon and heavy metal, such as platinum. The coating is thin enough to replicate details of the exposed specimen surface. The metal replica, not the specimen itself, is used for micrographs.

a Freeze-fracturing and freeze-etching. The micrograph shows part of a replica of a red blood cell that was prepared by the techniques described here.

5.3 DIFFUSION

Concentration Gradients and Diffusion

Cellular life depends on the energy inherent in molecules (or ions), which keeps them in constant motion. It depends also on **concentration gradients**. "Concentration" refers to the number of molecules of a substance in a specified volume of fluid. Add the word "gradient," and this means one region of the fluid contains more molecules than a neighboring region.

In the absence of other forces, molecules move down their concentration gradient. They do so because they constantly collide with one another, millions of times a second. Random collisions send the molecules back and forth, but the *net* movement is away from the place of greater concentration.

The net movement of like molecules down a concentration gradient is called **diffusion**. Diffusion is a key factor in the movement of substances across cell membranes and through fluid parts of the cytoplasm. In multicelled organisms, it is a key factor in the exchange of substances between cells and tissue fluids.

Suppose more than one substance is present in the same fluid. This makes no difference. Each substance still diffuses in some direction according to its *own* concentration gradient. Put a drop of dye at one side of a bowl of water. Dye molecules diffuse in one direction—to the region where they are less concentrated. And water molecules move in the opposite direction, to the region where *they* are less concentrated (Figure 5.4).

Diffusion is the net movement of like molecules or ions down their concentration gradient. They show a net outward movement from a region where they are most concentrated to a neighboring region where they are less concentrated.

Factors Influencing the Rate and Direction of Diffusion

The rate at which ions or molecules move down their concentration gradient depends on several factors. When the concentration gradient is steep, diffusion is faster. Then, far more molecules are moving outward, compared to the number moving in. As the gradient decreases, the difference in the number of molecules moving one way or the other becomes less pronounced, and diffusion is slower. When the gradient is gone, individual molecules are still in motion. But the total number moving one way or the other is just about the same at any given time. When the net distribution of molecules is just about uniform through the two adjoining regions, we call this "dynamic equilibrium."

Diffusion is faster at higher temperatures, because heat energy causes molecules to move more rapidly—hence to collide more frequently. And molecular size affects diffusion rates. Other factors being equal, smaller molecules move faster than large ones do.

Besides this, the rate and direction of diffusion may be modified by an **electric gradient**—a difference in charge between two adjoining regions. For example, a variety of dissolved ions and charged particles contribute to the overall electric charge near each surface of a cell membrane. Opposite charges attract. So the side of the membrane that is more *negatively* charged will exert the most pull on, say, sodium ions—which are *positively* charged. The pull can enhance the flow of sodium across the membrane. Many processes, such as the flow of information through your nervous system, depend on the combined force of electric and concentration gradients to attract specific substances across cell membranes.

As you will see shortly, a **pressure gradient**—a difference in pressure between two adjoining regions—also can influence the rate and direction of diffusion.

5.4 OSMOSIS

Some small molecules readily diffuse through the lipid bilayer of a cell membrane or through protein channels across it. Glucose and other molecules cannot do this; membrane proteins must pump them across. Because the membrane shows this selective permeability, concentrations of dissolved substances (that is, solutes) can increase on one side of the membrane and not the other. The resulting solute concentration gradients affect the movement of water into and out of cells.

Water by itself cannot get more or less concentrated. (Hydrogen bonds keep water molecules from crowding closer together or drifting apart.) Water becomes "less concentrated" only when solutes are dissolved in it.

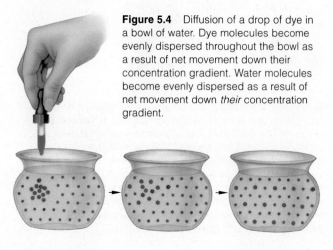

Figure 5.4 Diffusion of a drop of dye in a bowl of water. Dye molecules become evenly dispersed throughout the bowl as a result of net movement down their concentration gradient. Water molecules become evenly dispersed as a result of net movement down *their* concentration gradient.

Said another way, *a water concentration gradient is influenced by the number of molecules of all the solutes that are present on both sides of the membrane.*

Thus, the direction in which water moves across a membrane is influenced by **tonicity**—that is, the relative concentrations of solutes in two fluids. As Figures 5.5 and 5.6 show, water tends to move where solute concentrations are greater. When a membrane separates two *iso*tonic fluids, solute concentrations are equal—so there is no net movement of water in either direction. When solute concentrations are not equal, water tends to move from the *hypo*tonic solution (less solutes) to the *hyper*tonic solution (more solutes).

Unless cells have built-in mechanisms for adjusting to differences in tonicity between cytoplasmic fluid and the surroundings, they can shrivel or burst. As the *Focus* essay on the next page makes clear, some cells are better than others at doing this.

Bear in mind, other factors can modify the direction of water movement. For example, water also tends to move from regions of higher to lower pressure. Think of the fluid pressure generated by your beating heart. In your kidneys, the pressure forces some water and small solutes to leave blood capillaries and enter the surrounding tissue fluid. From there, water and solutes cross the membranes of kidney cells. Normally, tissue fluid is about three times more dilute than plasma (the fluid portion of blood), yet the pressure forces water to move against the gradient. The movement of molecules of water and solutes in the same direction in response to a pressure gradient is called **bulk flow**.

There is a special name for the movement of water across membranes in response to solute concentration gradients, fluid pressure, or both. It is called **osmosis**.

 Osmosis is the movement of water across a selectively permeable membrane in response to concentration gradients, fluid pressure, or both.

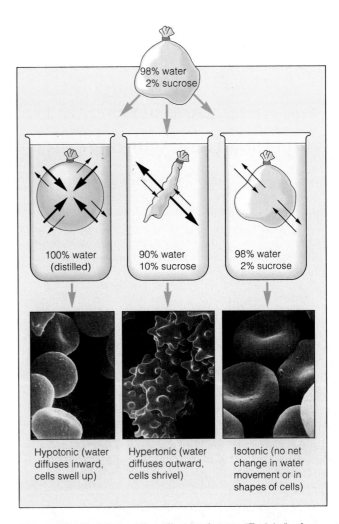

Figure 5.5 Tonicity and the diffusion of water. "Tonicity" refers to the relative concentrations of solutes in two fluids. In the sketches, membrane-like bags through which water but not sucrose can move are placed in three different solutions having different solute levels. Arrow widths indicate the relative amounts of water movement in each container.

The micrographs correspond to the sketches. They show the kinds of shapes that might be seen in human red blood cells placed in solutions that are hypotonic (water will move into cells), hypertonic (water will move out), or isotonic (no net movement). Solute levels of the fluids inside and outside red blood cells normally are in balance. These cells cannot make large adjustments in the concentration of cytoplasmic water. Drastic decreases in solutes in their fluid surroundings would make them burst; increases would make them shrivel.

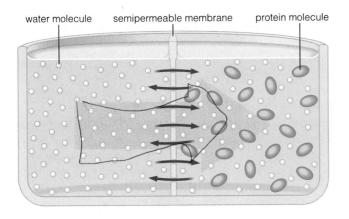

water molecule   semipermeable membrane   protein molecule

Figure 5.6 Osmosis. In this diagram, a membrane-like barrier divides a container into two compartments. Pure water was poured into the left compartment. A protein-rich solution of equal volume was poured into the one on the right. Water molecules can move across this membrane. The larger protein molecules cannot, so they occupy some of the available space in their compartment on an ongoing basis. Thus, the right compartment has fewer water molecules than the left. Water molecules follow this water concentration gradient. There is a net osmotic movement from left to right (*large blue arrow*).

Wilting Plants and Squirting Cells

When a plant with soft green leaves is growing well, you can safely bet that water in the surrounding soil is dilute (hypotonic), compared to water inside the plant's living cells. Plant cells, recall, have fairly rigid walls. As water moves into them, fluid pressure against the wall increases. "Turgor pressure" is the name for this internal fluid pressure on a cell wall.

Fluid pressure builds up against any walled cell when water moves inward by osmosis. But water will also be squeezed back out when the pressure is great enough to counter the attractive force of cytoplasmic fluid (which has more solutes than water from the soil). Both forces have the potential to cause the directional movement of water. The sum of these two opposing forces is called the "water potential."

When the soft parts of a plant remain erect, the movements of water into and out of its cells are equal. The constant turgor pressure keeps the cell walls plumped. But suppose the soil dries out. Suppose the inward movement of water dwindles or stops. Then, water will move out of cells and the soft parts of the plant body wilt.

Plants also wilt when the soil has too many solutes. A simple experiment will allow you to observe the wilting effect for yourself. Put 10 grams of table salt (NaCl) into 60 milliliters of water, then pour this salty solution into the soil around a potted tomato plant. As indicated by the clocks in the Figure *a* photographs, the plant will start to collapse after about five minutes. In less than thirty minutes, wilting will be severe. The sketches corresponding to the photographs show progressive plasmolysis—a shrinking of cytoplasm away from the cell wall.

Paramecium is not as vulnerable. This single-celled organism lives in freshwater. Because its cytoplasm is hypertonic relative to the surroundings, water tends to move into the cell by osmosis. If unchecked, the influx would bloat the cell and rupture its plasma membrane. *Paramecium* expels the excess through an energy-requiring mechanism involving contractile vacuoles (Figure *b*). Tubelike extensions of this type of organelle extend through the cytoplasm. Water enters the extensions and collects in a central vacuolar space. When filled, the vacuole contracts—squirting the excess water out through a small pore that opens to the environment.

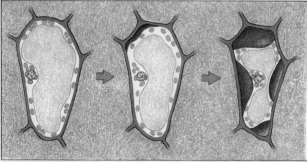

a

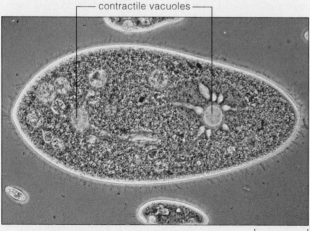

contractile vacuoles

b   10 μm

5.5 ROUTES ACROSS CELL MEMBRANES

We turn now to the mechanisms available to move solutes across cell membranes. As Figure 5.7 shows, the mechanisms range from unassisted diffusion, facilitated diffusion, and active transport to endocytosis and exocytosis. Through a combination of these mechanisms, cells or organelles are supplied with numerous raw materials and they are rid of wastes, at controlled rates. These mechanisms help maintain pH and volume inside the cell or organelle within functional ranges.

Oxygen, carbon dioxide, and other small molecules with no net charge readily diffuse across the membrane's lipid bilayer:

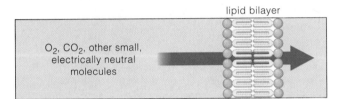

Glucose and other large, water-soluble molecules with no net charge almost never diffuse freely across the bilayer. Neither do positive or negative ions:

These substances must cross the membrane through the interior of transport proteins that span the bilayer.

With **passive transport**, water-soluble substances diffuse through the interior of channel proteins and many types of carrier proteins. With **active transport**, the membrane crossing occurs only when some type of carrier protein receives an energy boost, usually from an ATP molecule. Let's now take a closer look at how these mechanisms operate.

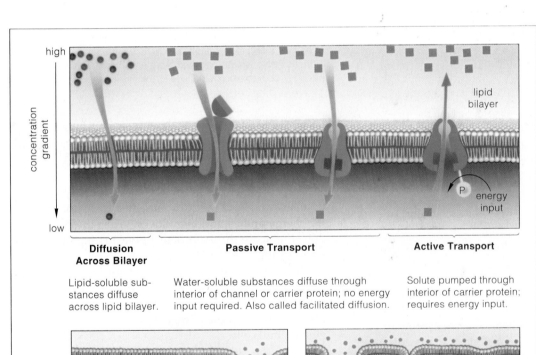

Figure 5.7 Overview of mechanisms by which solutes move across cell membranes. Exocytosis and endocytosis proceed at the plasma membrane only.

Diffusion Across Bilayer

Lipid-soluble substances diffuse across lipid bilayer.

Passive Transport

Water-soluble substances diffuse through interior of channel or carrier protein; no energy input required. Also called facilitated diffusion.

Active Transport

Solute pumped through interior of carrier protein; requires energy input.

Exocytosis

Vesicle moves to plasma membrane and fuses with it; contents released outside.

Endocytosis

Vesicle forms at surface of plasma membrane, then moves into cytoplasm.

5.6 PROTEIN-MEDIATED TRANSPORT

Characteristics of Transport Proteins

The type and number of transport proteins that span a membrane influence whether a given substance will move into or out of a given cell type, and how fast.

The transport proteins that serve as channels differ from one kind of organism to the next. For example, channel proteins in the plasma membrane of bacteria are like large pores; they are rather nonselective about which solutes travel through them. By contrast, channel proteins in the plasma membranes of animal and plant cells are chemically picky; they permit some solutes but not others to cross.

Cells also differ in their types and numbers of carrier proteins. Each type of carrier protein is highly selective, with a binding site that attracts a particular substance or a number of related substances. The carrier that transports amino acids, for instance, will not transport glucose.

To understand how the selective transport proteins work, you have to know that they are not rigid blobs of atoms. When they interact with molecules or ions, the proteins change from one shape into another shape, then back again.

Before a transport protein will allow passage of a solute, the solute must become weakly bound to it at a specific site somewhere on the protein surface. Binding leads to changes in the protein shape. Part of the protein closes in behind the bound solute—and part opens up to the opposite side of the membrane. There, the solute dissociates from the site. Think of the solute as hopping onto the protein on one side of the membrane, then hopping off on the other side.

Passive Transport

All channel proteins and many types of carrier proteins permit the passive diffusion of solutes across the membrane, down their concentration gradient. That is why the mechanism of passive transport is sometimes called "facilitated diffusion."

Consider the manner in which a carrier protein functions. It can move molecules (or ions) of a solute *both ways* across the membrane, depending on which way the carrier's binding site faces (Figure 5.8). At any given time, the *net* direction of movement depends on how many solute molecules are actually making random contact with vacant binding sites. Where the solute is more concentrated, binding and transport proceed more often. Where solute molecules are less concentrated, they still bind to the sites—but they do so at a lower rate. (With fewer molecules around, the random encounters with binding sites are not as frequent.)

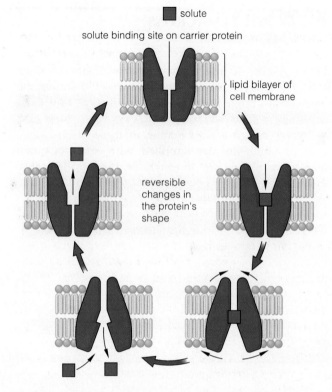

Figure 5.8 Passive transport across a cell membrane. Carrier proteins can move the ions or molecules of a solute in both directions. The *net* movement will be down the gradient (from higher to lower solute concentration) until concentrations become the same on both sides of the membrane.

By itself, the passive two-way transport would continue until solute concentrations became equal on both sides of the membrane. In most cases, however, other processes influence the outcome.

For example, the bloodstream delivers glucose to tissues throughout the body. Nearly all cells use glucose for energy and as building blocks. When the concentration of glucose in blood (and tissues) is high, the cells take up glucose. However, as fast as glucose is passively transported into the cells, glucose usually is being withdrawn from the cytoplasm for metabolic reactions. Thus, glucose metabolism helps maintain the concentration gradient that favors the movement of glucose into cells.

---

1. In passive transport, a transport protein moves molecules (or ions) of a specific solute across a cell membrane by undergoing a reversible change in shape.

2. Passive transport is based on diffusion. Barring other influences, it will continue until solute concentrations are the same on both sides of the membrane.

Active Transport

Carrier proteins that function in active transport pump a solute across the cell membrane, *against* its concentration gradient. Unlike passive transport (which will continue until concentrations are the same on both sides of the membrane), active transport continues until the solute becomes *more* concentrated on the side of the membrane where it is being pumped.

Active transport does not proceed spontaneously; it is an uphill process, so to speak. As Figure 5.9 shows, the carrier requires an input of energy, most often from ATP. When ATP donates energy to the carrier protein, the solute binding site becomes altered. Molecules (or ions) of the solute bind to the site more easily when it is in its altered configuration. After a molecule has been moved to the opposite side of the membrane, it dissociates from the binding site—which thereupon reverts to its less attractive configuration. At any given time, fewer molecules can make the return trip. So the net movement is to the side of the membrane where the solute is more concentrated.

By analogy, imagine skiers rapidly hopping on a ski lift that moves them up a mountain. At the top, the lift chairs tilt in a way that is not very inviting for skiers who want to hop on for a downhill ride. More skiers will be transported uphill than downhill.

One active transport system, the **calcium pump**, helps keep the calcium concentration in cells at least a thousand times lower than the concentration outside.

The **sodium-potassium pump** is one of the major cotransport systems. In this particular case, the binding of sodium ions to an energized carrier protein alters a binding site so that it becomes more easily occupied by potassium—the solute. Operation of many sodium-potassium pumps sets up concentration and electric gradients across a plasma membrane. And energy that is inherent in these specific gradients can be used to drive many cell activities.

We will return to active transport mechanisms in chapters that describe muscle contraction, information flow through nervous systems, and other physiological processes. For now, keep these points in mind:

---

1. In active transport, carrier proteins pump molecules or ions of a specific solute across a cell membrane, against their concentration gradient.

2. Active transport is not spontaneous; it requires an energy input, most often from ATP.

3. The energy input changes the solute binding site in a way that favors the net movement of the solute in the direction where it is most concentrated.

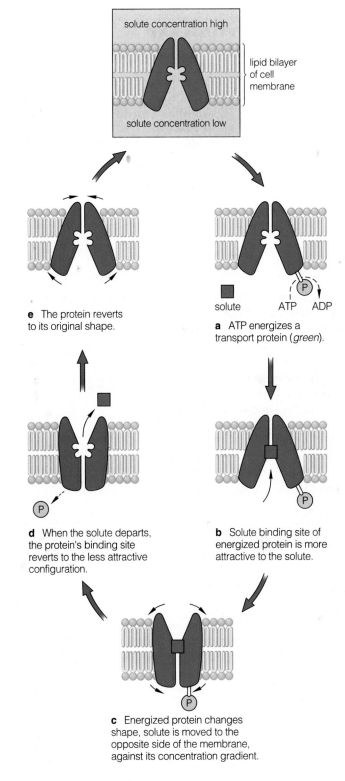

e The protein reverts to its original shape.

a ATP energizes a transport protein (*green*).

d When the solute departs, the protein's binding site reverts to the less attractive configuration.

b Solute binding site of energized protein is more attractive to the solute.

c Energized protein changes shape, solute is moved to the opposite side of the membrane, against its concentration gradient.

Figure 5.9 Active transport across a cell membrane. This carrier protein moves molecules of a specific solute in both directions. However, with an energy input from ATP, the chemical fit between the molecules and their binding site improves. Once a bound molecule is released, the site reverts to its unaltered, less attractive state. Even at low concentrations, molecules occupy the altered sites and move across the membrane faster. That is why there can be a greater *net* movement of molecules against the concentration gradient.

5.7 EXOCYTOSIS AND ENDOCYTOSIS

Protein channels and carriers can deal with one to several ions and molecules at a time. When it comes to moving materials in bulk across membranes, cells rely on exocytosis and endocytosis.

In **exocytosis**, a cytoplasmic vesicle moves to the plasma membrane and fuses with it. As shown in Figure 5.10a, as the vesicle membrane becomes incorporated into the plasma membrane, its contents end up fully exposed to the surroundings.

In **endocytosis**, part of the plasma membrane sinks inward and balloons around particles, fluid, or tiny prey. It seals on itself to form a vesicle, which transports its contents or stores them in the cytoplasm (Figure 5.10b).

Free-living phagocytic cells, including amoebas, trap bacterial cells and other foreign items by extending pseudopods around them. (Phagocyte literally means "cell eater.") As Figure 5.11 shows, these lobes of cytoplasm wrap around the trapped item and seal together, so that part of the plasma membrane forms a vesicle. After these particular vesicles move deeper into the cytoplasm, they fuse with lysosomes. Lysosomes, recall, are filled with digestive enzymes that can break down just about all biological molecules. The enzymes digest the trapped item into fragments and molecules. These are used in a variety of ways, depending on the cell type.

Animal cells also take up liquid droplets by endocytosis. Sometimes this transport process is called pinocytosis (which means "cell drinking"). A depression forms at the surface of the plasma membrane and dimples inward around solute-rich extracellular fluid. Here again, an endocytic vesicle forms and moves inside the cytoplasm.

In many cases, endocytosis is *receptor-mediated*. Specific types of particles or molecules bind to specific receptors at the plasma membrane. These receptors are clustered in shallow depressions in the membrane, appropriately called "coated pits." The coated pit shown in the Figure 5.12 micrographs is lined with membrane receptors that are specific for lipoprotein particles. All cells that metabolize or store cholesterol take up these particles. When lipoproteins bind to the receptors, the pit sinks into the cytoplasm, forming an endocytic vesicle.

Figure 5.13 summarizes the possible destinations of the vesicles formed by exocytosis and endocytosis.

Exocytosis and endocytosis are transport processes that move materials in bulk across the plasma membrane.

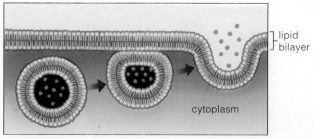

a Exocytosis

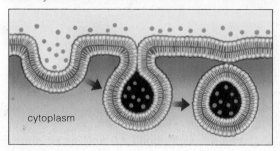

b Endocytosis

Figure 5.10 (**a**) Exocytosis. The membrane of a vesicle fuses with the plasma membrane, and its contents are released outside the cell. (**b**) Endocytosis. Part of the plasma membrane balloons inward and seals back on itself, forming a separate vesicle.

Figure 5.11 Phagocytosis, the means by which certain cells, including predatory amoebas and macrophages (infection-fighting white blood cells), engulf their targets. A vesicle forms around the target and moves into the cytoplasm, where it fuses with lysosomes—those sacs of digestive enzymes described on page 66.

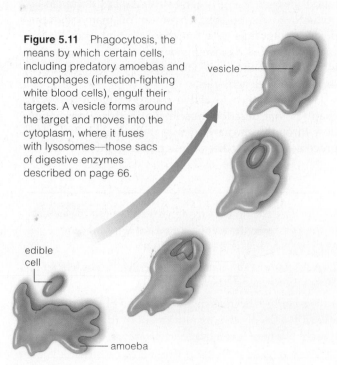

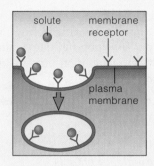
solute   membrane
         receptor

plasma
membrane

Figure 5.12   Example of receptor-mediated endocytosis, a transport process that begins when specific molecules bind with receptors at the surface of a plasma membrane. The electron micrographs show part of the plasma membrane from an immature chicken egg. Before it was plucked from the chicken, the egg was taking up lipoproteins and other substances necessary for its growth and development.

The shallow indentation in (**a**) is an example of a "coated pit." On the cytoplasmic side of the pit, molecules of clathrin, a type of protein, form a dense lattice. (**b**) Receptors at the membrane's outer surface bind lipoprotein particles. (**c**) The pit deepens and rounds out. (**d**) The fully formed coated vesicle encloses lipoproteins, which will be used or stored by the cell. Receptor-laden membrane that was appropriated by the vesicle will be recycled.

indentation on extracellular
side of plasma membrane

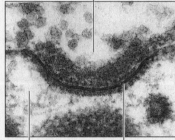 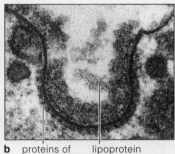

a   cytoplasm   plasma membrane

b   proteins of cytoskeleton   lipoprotein particles bound to membrane receptors

c   0.1μm

d   vesicle completely formed, moving into cytoplasm

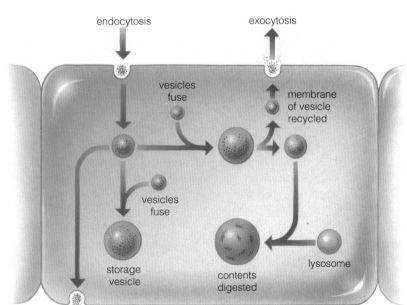

endocytosis

exocytosis

vesicles fuse

membrane of vesicle recycled

vesicles fuse

storage vesicle

contents digested

lysosome

substance released at opposing cell surface

Figure 5.13   Examples of the use and cycling of membrane lipids and proteins. Besides incorporating new membrane from some of the vesicles that budded earlier from ER membranes and Golgi bodies (page 64), the plasma membrane gives up bits of itself to endocytic vesicles and gets membrane bits back from exocytic vesicles.

SUMMARY

Membrane Structure and Function

1. Fluid of one sort or another bathes living cells; the cytoplasm of which is largely fluid, also. The cell's plasma membrane is the structural and functional boundary between the two fluids. In eukaryotic cells, internal cell membranes divide the fluid portion of cytoplasm into functionally diverse organelles.

2. Each cell membrane is two layers thick. Lipid molecules—phospholipids especially—make up this bilayer, which has many proteins embedded in it or positioned at one of its surfaces.

    a. The lipids are arranged with all hydrophobic tails sandwiched in between the hydrophilic heads.

    b. The lipid bilayer gives the membrane its water-impermeable structure. The proteins carry out most membrane functions.

3. These are the main features of the fluid mosaic model of membrane structure:

    a. Cell membranes are fluid structures, mainly because their lipids twist, move laterally, and flex their tails. Also, lipids with short tails, kinked tails, or ring structures disrupt what might otherwise be tight packing among the fatty acid tails.

    b. Cell membranes have a mosaic quality, in that they are a composite of diverse lipids, integrated proteins, and surface-bound proteins.

    c. The "mosaic" is asymmetrical, in that the two sides of a given membrane differ in the number, kind, and arrangement of lipids and proteins.

Functions of Membrane Proteins

1. Cells differ in the number and type of membrane proteins. The differences affect a cell's metabolism, volume, and pH, its responsiveness to extracellular substances, and its interactions with other cells.

2. All cell membranes incorporate transport proteins. Plasma membranes also incorporate receptor proteins, recognition proteins, and adhesion proteins.

3. Transport proteins function as channels and carriers.

    a. The interior of transport proteins opens to both sides of the membrane, and solutes move through them. Some channel proteins have molecular gates that can be closed and opened in controlled ways.

    b. Many carrier proteins function in active transport; they move specific solutes across the membrane only when they receive an energy boost, as from ATP.

    c. Solutes do not diffuse freely through most transport proteins. They bind to an exposed site on the protein, then the protein's shape changes, exposing the bound solute to the opposite side of the membrane.

4. Receptor proteins bind hormones and other extracellular substances that can trigger alterations in cell behavior or metabolism.

5. Recognition proteins are molecular fingerprints at the surface of each cell type. Infection-fighting white blood cells distinguish between recognition proteins of the body's own cells and those of invaders.

6. Adhesion proteins help cells adhere to one another and form cell-to-cell junctions.

Movement of Substances Into and Out of Cells

1. Specific solutes move across cell membranes by the following mechanisms:

    a. Diffusion is the unassisted movement of solutes from a region of higher to lower concentration.

    b. Osmosis is the movement of water across a selectively permeable membrane in response to solute concentration gradients, fluid pressure, or both.

    c. Passive transport is the movement of a solute through the interior of a channel protein or a carrier protein that does not require an energy input to operate.

    d. Active transport is an energy-requiring pumping of a solute through a carrier protein, against its concentration gradient.

2. Oxygen, carbon dioxide, and other small molecules with no net charge diffuse across the lipid bilayer. Ions and large, water-soluble molecules such as glucose are actively or passively transported across.

3. Tonicity refers to the relative concentrations of solutes of two fluids. When conditions are isotonic (equal concentrations), there is no osmotic movement in either direction. Water tends to move from hypotonic fluids (lower concentration) to hypertonic fluids (higher concentration).

4. Through exocytosis, cells eject larger volumes of dissolved substances across the plasma membrane, compared to membrane transport processes.

5. Through mechanisms of endocytosis, cells take in large particles or fluid droplets. Amoebas and other phagocytic cells use endocytosis when they trap their targets.

Review Questions

1. Describe the fluid mosaic model of plasma membranes. What makes the membrane fluid? What parts constitute the mosaic? *78–79*

2. List the structural features that all cell membranes have in common. *78–79*

3. Describe some features of membrane proteins. *80*

4. How does diffusion work? *82*

5. What is osmosis and what causes it? *82–83*

6. Explain the difference between active and passive transport mechanisms. *85*

7. What types of substances can readily diffuse across the lipid bilayer of a cell membrane? What types of substances must be actively or passively transported across? *85–87*

8. Can you explain the difference between exocytosis and endocytosis? What is receptor-mediated endocytosis? *88*

Self-Quiz *(Answers in Appendix IV)*

1. Cell membranes consist mainly of a _____ .
   a. carbohydrate bilayer and proteins
   b. protein bilayer and phospholipids
   c. phospholipid bilayer and proteins
   d. none of the above are correct

2. The most abundant components of cell membranes are _____ .
   a. phospholipids   c. proteins
   b. cholesterols   d. a and c are correct

3. In a lipid bilayer, the _____ of lipid molecules are sandwiched between their _____ .
   a. hydrophilic tails, hydrophobic heads
   b. hydrophilic heads, hydrophilic tails
   c. hydrophobic tails, hydrophilic heads
   d. hydrophobic heads, hydrophilic tails

4. Most membrane functions are carried out by _____ .
   a. proteins   c. nucleic acids
   b. phospholipids   d. hormones

5. Internal cell membranes as well as the plasma membrane incorporate _____ .
   a. transport proteins   d. recognition proteins
   b. receptor proteins   e. all are correct
   c. adhesion proteins

6. When a cell is placed in a hypotonic solution, water tends to _____ .
   a. move into the cell   c. show no net movement
   b. move out of the cell   d. move by endocytosis

7. A _____ is a device that spins test tubes and separates cell components according to their relative densities.

8. The direction in which a solute diffuses from one region to another will *not* be influenced by _____ .
   a. how steep the concentration gradient is
   b. an electric gradient
   c. a pressure gradient
   d. the presence of other solutes
   e. all of these factors influence diffusion

9. _____ can diffuse across a lipid bilayer.
   a. glucose   c. carbon dioxide
   b. oxygen   d. b and c are correct

10. Sodium ions move across a membrane, through the interior of a protein that has received an energy boost. This is an example of _____ .
    a. passive transport   c. facilitated diffusion
    b. active transport   d. a and c are correct

11. Match each event or condition with the appropriate description.
    ___ movement of water across a membrane in response to solute concentration or pressure gradients
    ___ the energy-requiring pumping of a solute against its concentration gradient
    ___ exocytosis
    ___ isotonic, hypotonic, hypertonic
    ___ unassisted movement of solutes down a concentration gradient
    ___ endocytosis

    a. refers to tonicity
    b. diffusion
    c. cells take in large particles or fluid droplets
    d. active transport
    e. vesicle fuses with plasma membrane, contents released to outside
    f. osmosis

Selected Key Terms

active transport *85*   lipid bilayer *78*
adhesion protein *80*   osmosis *83*
bulk flow *83*   passive transport *85*
calcium pump *87*   phospholipid *78*
centrifuge *80*   pressure gradient *82*
concentration gradient *82*   receptor protein *80*
diffusion *82*   recognition protein *80*
electric gradient *82*   sodium-potassium pump *87*
endocytosis *88*   tonicity *83*
exocytosis *88*   transport protein *80*
fluid mosaic model *79*

Readings

Bretscher, M. October 1985. "The Molecules of the Cell Membrane." *Scientific American* 253(4):100–108.

Dautry-Varsat, A., and H. Lodish. May 1984. "How Receptors Bring Proteins and Particles Into Cells." *Scientific American* 250(5):52–58. Describes receptor-mediated endocytosis.

Singer, S., and G. Nicolson. 1972. "The Fluid Mosaic Model of the Structure of Cell Membranes." *Science* 175:720–731.

Unwin, N., and R. Henderson. February 1984. "The Structure of Proteins in Biological Membranes." *Scientific American* 250(2): 78–94.

6 GROUND RULES OF METABOLISM

Growing Old With Molecular Mayhem

Somewhere in those slender strands of DNA in your cells are snippets of instructions for constructing two wonderful proteins. Those proteins go by the names superoxide dismutase and catalase. Both are enzymes—they make metabolic reactions proceed much, much faster than they would on their own. And both help keep you from growing old before your time.

The two enzymes help your cells clean house, so to speak. Together, they produce and then hack up hydrogen peroxide (H_2O_2), a normal but toxic outcome of certain oxygen-requiring reactions. Oxygen (O_2) picks up electrons from those reactions, and sometimes it does not get quite as many as it is supposed to. So it takes on a negative charge (O_2^-).

Like other unbound, molecular fragments with the wrong number of electrons, O_2^- is a "free radical." Free radicals are *so* reactive, they even attach to molecules that usually do not take part in random reactions—molecules like DNA.

Enter superoxide dismutase. Under its prodding, two of the rogue oxygen molecules combine with hydrogen ions, forming H_2O_2 and O_2.

Enter catalase. Under *its* prodding, two molecules of hydrogen peroxide react and split into ordinary water and ordinary oxygen: $2H_2O_2 \longrightarrow 2H_2O + O_2$.

As people age, their capacity to produce functional proteins—including enzymes—begins to falter. Among those enzymes are superoxide dismutase and catalase.

a

b

Figure 6.1 (**a**) Owner of a good supply of functional enzymes that keep free radicals in check. (**b**) Owner of skin with age spots—visible evidence of free radicals on the loose.

They are produced in diminishing numbers, crippled form, or both. Now free radicals and hydrogen peroxide can accumulate. Like loose cannons, they career through cells with tiny blasts at the structural integrity of proteins, DNA, membranes, and other vital parts. Cells suffer or die outright. Those brown "age spots" on an older person's skin are evidence of assaults by free radicals (Figure 6.1). The spots are masses of brown pigment molecules that build up in cells when free radicals take over—all for the want of two enzymes.

With this chapter, we start to examine activities that keep cells alive and functioning smoothly. Sometimes the topics may seem remote from the world of your interests. But they help define who *you* are and who you will become, age spots and all.

KEY CONCEPTS

1. Cells trap and use energy for building, stockpiling, breaking apart, and eliminating substances in ways that help them survive and reproduce. These activities are called metabolism.

2. Metabolism proceeds as long as cells acquire and use energy. To stay alive, then, cells must replace the energy they inevitably lose during each metabolic reaction. The sun is the original source of energy replacements through most of the biosphere.

3. Different metabolic pathways, operating in coordinated ways, help maintain, increase, or decrease the relative amounts of substances inside cells.

4. Enzymes increase the rate of specific reactions. They take part in nearly all metabolic pathways. So does ATP, an organic compound that transfers energy from one reaction site to another in cells.

Find a light microscope somewhere and use it to peer down on a living cell suspended in a water droplet. The image is eerie—something that small is practically pulsating with movements. Even as you watch it, the cell takes in energy-rich solutes from the water. It busily builds membranes, stores things, checks out the DNA, and replenishes pools of enzymes. It is alive; it is growing; it may divide in two. Multiply these proceedings by 65 trillion cells and you have an idea of what goes on in your own body as you sit quietly, observing that energetic single cell!

Metabolism is the somewhat dreary name for this dynamic activity. It refers to the cell's capacity to acquire energy and use it to build, store, break apart, and eliminate substances in controlled ways. Metabolism is the basis for survival and reproduction—and it all begins with energy.

Energy is a capacity to make things happen, to do work. You use energy to wax a car, even to watch a movie. In both cases, cells of your muscles, brain, and other body parts are being put to work. Energy in muscle cells drives the contractions that move your body parts and hold them in various positions. Energy allows your brain cells to chatter among themselves and guide what you are doing. Gain insight into how cells secure, use, and lose energy, and you gain deeper understanding into the nature of life.

ENERGY AND LIFE

How Much Energy Is Available?

You, like cells, cannot create your own energy from scratch. You must get it from someplace else. That is the message of the **first law of thermodynamics:**

The total amount of energy in the universe remains constant. More energy cannot be created, and existing energy cannot be destroyed. It can only be converted from one form to another.

Consider what this law means. The universe has only so much energy, distributed in a variety of forms. One form can be converted to another, as when corn plants absorb sunlight energy and convert it to the chemical energy of starch. By eating corn, you can extract and convert some of the energy in starch to other forms, such as mechanical energy for your movements. With each conversion, a little energy escapes to the surroundings as heat. Even when you are "doing nothing," your body steadily gives off about as much heat as a 100-watt light bulb because of ongoing conversions in your cells. When released energy "heats up" your surroundings, it really is being transferred to the rather disorderly array of molecules that make up the air (Figure 6.2). It increases the number of ongoing, random collisions among those molecules, and with each encounter, a bit more energy spreads outward. However, *none of the energy has vanished*. It simply has become dissipated, in a random, disorderly way, among molecules in the air.

The One-Way Flow of Energy

In the world of life, most of the energy available for energy conversions is stored in covalent bonds. Glucose, starch, glycogen, fatty acids, and other biological molecules with a number of covalent bonds have a high energy content. When reactions break a bond or rearrange atoms around it, atoms move—and add to the molecular commotion (heat energy) of the surroundings. Cells generally cannot recapture energy that has been lost as heat. They can only direct atoms into different bonds.

For example, most cells can break apart glucose until all that remains is six molecules of carbon dioxide and six of water. The breakdown reactions are energetically favored; the carbon and hydrogen atoms of glucose form more stable arrangements when bonded to oxygen, which the cells readily secure from the atmosphere. The total bond energy of glucose is many times greater than the total bond energy of carbon dioxide molecules formed when glucose is broken down. The differences reflect the amount of energy lost as heat during many steps of the breakdown reactions. (We

Figure 6.2 Example of how the total energy content of a system *together with its surroundings* remains constant.

"System" means all matter in a specific region, such as a human body, a plant, a DNA molecule, or a galaxy.

"Surroundings" can be a small region in contact with the system or as vast as the entire universe. The system shown here is giving off heat to the surroundings by evaporative water loss from sweat. What one loses, the other gains, so the total energy content of both doesn't change.

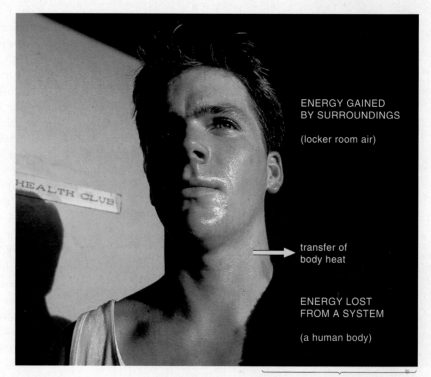

ENERGY GAINED
BY SURROUNDINGS

(locker room air)

transfer of
body heat

ENERGY LOST
FROM A SYSTEM

(a human body)

net energy change = 0

measure such differences in terms of kilocalories per mole. A *kilocalorie* is the same thing as a thousand calories—the amount of energy that will heat 1,000 grams of water from 14.5°C to 15.5°C at standard pressure.)

With their high energy content, glucose and other molecules of life are "high-quality" forms of energy. By contrast, a small amount of heat energy spread out in the air is "low quality," because it doesn't lend itself to conversions in cells.

Bad news for cells of the remote future: The amount of low-quality energy in the universe is increasing. Because no energy conversion can ever be 100 percent efficient, the total amount of energy in the universe is spontaneously moving from forms of higher to lower energy content. That, basically, is the point to remember about the **second law of thermodynamics:**

Taken as a whole, the amount of energy in the universe is spontaneously flowing from forms of higher to lower energy content.

Without energy to maintain it, any organized system tends to become disorganized over time. **Entropy** is a measure of the degree of a system's disorder. Think about the Egyptian pyramids—originally organized, presently crumbling, and many thousands of years from now, dust. The ultimate destination of the pyramids and everything else in the universe is a state of maximum entropy. Billions of years from now, all of the energy available for conversions will be dissipated, and nothing we know of will pull it together again.

Can it be that life is one glorious pocket of resistance to the depressing flow toward maximum entropy? After all, every time a new organism grows, atoms become linked in precise arrays, and energy becomes more concentrated and organized, not less so! Yet a simple example will show that the second law does indeed apply to life on earth.

The primary energy source for life on earth is the sun, which is steadily losing energy. Plants capture some sunlight energy, convert it in various ways, then lose energy to other organisms that feed, directly or indirectly, on plants. At each energy transfer along the way, some energy is lost, usually as heat that joins the universal pool. Overall, energy still flows in one direction. The world of life maintains a high degree of organization only because it is being resupplied with energy lost from someplace else (Figure 6.3).

A steady flow of sunlight energy into the interconnected web of life compensates for the steady flow of energy leaving it.

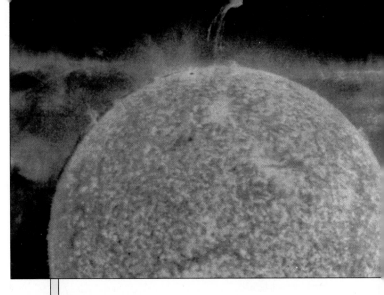

ENERGY LOST
one-way flow of energy away from the sun

ENERGY GAINED
one-way flow of energy from the sun into organisms

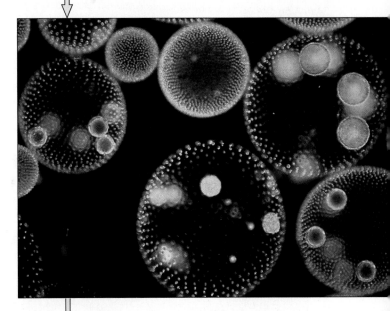

ENERGY LOST
one-way flow of energy from organisms to the surroundings

Figure 6.3 The one-way flow of energy into the world of life that compensates for the one-way flow out of it.

Living cells capture some of the energy being lost from the sun and convert it to useful forms. Energy is released during each conversion and lost to the surroundings, mostly as heat.

Green "dots" in the lower photograph are photosynthetic cells (*Volvox*), organized in tiny, spherical colonies. Orange "dots" are cells set aside for reproduction. They use energy to form new colonies inside the parent sphere.

6.2 ENERGY AND THE DIRECTION OF METABOLIC REACTIONS

Most of the one-way flow of energy through the world of life proceeds in cells that engage in photosynthesis and aerobic respiration. The following sections will help you make sense of these activities, both of which proceed through a step-by-step series of reactions.

Energy Losses and Energy Gains

When cells completely break down a glucose molecule, the covalent bonds of the resulting carbon dioxide and water molecules are more stable, in that it takes more energy to break them. But the total bond energies are lower than they were in glucose. This is an example of a reaction that ends with a net *loss* in energy (Figure 6.4a). It is an *exergonic* reaction, a word that means "energy out." Every day in an adult human of average size, the breakdown of glucose and other molecules by such reactions provides the body with an average of 1,200 to 2,800 kilocalories of energy.

How can a cell build starch, fatty acids, and other energy-rich molecules from smaller, energy-poor ones? Doesn't the overall synthesis of such molecules run counter to the spontaneous direction of energy flow? It does indeed. However, *inputs of energy*—as from ATP molecules—can drive reactions in an energetically unfa-

vorable direction. Such reactions typically end with a *net gain* in energy (Figure 6.4b). They are said to be *endergonic*, a word that means "energy in."

Reversible Reactions

Most reactions are reversible. They proceed in the forward direction, from starting substances (reactants) to products. They also proceed in reverse—with products being converted back to reactants. Figure 6.5 is an example of a reversible reaction, as the two opposing arrows signify.

Whether a reversible reaction runs more strongly in the forward or reverse direction depends partly on the ratio of reactant to product molecules. Remember, molecules are in constant random motion that puts them on collision courses with one another. The more concentrated they are, the more often they will collide. And energy released during a collision may be enough to make molecules combine, change shape, or split into smaller parts.

With a high concentration of reactant molecules, the reaction proceeds strongly in the forward direction. When the product concentration increases enough, a greater number of product molecules will be available to revert to reactants.

A reversible reaction tends spontaneously toward **chemical equilibrium**, at which time it proceeds at about the same pace in the forward and reverse direc-

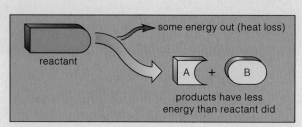

a The reactants of some reactions have *more* energy than the products. Many such reactions release energy that cells can use, as happens during aerobic respiration.

Figure 6.4 Energy changes in metabolic reactions.

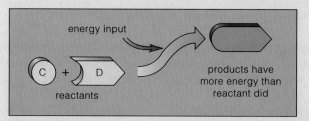

b The reactants of other reactions have *less* energy than the products. Energy inputs drive such reactions, as when an energy input from the sun drives photosynthesis.

c Energy inputs drive reactions by which cells build energy-rich molecules. When such molecules are split apart, some energy is released and can be harnessed to do cellular work.

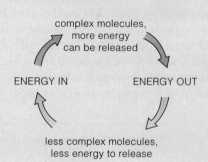

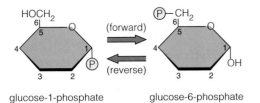

HOCH₂ ... glucose-1-phosphate   glucose-6-phosphate

Figure 6.5 A reversible reaction. Glucose is primed to enter reactions when a phosphate group becomes attached to it. With high concentrations of glucose-1-phosphate, the reaction tends to run in the forward direction. With high concentrations of glucose-6-phosphate, it runs in reverse. (The 1 and 6 of these names simply identify which particular carbon atom of the glucose ring has a phosphate group attached to it.)

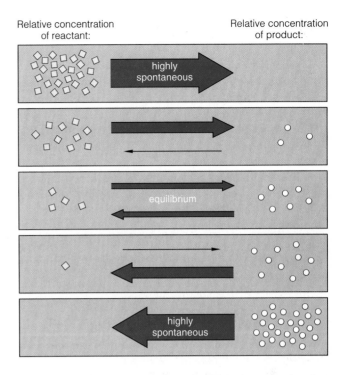

Relative concentration of reactant: Relative concentration of product:

highly spontaneous

equilibrium

highly spontaneous

Figure 6.6 Chemical equilibrium. With a high concentration of reactant molecules, a reaction tends to proceed most strongly in the forward direction (to product molecules). With a high concentration of product molecules, it proceeds most strongly in reverse. At equilibrium, the rates of the forward and reverse reactions are the same.

tions (Figure 6.6). The *amounts* of reactant and product molecules may or may not be the same at equilibrium. Think of a party with as many people wandering in as wandering out of two adjoining rooms. The total number in each room stays the same—say, thirty in one and ten in the other—even though the mix of people in each room continually changes.

Every reaction has its own characteristic ratio of products to reactants at equilibrium. For example, in one reaction, glucose-1-phosphate is rearranged into glucose-6-phosphate. (As Figure 6.5 shows, these are simply glucose molecules with a phosphate group attached to one of their six carbon atoms.) Suppose we allow the reaction to proceed when the concentrations of both substances are the same. At that point, the forward reaction proceeds 19 times faster than the reverse reaction does. Said another way, the forward reaction is producing more molecules in the same amount of time. The forward and reverse reactions eventually proceed at the same rate—but only when there are nineteen molecules of glucose-6-phosphate for every glucose-1-phosphate molecule. For this reaction, the ratio of products to reactants at equilibrium is 19:1.

Cells never stop building and tearing down molecules and moving them about in specific directions until they die. What does their great juggling act accomplish? Think it through. Cells can use only so many molecules at a given time and have only so much internal space to hold any excess. If they produce more of a substance than they can use, put into storage, or secrete, the excess might cause problems.

Consider *phenylketonuria* (PKU). People affected by this genetic (heritable) disorder produce a defective enzyme that prevents cells from using phenylalanine, one of the amino acids. When this amino acid accumu-

lates, the excess enters reactions that produce phenylketones. An accumulation of *these* product molecules can damage the brain in less than a few months. In most developed countries, routine screening programs identify affected newborns, who can grow up symptom-free on a phenylalanine-restricted diet.

---

1. The substances present at the conclusion of a reaction—the end products—may have less or more energy than the starting substances (the reactants).

2. Most reactions are reversible. Besides proceeding in the forward direction, from reactants to end products, they can proceed in the reverse direction—from end products back to reactants.

3. A reversible reaction moves spontaneously toward equilibrium, a state in which it proceeds at about the same rate in both directions.

6.3 METABOLIC PATHWAYS

All cells maintain, increase, and decrease the concentrations of substances by coordinating different metabolic pathways. A **metabolic pathway** is an orderly sequence of reactions, with specific enzymes acting at each step. The pathways are linear or circular (Figure 6.7). Often they are coupled, with products of one serving as reactants for others.

The main metabolic pathways are biosynthetic or degradative, overall. In *biosynthetic* pathways, small molecules are assembled into proteins, lipids, and other large molecules of higher energy content. In *degradative* pathways, large molecules are broken down to products of lower energy content. Both types have these participants:

Substrates: Substances able to enter a reaction; also called reactants or precursors.

Intermediates: Compounds formed between the start and the end of a metabolic pathway.

Enzymes: Proteins that catalyze (speed up) reactions.

Cofactors: Organic molecules or metal ions that assist enzymes or carry atoms or electrons from one reaction site to another.

Energy Carriers: Mainly ATP, which readily donates energy to diverse reactions.

End Products: Substances present at the end of a metabolic reaction or pathway.

Let's take a quick look at the roles of these substances before turning to the chapters on the main metabolic pathways.

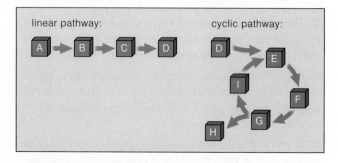

Figure 6.7 Example of a linear metabolic pathway that is coupled to a cyclic (circular) metabolic pathway.

6.4 ENZYMES

Characteristics of Enzymes

Enzymes are catalytic molecules, meaning that they greatly speed up specific reactions. They enhance the rate at which the reactions approach equilibrium. With few exceptions, enzymes are proteins.

Enzymes have four features in common. *First*, they do not make anything happen that could not happen on its own. But they usually make it happen at least a million times faster. *Second*, enzymes are not altered permanently or used up in the reactions they mediate; the same enzyme may be used over and over again. *Third*, the same enzyme works for the forward and reverse directions of a reaction. *Fourth*, each type of enzyme is highly selective about its substrates.

Substrates are molecules that a specific enzyme can chemically recognize, bind, and modify in a specific way. For example, thrombin, an enzyme involved in blood clotting, only recognizes and splits a peptide bond between two amino acids, arginine and glycine:

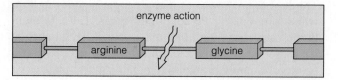

enzyme action

arginine · glycine

Enzyme-Substrate Interactions

An enzyme has one or more crevices in its surface where substrates interact with it. Such a crevice is an **active site**, the place where a specific reaction is catalyzed (Figure 6.8).

According to Daniel Koshland's **induced-fit model**, each substrate has a surface region that almost *but not quite* matches chemical groups in an active site. When substrates first settle into the site, the contact strains some of their bonds. Strained bonds are easier to break, and this helps pave the way for new bonds (within the products). Also, interaction with charged or polar groups in the site favors a redistribution of electric charge that primes the substrates for conversion to an activated state.

Substrates reach an activated, transition state when they fit most precisely in the active site (Figure 6.9). The enzyme induces the fit in several ways. For instance, weak but extensive bonding at the active site puts substrates in positions that make them collide and so promotes reaction. If those same molecules were colliding on their own, they would do so from random directions. The reaction rate would not be very impressive, because mutually attractive chemical groups would meet up less frequently.

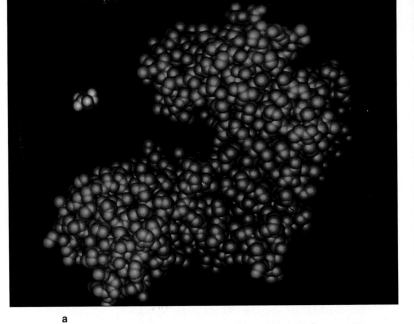
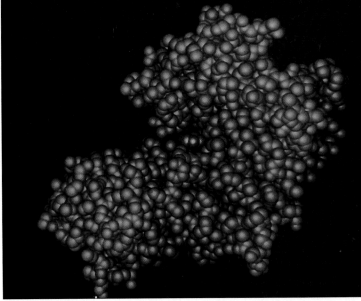

a

b

Figure 6.8 An enzyme at work. (**a**) Model of the enzyme hexokinase (*green shading*) and its substrate, a glucose molecule (*red*). The glucose is heading toward the active site, a cleft in the enzyme. (**b**) When glucose contacts the site, the upper and lower parts of the enzyme temporarily close in around it and prod the molecule to enter a specific reaction.

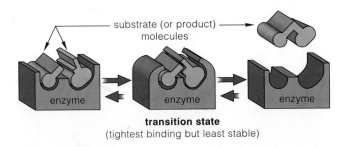

substrate (or product) molecules

enzyme   enzyme   enzyme

transition state
(tightest binding but least stable)

Figure 6.9 Induced-fit model of enzyme-substrate interactions. Only when the substrate is bound in place is an enzyme's active site complementary to it. The fit is most precise during a transition state of a reaction. The enzyme-substrate complex is short-lived, partly because the bonds holding it together are usually weak.

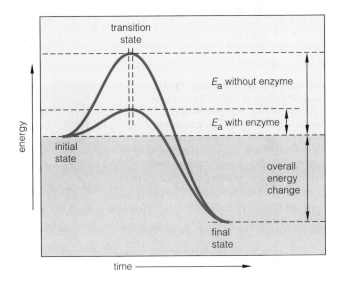

Figure 6.10 Energy hill diagram showing the effect of enzyme action. An enzyme greatly enhances the rate at which a given number of molecules complete a reaction because it lowers the required activation energy (E_a). It takes less energy to boost reactants to the crest (transition state) of a lower energy hill.

Once substrates are in the transition state, they react spontaneously, just as a boulder pushed up and over the crest of a hill rolls down on its own. However, they simply won't reach that state unless they collide with some minimum amount of energy. That amount, the **activation energy**, is like an "energy hill" that must be surmounted (Figure 6.10). By putting its substrates on a precise collision course, an enzyme makes the energy hill smaller, so to speak. And this means the reaction will proceed more rapidly.

An enzyme enhances the rate at which a reaction proceeds. It does this by lowering the amount of activation energy necessary to make its substrates react.

Effects of Temperature and pH on Enzyme Activity

Each type of enzyme functions best within a certain temperature range. As Figure 6.11 indicates, reaction rates decrease sharply when temperatures become too high. The increase in heat energy disrupts weak bonds holding the enzyme in its three-dimensional shape. This alters the active site, and substrates cannot bind to it. Often, exposure to temperatures that are much higher than an organism typically encounters destroys enzymes—and disrupts metabolism. This happens during a dangerously high *fever*. Humans usually die when internal body temperature reaches 44°C (112°F).

Each enzyme also functions best within a certain pH range. Higher or lower pH values generally disrupt its structure and function (Figure 6.12). Most enzymes function best in neutral solutions (pH 7). Pepsin is one of the exceptions. This protein-digesting enzyme performs its task in gastric fluid, which is extremely acidic.

Enzymes function best within limited ranges of temperature and pH.

Control of Enzyme Activity

Earlier you read that cells maintain, increase, and decrease concentrations of substances by coordinating different metabolic pathways. They do this mostly by controlling enzymes. Some controls govern enzyme synthesis and so cut down or beef up the number of enzyme molecules that are operating at any time at a key step in a pathway. Other controls stimulate or inhibit enzymes that are already formed.

Internal controls come into play when concentrations of substances change within the cell itself. Think of a bacterium, busily synthesizing tryptophan and other amino acids necessary to construct its proteins.

After a bit, protein synthesis slows and no more tryptophan is needed. But the tryptophan pathway is still in full swing, so the cellular concentration of its end product rises. Now a control mechanism called **feedback inhibition** operates. By this mechanism, when production of a substance triggers a cellular change, the substance itself shuts down its further production (Figure 6.13). In this case, unused molecules of tryptophan inhibit a key enzyme in the pathway.

This enzyme happens to be governed by *allosteric* control. Besides the active site, allosteric enzymes have

Figure 6.11 (**a**) Effect of increases in temperature on the activity of one kind of enzyme. (**b**) Observable evidence of temperature effects on an enzyme that influences the coat color of Siamese cats. Hairs on the ears and paws of this cat contain more dark-brown pigment (melanin) than hairs on the rest of the body. A heat-sensitive enzyme that controls melanin production is less active in warmer body regions—which have lighter hair as a result.

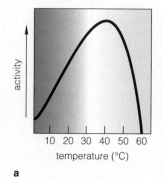

a

b

Figure 6.12 Effect of pH on the activity of three different enzymes. One enzyme (*black*) functions best in neutral solutions. Another kind of enzyme (*orange*) functions best in basic solutions, and another (*blue*), in acidic solutions.

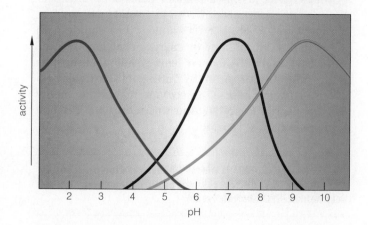

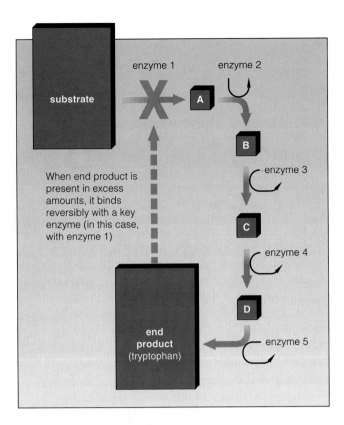

Figure 6.13 Feedback inhibition of a metabolic pathway. In this example, five enzymes act in sequence to convert a substrate into tryptophan, the end product. When an excessive number of end-product molecules accumulate, some of them bind to the first enzyme in the sequence and so block the entire pathway.

control sites where specific substances can bind and alter enzyme activity.

When the pathway is blocked, fewer tryptophan molecules are around to inhibit the key enzyme—so production rises. In such ways, feedback inhibition quickly adjusts concentrations of substances in cells.

In humans and other multicelled organisms, control of enzyme activity is just amazing. Cells not only work to keep themselves alive, they work with other cells in coordinated ways that benefit the whole body! Hormones are signaling agents in this vast enterprise. Specialized cells release these chemical signals into the bloodstream. Any cell having receptors for a particular hormone will take it up, and its program for constructing a particular protein or some other internal activity will be altered. The hormone trips internal control agents into action, and the activities of specific enzymes change.

Enzymes act only on specific substrates, and controls over their activity are central to the directed flow of substrates into, through, and out of the cell.

6.5 ENZYME HELPERS

During a metabolic reaction, enzymes speed the transfer of one or more electrons, atoms, or functional groups from one substrate to another. "Cofactors" help catalyze the reaction or serve briefly as transfer agents.

Some of the cofactors are specific types of organic compounds. Others are nothing more than metal ions that associate with the enzyme.

Coenzymes

The enzyme helpers called **coenzymes** are complex organic molecules, many of which are derived from vitamins. The coenzymes abbreviated **NAD⁺** (short for "nicotinamide adenine dinucleotide") and **FAD** (flavin adenine dinucleotide) are examples.

Both of these coenzymes can pick up hydrogen atoms that are liberated during glucose breakdown. They transfer these unbound protons (H^+) to other reaction sites. Electrons, being attracted to the opposite charge of the protons, go along for the ride. When carrying their cargo, the two coenzymes are abbreviated NADH and $FADH_2$, respectively.

The coenzyme **NADP⁺** (for nicotinamide adenine dinucleotide phosphate) functions in photosynthesis. It also transfers protons and electrons from certain degradative pathways to biosynthetic pathways, such as the ones by which fatty acids are assembled. Discerning enzymes recognize its phosphate "tag" (the P^+ attached to the molecule) and accept its proton and electron cargo. Like NAD⁺, this coenzyme is derived from the vitamin niacin. When carrying its cargo, it is abbreviated NADPH.

Metal Ions

The metal ions that serve as cofactors include ferrous iron (Fe^{++}). This metal ion is a component of cytochrome molecules. The cytochromes are carrier proteins that show enzyme activity. They are embedded in cell membranes, such as the membranes of chloroplasts and mitochondria.

a cytochrome

Coenzymes and metal ions assist enzymes in the transfer of electrons, atoms, and functional groups from one substrate to another.

6.6 ELECTRON TRANSFERS IN METABOLIC PATHWAYS

If you were to throw some glucose into a wood fire, its atoms would quickly let go of one another and combine with oxygen in the atmosphere, forming CO_2 and H_2O. Of course, all of the released energy would be lost as heat to the surroundings.

Cells do not "burn" glucose all at once. Doing so would waste its stored energy. Instead, enzymes pluck atoms away from the molecule during controlled steps of a degradative pathway, so that intermediate molecules form along the route. Thus, only some of the bond energy is released at each enzyme-mediated step.

Recall that chemical bonds are interactions between electrons of different atoms. Breaking those bonds puts electrons up for grabs, so to speak. And the liberated electrons can be sent through **electron transport systems**. As microscopists discovered, such systems are bound in the internal membrane systems of chloroplasts and mitochondria. They actually are organized arrays of enzymes and coenzymes that transfer elec-trons in a highly organized sequence. One molecule donates electrons, and the next in line accepts them.

You may hear of *oxidation-reduction* reactions. This rather obtuse name simply refers to electron transfers. A donor that gives up electrons is said to be oxidized. An acceptor that acquires electrons is reduced.

Electron transport systems operate by accepting electrons at higher energy levels, releasing them at lower energy levels, and making use of the energy they release. Think of a transport system as a staircase (Figure 6.14). Excited electrons at the top step have the most energy. They drop down, one step at a time. At certain steps, energy being released is harnessed to do work—for instance, to move ions in ways that set up ion concentration and electric gradients across a membrane. Such gradients, you will discover, are central to the formation of ATP.

Some transfers proceed after atoms or molecules absorb enough energy to "excite" electrons—that is, to boost the electrons farther from the nucleus. When an excited electron returns to the lowest energy level available, it gives off energy. The *Focus* essay describes some interesting evidence of this energy release.

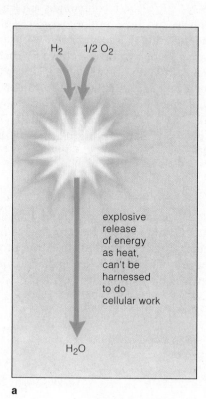

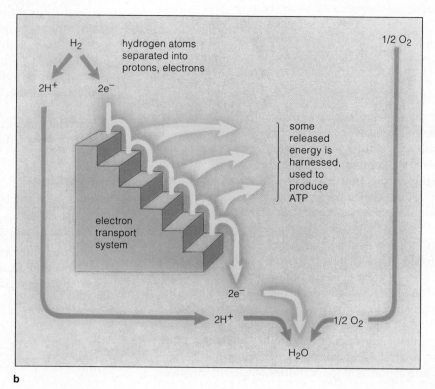

Figure 6.14 Controlled release of energy in metabolic reactions. **(a)** Hydrogen and oxygen exposed to an electric spark will react and release energy all at once. **(b)** In cells, the same type of reaction proceeds in many small steps that allow some of the released energy to be harnessed in useful forms. The "steps" are electron transfers, often between molecules that operate together as an electron transport system.

a

You Light Up My Life—Visible Effects of Excited Electrons

At night, during certain metabolic reactions, fireflies, various beetles, and some other organisms flash with light. These displays of **bioluminescence** take place when enzymes called luciferases excite the electrons of luciferins, a class of highly fluorescent substances. When the excited electrons quickly return to a lower energy level, they release energy—in the form of light.

In the tropical forests of Jamaica, click beetles, known locally as kittyboos, fly about at night and give startling displays of bioluminescence (Figure *a–c*). These particular beetles belong to the genus *Pyrophorus*, and different varieties emit green, greenish-yellow, yellow, or orange flashes.

Molecular biologists have identified the kittyboo genes responsible for the flashes. More than this, they have managed to insert copies of the genes into other organisms, including bacteria (Figure *d*). How they do these gene transfers is a topic of Chapter 16.

Besides being fun to think about, transfers of the genes for bioluminescence have practical applications. Each year, for example, 3 million people die from a lung disease caused by *Myobacterium tuberculosis*. Different strains of this bacterium are resistant to different antibiotics. A patient cannot be treated effectively until the strain causing her or his particular infection is identified. Before 1993, the tests used for identification took weeks to complete.

Now, bacterial cells from a patient are rapidly cultured and exposed to luciferase genes. Some cells take up the genes, which may become inserted into the bacterial DNA. Clinicians then expose colonies of the genetically modified bacteria to different antibiotics. If an antibiotic has no effect, the cells churn out gene products—*including luciferase*—and the colony glows. If a colony doesn't glow, this is evidence that the antibiotic has stopped the infectious cells in their metabolic tracks.

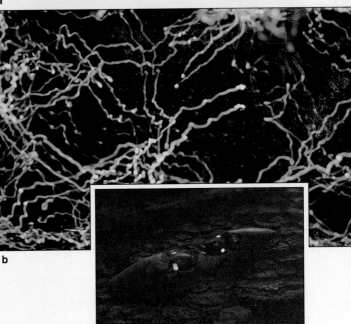

b

c

(**a**) One of the Jamaican kittyboo beetles (*Pyrophorus noctilucus*). The time-lapse photograph in (**b**) reveals the random paths of kittyboos on the wing, as they light up the night with bioluminescent flashes. The flashes help potential mates find each other in the dark (**c**). The micrograph in (**d**) shows four colonies of bacterial cells that have taken up kittyboo genes responsible for bioluminescence.

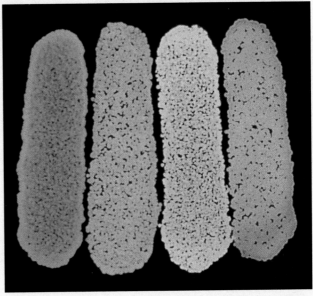

d

103

6.7 ATP—THE MAIN ENERGY CARRIER

Structure and Function of ATP

Photosynthetic cells do not run directly on sunlight. First they convert sunlight energy to the chemical energy of **ATP**. This is the abbreviation for adenosine triphosphate, a type of nucleotide.

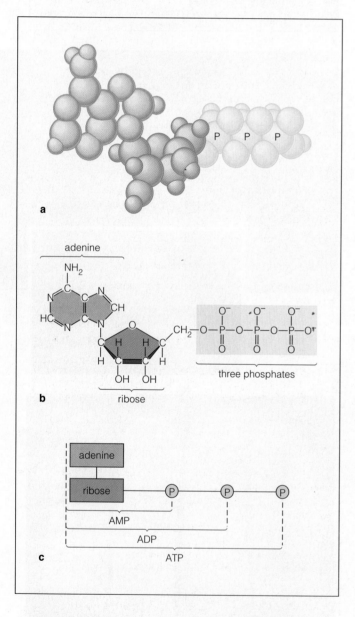

Figure 6.15 ATP—adenosine triphosphate, the main energy carrier in cells. (**a**) Three-dimensional model depicting its component atoms. (**b**) The structural formula for ATP. (**c**) Adenosine diphosphate (ADP) forms when ATP gives up one phosphate group. Adenosine monophosphate (AMP) forms when another phosphate group is removed.

Besides this, no cell whatsoever *directly* uses energy released during glucose breakdown. First the energy is harnessed and converted to ATP energy.

As Figure 6.15 shows, an ATP molecule consists of adenine (a nitrogen-containing compound), the five-carbon sugar ribose, and a string of three phosphate groups. Covalent bonds hold these components together—and the bonding arrangement is not that stable. Many hundreds of different enzymes can easily split off the outermost phosphate group by hydrolysis reactions. When they do, a great deal of the energy that is released can be tapped for hundreds of different tasks—such as breaking apart molecules, actively transporting substances across cell membranes, and triggering muscle contraction.

Thus ATP molecules are like coins of a nation—they are the cell's common currency of energy.

The ATP/ADP Cycle

Given the central role of ATP in metabolism, it comes as no surprise that cells have mechanisms for renewing the molecule after it gives up phosphate. In the **ATP/ADP cycle**, an energy input drives the binding of adenosine diphosphate (ADP) to a phosphate group or to unbound phosphate (P_i), forming ATP. Then the ATP is ready to donate a phosphate group elsewhere and revert back to ADP (Figure 6.16).

The attachment of phosphate to a molecule is called **phosphorylation**. Keep this molecular event in mind in chapters to follow. Generally speaking, when a molecule becomes phosphorylated, as by ATP, its store of energy increases *and it becomes primed to enter a specific reaction.*

ATP delivers energy to or picks up energy from almost all metabolic pathways.

With the ATP/ADP cycle, cells have a renewable means of conserving energy and transferring it to specific reactions.

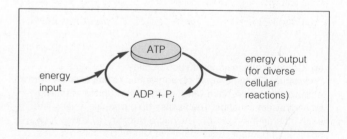

Figure 6.16 The ATP/ADP cycle.

SUMMARY

1. Cells acquire and use energy to build, store, break down, and rid themselves of substances. These activities are called metabolism. They underlie growth, maintenance, and reproduction of all organisms.

2. Energy flows in one direction through the world of life. The sun is the primary energy source. It replaces metabolically generated energy that all cells inevitably lose to their surroundings (as heat, mostly).

3. A metabolic pathway is a stepwise sequence of reactions in cells. In biosynthetic pathways, large molecules are assembled, and energy becomes stored in them. In degradative pathways, large molecules are broken down to smaller ones, and energy is released.

4. These substances take part in metabolic reactions:

   a. Substrates (or reactants): the substances that enter a specific reaction.

   b. Enzymes: proteins that serve as catalysts (they speed up reactions).

   c. Cofactors: coenzymes (including NAD^+) and metal ions that help catalyze reactions or carry electrons, hydrogen, or functional groups stripped from substrates.

   d. Energy carriers: mainly ATP, which readily donates energy to other molecules. Most biosynthetic pathways are driven by ATP energy.

   e. End products: the substances formed at the end of a metabolic pathway.

5. Control mechanisms stimulate or inhibit the activity of enzymes at key steps in metabolic pathways. They help coordinate the flow of substances into, through, and out of cells.

6. Electron transport systems are organized sequences of enzymes and coenzymes. They are built into cell membranes, such as those of chloroplasts and mitochondria. Electrons stripped from substrates are transferred through these systems. During certain transfers, energy is released that can be used to do work—for example, to make ATP.

Review Questions

1. State the first and second laws of thermodynamics. Does life violate the second law? *94–95*

2. In metabolic reactions, does equilibrium imply equal amounts of reactants and products? Think of a cellular activity that might keep a reaction from approaching equilibrium. *96–97*

3. Describe an enzyme and its role in metabolic reactions. *98–99*

4. What are the three components of ATP? Why is phosphorylation of a molecule by ATP so important? *104*

Self-Quiz *(Answers in Appendix IV)*

1. A cell's capacity to acquire and use energy for building and breaking apart molecules is called **metabolism**

2. Two laws of **Therm.** govern how cells acquire, convert, and transfer energy during metabolic reactions.

3. **Sun** is the primary source of energy for life on earth.
   a. Food   c. The sun
   b. Water   d. ATP

4. Which is *not* true of chemical equilibrium?
   a. Product and reactant concentrations are always equal.
   b. The rates of the forward and reverse reactions are the same.
   c. There is no further net change in product and reactant concentrations.

5. In a biosynthetic pathway, _____ .
   a. large molecules are broken down to smaller ones
   b. energy is not required for the reactions
   c. large molecules are assembled from simpler ones
   d. both a and b

6. Enzymes _____ .
   a. enhance reaction rates   c. act on specific substrates
   b. are affected by pH   d. all of the above

7. Electron transport systems involve _____ .
   a. enzymes and cofactors   c. cell membranes
   b. electron transfers and   d. all of the above
      released energy

8. The main energy carriers in cells are _____ .
   a. NAD^+ molecules   c. ATP molecules
   b. cofactors   d. enzymes

9. Match each substance with its correct description.
   c_ a coenzyme or metal ion   a. substrate
   d_ mainly ATP   b. enzyme
   a_ substance entering a reaction c. cofactor
   b_ protein that catalyzes a reaction d. energy carrier

Selected Key Terms

activation energy *99*
active site *98*
ATP (adenosine triphosphate) *104*
ATP/ADP cycle *104*
bioluminescence *103*
chemical equilibrium *96*
coenzyme *101*
cofactor *98*
electron transport system *102*
end product *98*
energy *93*
energy carrier *98*
entropy *95*
enzyme *98*
FAD (flavin adenine dinucleotide) *101*
feedback inhibition *100*
first law of thermodynamics *94*
induced-fit model *98*
intermediate *98*
metabolic pathway *98*
metabolism *93*
NAD^+ *101*
$NADP^+$ *101*
phosphorylation *104*
second law of thermodynamics *95*
substrate *98*

Readings

Fenn, J. 1982. *Engines, Energy, and Entropy.* New York: Freeman. Deceptively simple paperback.

Rusting, R. December 1992. "Why Do We Age?" *Scientific American* 267(6):131–141.

7 ENERGY-ACQUIRING PATHWAYS

Sun, Rain, and Survival

Just before dawn in the Midwest, the air is dry and motionless. The heat that has scorched the land for weeks still rises from the earth and hangs in the air of a new day. There are no clouds in sight. There is no promise of rain. For hundreds of miles, crops stretch out, withered and nearly dead. All the marvels of modern agriculture can't save them. In the absence of one vital substance—water—life in each cell of those many hundreds of thousands of plants has ceased.

In Los Angeles, a student wonders if the Midwest drought will bump up food prices. In Washington, D.C., economists analyze crop failures in terms of tonnage available for domestic consumption and export.

Thousands of kilometers away, in the vast Sahel Desert of Africa, grasses and cattle are dying after a similar unrelenting drought. Children with bloated bellies and spindly legs wait passively for death. Deprived of nourishment for too long, cells of their bodies will never function normally again.

You are about to explore pathways by which cells trap and use energy. At first these pathways may seem to be far removed from your everyday world. *Yet the food that nourishes you and nearly all other organisms cannot be produced or used without them.*

We will return to this point in later chapters, when we address major concerns such as human population

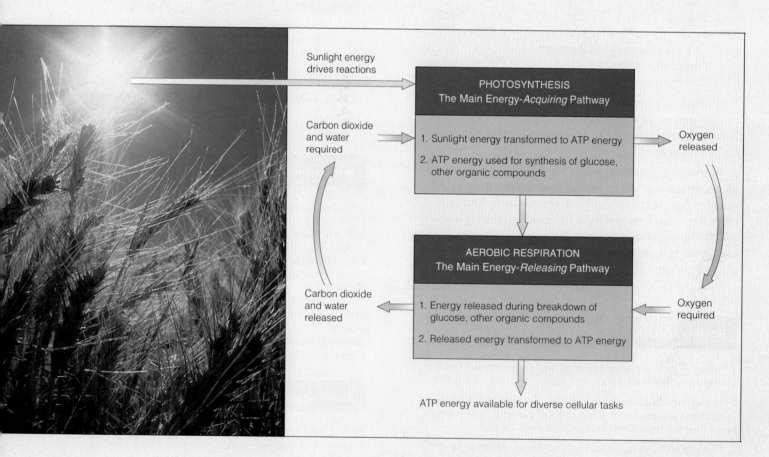

Figure 7.1 Links between photosynthesis and aerobic respiration, the main energy-acquiring and energy-releasing pathways in the world of life.

growth, nutrition, limits on agriculture, genetic engineering of plants, and effects of pollution on crops. Here, our point of departure is the *source* of food—which isn't a farm or a supermarket or a refrigerator. What we call "food" was put together somewhere in the world by living cells from glucose and other organic compounds. Such compounds are built on a framework of carbon atoms, so the questions become these:

1. Where does the carbon come from in the first place?

2. Where does the energy come from to drive the assembly of carbon-based compounds?

Answers to these questions depend on an organism's mode of nutrition.

Organisms classified as **autotrophs** are "self-nourishing," which is what *autotroph* means. Their carbon source is carbon dioxide (CO_2), a gaseous substance all around us in the air and dissolved in water. The world's plants, certain protistans, and certain bacteria are *photo*autotrophs; sunlight is their energy source. A few bacteria are *chemo*autotrophs; they can extract energy from an inorganic substance, such as sulfur.

Most organisms classified as **heterotrophs** get their carbon *and* energy from organic compounds that other organisms have already put together. They eat autotrophs, each other, and organic wastes. (*Hetero-* means other, as in "nourished by other organisms.") That is how animals, fungi, many of the protistans, and most bacteria stay alive.

When you think about this grand pattern of who eats whom, one thing becomes clear. *Survival of nearly all organisms ultimately depends on photosynthesis, the main pathway by which carbon and energy enter the world of life.*

Once glucose and other organic compounds are assembled, cells put them in storage or use them as building blocks. When cells require energy, they can break such compounds apart by several pathways. However, *the predominant energy-releasing pathway is called aerobic respiration.* Figure 7.1 provides you with a preview of the links between photosynthesis and aerobic respiration—the focus of this chapter and the next.

KEY CONCEPTS

1. Carbon-based compounds are the building blocks and energy stores of life. Plants assemble these compounds by photosynthesis. First they trap sunlight energy and convert it to chemical energy (in the form of bonds in ATP molecules). Then ATP delivers energy to reactions in which glucose is put together from carbon dioxide and water. Finally, glucose subunits are combined to form starch and other molecules.

2. Photosynthesis is the biosynthetic pathway by which most carbon and energy enter the web of life.

3. In plant cells, photosynthesis proceeds in organelles called chloroplasts. The pathway starts at a membrane system inside the chloroplast. Its machinery includes light-absorbing pigments, enzymes, and a coenzyme ($NADP^+$), which delivers hydrogen and electrons to the synthesis reactions.

4. Photosynthesis is often summarized this way:

$$12H_2O + 6CO_2 \xrightarrow{\text{sunlight}} 6O_2 + C_6H_{12}O_6 + 6H_2O$$

7.1 PHOTOSYNTHESIS: AN OVERVIEW

Energy and Materials for the Reactions

Photosynthesis is an ancient pathway, and it evolved in distinct ways in different organisms. To keep things simple, let's focus on what goes on in lettuce, weeds, and other leafy plants.

The pathway consists of two stages, each with its own set of reactions. In the *light-dependent* reactions, energy from sunlight is absorbed and converted to ATP energy. Water molecules are split, and the coenzyme NADP$^+$ picks up the liberated hydrogen and electrons, thereby becoming NADPH.

In the *light-independent* reactions, ATP donates energy to sites where glucose ($C_6H_{12}O_6$) is put together from carbon, hydrogen, and oxygen. Carbon dioxide (CO_2) provides the carbon and oxygen. Water provides the hydrogen, as delivered by NADPH.

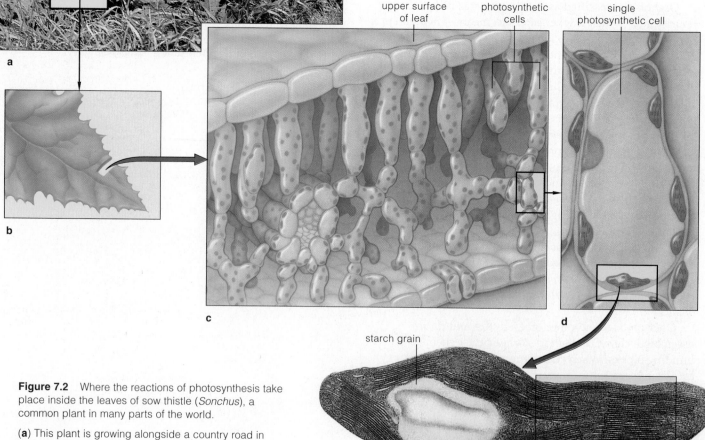

upper surface of leaf   photosynthetic cells   single photosynthetic cell

starch grain

Figure 7.2 Where the reactions of photosynthesis take place inside the leaves of sow thistle (*Sonchus*), a common plant in many parts of the world.

(**a**) This plant is growing alongside a country road in Germany. The subsequent photographs and diagrams take us inside one of the leaves.

(**b,c**) Here is a close-up of one leaf, then of a small section cut from the leaf. (**d**) Within that leaf section are numerous photosynthetic cells, one of which is illustrated. (**e**) Inside that cell are chloroplasts, the organelle of photosynthesis.

e chloroplast

Photosynthesis is often summarized in this manner:

$$12H_2O + 6CO_2 \xrightarrow{\text{sunlight}} 6O_2 + C_6H_{12}O_6 + 6H_2O$$

This summary equation shows glucose as an end product in order to keep the chemical bookkeeping simple. However, the reactions don't really stop with glucose. Glucose and other simple sugars combine at once to form sucrose, starch, and other carbohydrates—the true end products of photosynthesis.

Where the Reactions Take Place

The two stages of photosynthesis proceed at different sites inside the chloroplast. Only the photosynthetic cells of plants and a few protistans contain this type of organelle (page 68). Each chloroplast has two outer membranes wrapped around its interior, the **stroma**. An inner membrane weaves through the stroma, which is largely fluid. Often the membrane has the form of flattened channels and disks, which are arranged in stacks called grana (singular, granum). The first stage of photosynthesis proceeds at this inner membrane, which is the **thylakoid membrane system**. The spaces inside the disks and channels connect as a single compartment for hydrogen ions, which are used in ATP production. The second stage of photosynthesis—the reactions by which sugars are assembled—takes place in the stroma.

Figure 7.2 marches down through a chloroplast from sow thistle (*Sonchus*), a common weed. Two thousand of those chloroplasts, lined up single file, would be no wider than a dime. Imagine all the chloroplasts in just one weed or lettuce leaf, each a tiny factory for producing sugars and starch—and you get an idea of the magnitude of metabolic events required to feed you and all other organisms living together on this planet.

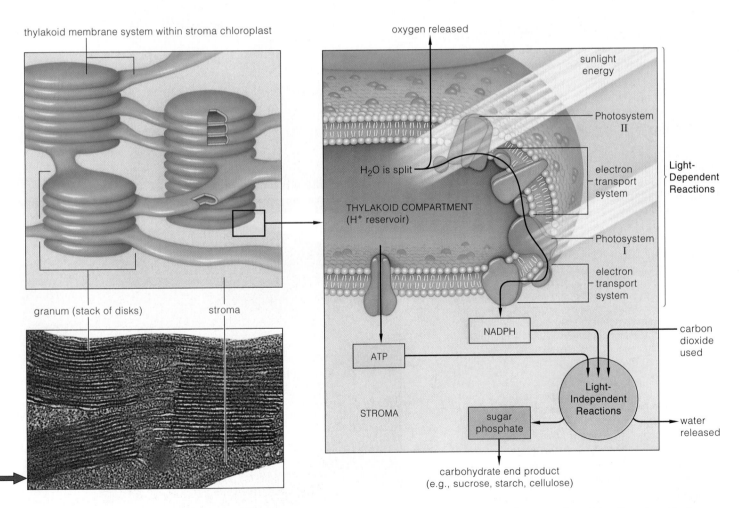

(**f**) Inside the photosynthetic cell, light-dependent reactions proceed at the thylakoid membrane system. Light-independent reactions proceed within the stroma.

(**g**) This diagram provides an overview of the sites where the key steps of both stages of reactions proceed.

7.2 LIGHT-TRAPPING PIGMENTS

Generally speaking, **pigments** are molecules that can absorb light. In animals, melanin and other pigments have roles in vision, coloration of body surfaces, and other functions. In photosynthetic organisms, a variety of pigment molecules trap photons from the sun.

Photons are packets of light energy that travel through space in undulating motion, rather like ocean waves. A photon's wavelength (the distance from one wave peak to the next) is related to its energy. The most energetic photons travel as short wavelengths, and the least energetic as long wavelengths. Humans and some other animals can perceive part of the spectrum of different wavelengths as different colors of light (Figure 7.3a).

In the thylakoid membranes of chloroplasts, clusters of pigments trap light of certain wavelengths. Just as some ocean waves are more exciting than others to surfers, certain wavelengths are more exciting to these pigments. For example, **chlorophylls** absorb violet-to-blue as well as red wavelengths (Figure 7.3b). They are key photosynthetic pigments in green algae and plants (Figure 7.4). They transmit green wavelengths, so that is why plant parts having an abundance of chlorophyll appear green to us. As another example, a variety of **carotenoids** absorb violet and blue wavelengths but transmit red, orange, and yellow.

In all plants, chlorophyll *a* is the main pigment of photosynthesis. Chlorophyll *b*, carotenoids, and other pigments absorb energy of different wavelengths. They don't use the energy themselves; they transfer it to the main pigment and so enhance its effectiveness.

In green leaves, the carotenoids are far less abundant than the chlorophylls. Often they become visible in autumn, when many plants stop producing chlorophyll (Figure 7.5). Several other pigments contribute to the distinctive coloration of various organisms. Among these are the red and blue **phycobilins**, the signature pigments of red algae and cyanobacteria.

Chlorophylls and other pigment molecules absorb different wavelengths of light, which correspond to different colors.

Figure 7.3 Wavelengths of visible light. These fall within a much larger range of wavelengths, called the electromagnetic spectrum (**a**). Various organisms use the wavelengths that range from about 400 to 750 nanometers for photosynthesis, vision, and other light-requiring processes. Shorter wavelengths, including ultraviolet light and x-rays, are energetic enough to break bonds in organic compounds. They can destroy cells.

(**b**) Wavelengths absorbed by photosynthetic pigments. Colors within the graph line for each kind of pigment correspond to the wavelengths that it absorbs. Three pigment classes are represented: the chlorophylls (*green* lines), carotenoids (*yellow-orange* line), and phycobilins (*purple* and *blue* lines). Peaks in the ranges of absorption correspond to the measured amount of energy absorbed and used in photosynthesis. The dashed line shows what would happen if we *combined* the amounts of energy that all the different pigments absorb. Taken together, photosynthetic pigments can absorb most of the wavelengths of visible light.

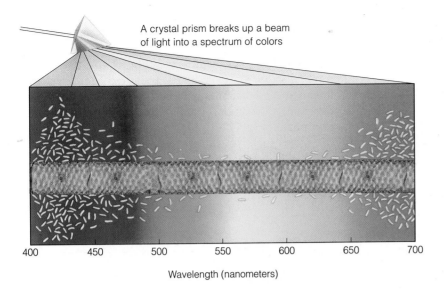

A crystal prism breaks up a beam of light into a spectrum of colors

Wavelength (nanometers)

Figure 7.4 T. Englemann's 1882 experiment that correlated photosynthetic activity of a strandlike green alga (*Spirogyra*) with certain wavelengths of light.

Oxygen is a by-product of photosynthesis. Many organisms use oxygen (for aerobic respiration). Among them are certain free-living bacteria that tumble about and so move through their aquatic habitat. Englemann reasoned that oxygen-requiring bacteria living in the same habitats as the alga would congregate where oxygen was being produced. He put a strand of the alga in a drop of water containing such bacteria, then mounted it on a microscope slide. He used a crystal prism to cast a spectrum of colors across the slide. Bacteria congregated mostly where violet and red wavelengths fell across the strand. Those wavelengths were most effective for photosynthesis (and oxygen production).

maple leaf in summer

maple leaf in autumn

Figure 7.5 Changes in leaf color in autumn.

Chloroplasts of mature leaves contain chlorophylls, carotenoids, and other pigments. (The carotenoids include the yellow carotenes and xanthophylls.) Intensely green leaves have an abundance of chlorophyll pigments that are masking the presence of other pigments. In many species, the gradual reduction in daylength in autumn and other factors trigger the breakdown of chlorophyll, and more colors show through.

Also in autumn, water-soluble anthocyanins accumulate in the central vacuoles of leaf cells. These pigments appear red when the plant fluids are slightly acidic, blue when the fluids are basic (alkaline), or colors in between when fluids are at intermediate pH levels. Soil conditions contribute to the differences in pH.

Birch, aspen, and other tree species always have the same characteristic color in autumn. The color of other species, including maple, ash, and sumac, varies around the country. Color even varies from one leaf to the next, depending on the pigment combinations.

7.3 LIGHT-DEPENDENT REACTIONS

Three events unfold during the **light-dependent reactions**, the first stage of photosynthesis. *First*, pigments absorb sunlight energy and give up electrons. *Second*, electron and hydrogen transfers lead to ATP and NADPH formation. *Third*, the pigments that gave up electrons in the first place get electron replacements.

Photosystems

Embedded in thylakoid membranes are light-trapping clusters called **photosystems**. There may be many thousands of them, and each includes 200 to 300 pigment molecules. Most of the pigments merely "harvest" sunlight. When they absorb a photon, one of their electrons gets boosted to a higher energy level:

Energy boost from absorbed photon raises an electron in a pigment to a higher energy level (next "shell" away from nucleus)

Electron gives off extra energy, returns to lower energy level

When the electron returns to a lower level, it quickly gives up the added energy. The energy bounces among pigments, and a bit is lost at each bounce (as heat). Soon the energy remaining corresponds to a certain wavelength that only a few special chlorophylls can trap. Only these chlorophylls give up the electrons used in photosynthesis. As you will see, they will transfer them to an electron-accepting molecule poised at the start of a neighboring transport system.

ATP and NADPH: Loading Up Energy, Hydrogen, and Electrons

Recall that **electron transport systems** are organized sequences of enzymes and other proteins bound in a cell membrane. When excited electrons are transferred through these systems, they release their extra energy, some of which is harnessed to drive specific reactions. In thylakoid membranes, two kinds of photosystems give up electrons to different transport systems. They allow plants to make ATP by two different pathways, one cyclic and the other noncyclic.

Cyclic Pathway One pathway starts at an electron-donating chlorophyll of a "type I" photosystem. In the **cyclic pathway of ATP formation**, electrons "cycle" from chlorophyll P700, through a transport system, then back to P700. Energy linked with the electron flow drives the formation of ATP from ADP and unbound phosphate:

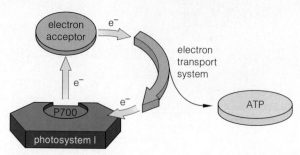

The cyclic pathway is probably the oldest means of ATP production. The first cells to use it were as tiny as existing bacteria, so their body-building programs were scarcely enormous. ATP alone would have provided enough energy to build their organic compounds. Building larger organisms requires far more organic compounds—*and vast amounts of hydrogen atoms and electrons*. Long ago, in the forerunners of multicelled plants, the cyclic pathway's machinery underwent expansion and became the basis of a more efficient ATP-forming pathway. Amazingly, this tiny bit of new machinery changed the course of evolution.

Noncyclic Pathway Today, the cyclic pathway still operates in trees, weeds, and other leafy members of the plant kingdom. But the **noncyclic pathway of ATP formation** dominates. Electrons are not cycled through this pathway. They depart (in NADPH), and electrons from *water molecules* replace them.

The pathway goes into operation when the sun's rays bombard a "type II" photosystem. Photon energy makes this photosystem's special chlorophyll (P680) give up electrons. It also triggers **photolysis**, a reaction sequence in which water molecules split into oxygen, hydrogen ions, and electrons. P680 attracts the rather unexcited electrons as replacements for the excited ones that got away (Figure 7.6).

Meanwhile, the excited electrons are transferred through a transport system—then to chlorophyll P700 of *photosystem I*. Electrons arriving at P700 haven't lost all of their extra energy. Because photons are also bombarding P700, the incoming energy boosts electrons to a higher energy level that allows them to enter a second transport system. One enzyme of this system has a helper—$NADP^+$. This coenzyme picks up two electrons and a hydrogen ion, and so becomes NADPH. It carts hydrogen and electrons to sites where organic compounds are built.

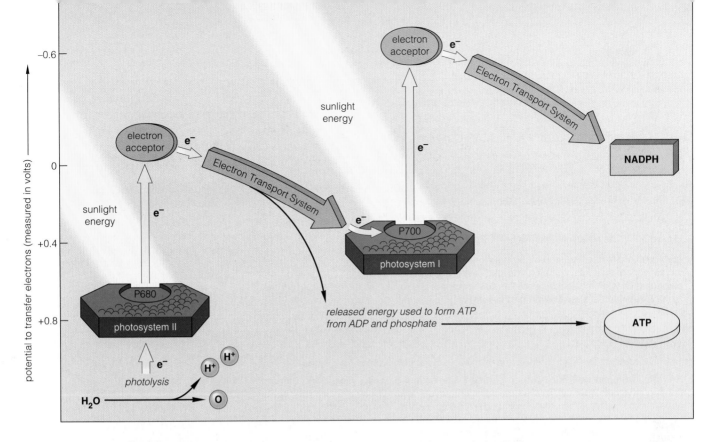

Figure 7.6 Noncyclic pathway of ATP formation, which also yields NADPH. Electrons derived from the splitting of water molecules (photolysis) travel through two photosystems. By working together, these boost electrons to an energy level that is high enough to lead to NADPH formation.

Figure 7.7 Visible evidence of photosynthesis—oxygen emerging from leaves of *Elodea*, an aquatic plant.

The Legacy—A New Atmosphere

On sunny days, on the surfaces of aquatic plants, you can see bubbles of oxygen. Figure 7.7 shows an example of this. The oxygen is a by-product of the noncyclic pathway of photosynthesis. As described in Chapter 21, this pathway may have evolved more than 2 billion years ago. At first, the oxygen simply dissolved in seas, lakes, wet mud, and other bacterial habitats. By about 1.5 billion years ago, however, large amounts of dissolved oxygen were escaping into what had been an oxygen-free atmosphere. Its accumulation changed the atmosphere forever. And it made possible aerobic respiration, the most efficient pathway for extracting energy

from organic compounds. The emergence of the noncyclic pathway ultimately allowed you and all other animals to be around today, breathing the oxygen that helps keep your cells alive.

In the light-dependent reactions, energy from the sun drives the formation of ATP (which carries energy) and NADPH (which carries hydrogen and electrons).

Oxygen, a by-product of photosynthesis, profoundly changed the early atmosphere and made aerobic respiration possible.

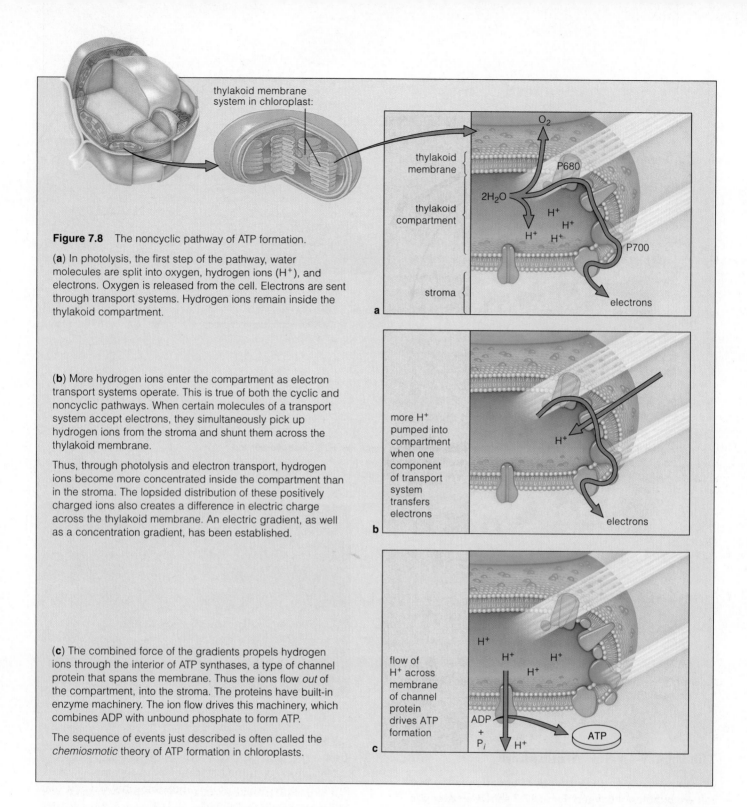

Figure 7.8 The noncyclic pathway of ATP formation.

(**a**) In photolysis, the first step of the pathway, water molecules are split into oxygen, hydrogen ions (H^+), and electrons. Oxygen is released from the cell. Electrons are sent through transport systems. Hydrogen ions remain inside the thylakoid compartment.

(**b**) More hydrogen ions enter the compartment as electron transport systems operate. This is true of both the cyclic and noncyclic pathways. When certain molecules of a transport system accept electrons, they simultaneously pick up hydrogen ions from the stroma and shunt them across the thylakoid membrane.

Thus, through photolysis and electron transport, hydrogen ions become more concentrated inside the compartment than in the stroma. The lopsided distribution of these positively charged ions also creates a difference in electric charge across the thylakoid membrane. An electric gradient, as well as a concentration gradient, has been established.

(**c**) The combined force of the gradients propels hydrogen ions through the interior of ATP synthases, a type of channel protein that spans the membrane. Thus the ions flow *out* of the compartment, into the stroma. The proteins have built-in enzyme machinery. The ion flow drives this machinery, which combines ADP with unbound phosphate to form ATP.

The sequence of events just described is often called the *chemiosmotic* theory of ATP formation in chloroplasts.

A Closer Look at ATP Formation in Chloroplasts

As you may have noticed, we've saved the trickiest question for last. *How*, exactly, does ATP form during the noncyclic (and cyclic) pathways?

As Figure 7.8 shows, when electrons flow through the membrane-bound transport systems, they pick up hydrogen ions (H^+) outside the membrane and dump them into the thylakoid compartment. This sets up H^+ concentration and electric gradients across the membrane. Hydrogen ions that were split away from water molecules increase the gradients. The ions respond by flowing out through the interior of channel proteins that span the membrane. Energy associated with the flow drives the binding of unbound phosphate to ADP, the result being ATP.

7.4 LIGHT-INDEPENDENT REACTIONS

The **light-independent reactions** are the "synthesis" part of photosynthesis. ATP molecules deliver the required energy for the reactions. NADPH molecules deliver the required hydrogen and electrons. Carbon dioxide (CO_2) in the air around photosynthetic cells provides the carbon and oxygen.

We say the reactions are light-independent because they don't depend directly on sunlight. They can proceed even in the dark, as long as ATP and NADPH are available.

Capturing Carbon

Let's track a CO_2 molecule that diffuses into the air spaces inside a sow thistle leaf and ends up next to a photosynthetic cell (Figure 7.2d). From there, it diffuses across the plasma membrane and on into the stroma of a chloroplast. The light-independent reactions start when the carbon atom of the CO_2 becomes attached to **RuBP** (ribulose bisphosphate), a molecule with a backbone of five carbon atoms. This is called **carbon dioxide fixation**.

Building the Glucose Subunits

Attaching carbon to RuBP is the first step of a cyclic pathway that produces a sugar phosphate molecule *and* regenerates the RuBP. The pathway is named the **Calvin-Benson cycle**, in honor of its discoverers. A specific enzyme catalyzes each step. For our purposes, we can simply focus on the carbon atoms of the pathway's substrates, intermediates, and end products. Figure 7.9 shows the carbon atoms as red circles.

The attachment of a carbon atom to RuBP produces an unstable six-carbon intermediate. This splits into two molecules of **PGA** (phosphoglycerate), each with a three-carbon backbone. ATP donates a phosphate group to each PGA. NADPH donates hydrogen and electrons to the resulting intermediate, thereby forming **PGAL** (phosphoglyceraldehyde).

The reaction steps just outlined proceed not once but *six* times. In other words, six CO_2 molecules are fixed, and twelve PGAL molecules are produced. Most of the PGAL becomes rearranged to form new RuBP molecules, which can be used to fix more carbon. But *two* of the PGAL combine, forming a six-carbon sugar phosphate. When a sugar has a phosphate group attached, it is primed for further reaction (refer to Figure 7.9).

The Calvin-Benson cycle produces enough RuBP molecules to replace the ones used in carbon dioxide fixation. The ADP, NADP$^+$, and phosphate leftovers

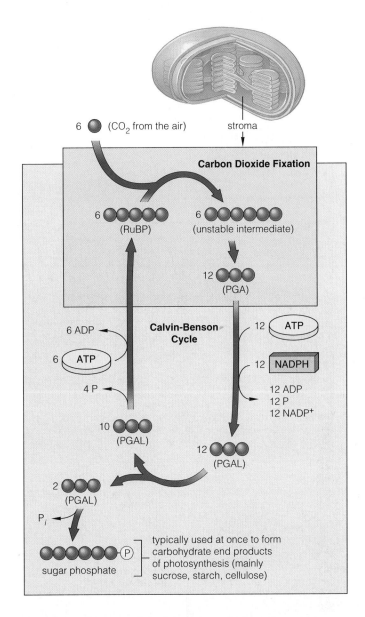

Figure 7.9 Summary of the light-independent reactions of photosynthesis. Carbon atoms of the key molecules are shown in red. All of the intermediates have one or two phosphate groups attached. However, for simplicity, only the phosphate on the resulting sugar phosphate is shown.

diffuse back to sites of the light-dependent reactions and can be converted once more to NADPH and ATP. The sugar phosphate formed in the cycle can serve as a building block for sucrose, starch, or cellulose—the plant's main carbohydrates. Synthesis of these large organic compounds by other pathways marks the conclusion of the light-independent reactions.

During the Calvin-Benson cycle, carbon is "captured" from carbon dioxide, a sugar phosphate forms in reactions that require ATP and NADPH, and RuBP (needed to capture the carbon) is regenerated.

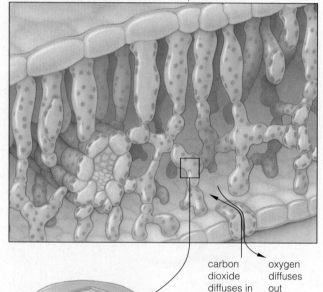

carbon dioxide diffuses in

oxygen diffuses out

one chloroplast from one photosynthetic cell within a leaf

7.5 THE REACTIONS, START TO FINISH

Be sure to look carefully at Figure 7.10. It summarizes the key reactants, intermediates, and products of both the light-dependent and light-independent reactions of photosynthesis.

During daylight hours, photosynthetic cells convert the newly formed sugar phosphates to sucrose or starch. Of all plant carbohydrates, sucrose is the most easily transportable, and starch is the most common storage form. The cells convert excess PGAL to starch

Figure 7.10 Summary of the main reactants, intermediates, and products of photosynthesis, corresponding to the equation:

$$12H_2O + 6CO_2 \xrightarrow{\text{sunlight}} 6O_2 + C_6H_{12}O_6 + 6H_2O$$

Starting with the light-dependent reactions, the splitting of twelve water molecules yields twenty-four electrons and six molecules of O_2. For every four electrons, three ATP and two NADPH form.

During the light-independent reactions, *each turn* of the Calvin-Benson cycle requires one CO_2, three ATP, and two NADPH. Each sugar phosphate that forms has a backbone of six carbon atoms—so its formation requires six turns of the cycle.

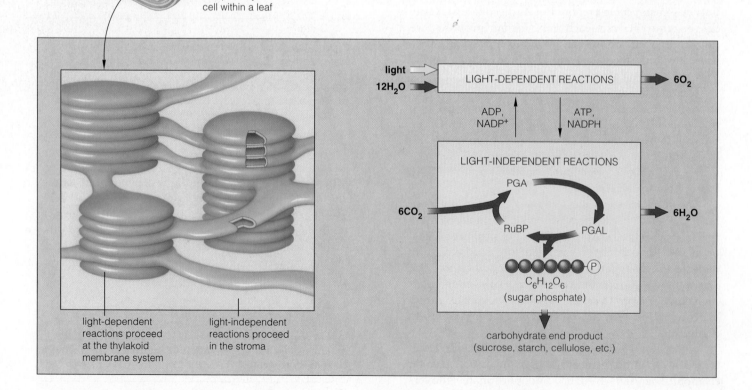

light-dependent reactions proceed at the thylakoid membrane system

light-independent reactions proceed in the stroma

light
$12H_2O$
LIGHT-DEPENDENT REACTIONS
$6O_2$

ADP, NADP⁺
ATP, NADPH

LIGHT-INDEPENDENT REACTIONS

PGA

$6CO_2$

RuBP
PGAL

$6H_2O$

$C_6H_{12}O_6$
(sugar phosphate)

carbohydrate end product
(sucrose, starch, cellulose, etc.)

Pastures of the Seas

Drifting through the surface waters of the world ocean are uncountable numbers of single cells. You can't see them without a microscope—a row of 7 million cells of one species would be less than a quarter-inch long. Yet are they abundant! In some parts of the world, a cup of seawater may hold 24 million cells of one species, and that doesn't even include all the cells of *other* aquatic species.

Many of the drifters are photosynthetic bacteria, protistans, and plants. Together they are the pastures of the seas, the food base for diverse consumers. The pastures "bloom" in spring, when waters become warmer and enriched with nutrients churned up from the deep by winter currents. Then, populations burgeon as cells divide again and again. Biologists had no idea that the number of cells and their distribution were so mind-boggling until satellites provided photographs from space. For example, the satellite image in Figure *b* shows a springtime bloom in the North Atlantic stretching from North Carolina all the way to Spain!

Those single cells have enormous impact on the world's climate. As they collectively photosynthesize, they sponge up nearly half the carbon dioxide we humans release each year (as when we burn fossil fuels and forests). Without them, carbon dioxide in the air would accumulate more rapidly and possibly accelerate global warming, as described in Chapter 49. If our planet warms too much, vast coastal regions may become submerged, and the current food-producing nations may be hit hard.

Amazingly, we daily dump industrial wastes, fertilizers, and raw sewage into the ocean and alter the living conditions for those drifting cells. How much of that noxious chemical brew can they tolerate?

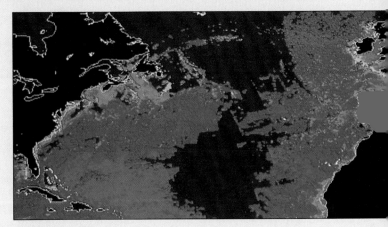

a Photosynthetic activity in winter. (In this color-enhanced image, red-orange shows where chlorophyll concentrations are greatest.)

b Photosynthetic activity in spring.

also. They briefly store the starch as grains in the stroma (you can see one of these in Figure 7.2*e*). After the sun goes down, the cells convert their starch to sucrose, for export to other living cells in leaves, stems, and roots. Ultimately, the products and intermediates of photosynthesis end up as energy sources and building blocks for all of the lipids, amino acids, and other organic compounds required for plant growth, survival, and reproduction.

When studying Figure 7.10, remind yourself of how photosynthesis "fits" in the world of living things. Think of the mind-boggling numbers of single-celled and multicelled photosynthetic organisms in which the reactions are proceeding at this very moment, on land and in the sunlit waters of the earth. The sheer volume of the reactant and product molecules that the photosynthesizers deal with may surprise you. The *Focus* essay above is a case in point.

FIXING CARBON— SO NEAR, YET SO FAR

If light intensity, air temperature, rainfall, and soil composition were uniform all year long, everywhere in the world, then maybe the photosynthetic reactions would proceed in exactly the same way in all plants. However, environments obviously differ—and so do the details of photosynthesis among plants that have evolved in different places. A brief comparison of two different carbon-fixing adaptations to stressful environmental conditions will be enough to emphasize this point.

Fixing Carbon Twice, In Two Cell Types

Think of a Kentucky bluegrass plant, with its narrow, blade-shaped leaves. As is true of all plants, its growth depends on CO_2 uptake. Athough CO_2 is plentiful in the air, it is not always abundantly available to photosynthetic cells *inside the leaves*. Leaves have a waxy cover that restricts water loss. Water escapes mainly through stomata (singular, stoma), which are tiny openings across the leaf surface. Most CO_2 diffuses into the leaves, and most O_2 diffuses out through these openings.

On hot, dry days, all plants close their stomata and so conserve water. But when they do, CO_2 can't diffuse into their leaves. Meanwhile, photosynthetic cells are busy, so oxygen builds up in each leaf. The stage is set for a wasteful process called "photorespiration." By this process, oxygen instead of CO_2 becomes attached to the RuBP used in the Calvin-Benson cycle, with different results:

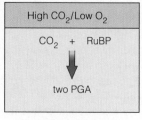
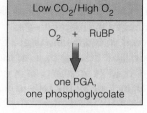

High CO_2/Low O_2	Low CO_2/High O_2
CO_2 + RuBP	O_2 + RuBP
↓	↓
two PGA	one PGA, one phosphoglycolate
Calvin-Benson cycle predominates	photorespiration predominates

Formation of sugar phosphates depends on PGA. When photorespiration wins out, less PGA forms—and the bluegrass plant's capacity for growth suffers.

Kentucky bluegrass and many other species are known as **C3 plants**, because the three-carbon PGA is the first intermediate formed by carbon fixation. By contrast, when corn, crabgrass, sugarcane, and many other plants fix carbon, the resulting intermediate is not PGA. It is the four-carbon oxaloacetate (Figure 7.11). Hence the name, **C4 plants**.

Figure 7.11 C4 pathway. (**a**) Internal structure of a leaf from corn (*Zea mays*), a typical C4 plant.

Photosynthetic mesophyll cells (*light green*) surround bundle-sheath cells (*dark green*). These in turn surround veins, the transport tubes that carry water into leaves and photosynthetic products away from them.

(**b**) A carbon-fixing system precedes the Calvin-Benson cycle in C4 plants.

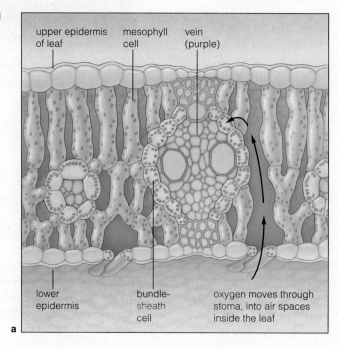
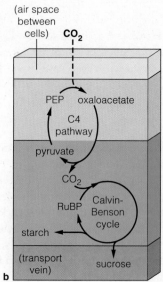

C4 plants maintain adequate amounts of CO_2 inside their leaves even though they, too, close stomata on hot, dry days. They fix CO_2 not once but twice, in two different types of photosynthetic cells. *Mesophyll* cells have first crack at the CO_2 in the leaf and use it to form oxaloacetate. It is a temporary fix. The oxaloacetate is quickly transferred to *bundle-sheath* cells that wrap around every vein in C4 leaves. In those cells, CO_2 is released and fixed again—in the Calvin-Benson cycle (Figure 7.11).

With their more efficient carbon-fixing mechanism, C4 species get by with tinier stomata—and so lose less water—than C3 species. Generally speaking, C4 plants are better adapted where temperatures are highest during the growing season.

For example, of all plant species that evolved in Florida, 80 percent are C4 plants—compared to 0 percent in Manitoba, Canada. Kentucky bluegrass and other C3 species have an advantage where temperatures drop below 25°C; they are less sensitive to cold. Mix C3 and C4 species from different regions in a garden, and one or the other kind will do better at least part of the year. That is why a lawn of Kentucky bluegrass does well during cool spring weather in San Diego, only to be overwhelmed during the hot summer by a C4 plant—crabgrass.

Figure 7.12 Giant saguaro (*Carnegiea gigantea*) and prickly pear (*Opuntia*) growing in the Sonoran desert of Arizona. Both kinds of cacti have reduced leaves, protective spines, and fleshy stems that function in photosynthesis as well as in water storage.

Fixing and Storing Carbon by Night, Using It by Day

We find a different carbon-fixing adaptation in deserts and other severely dry environments. Consider the **succulents**—plants with juicy, water-storing tissues and very thick surface layers that restrict water loss. They include cacti, such as the giant saguaro and prickly pear of the Sonoran desert (Figure 7.12). Like many flowering plants, they cannot open their stomata during the day without losing precious water. They open them and fix carbon dioxide only *at night*. Their cells store the resulting intermediate in the central vacuole, then use it in photosynthesis *the next day*—when stomata close.

Species that show this adaptation are called **CAM plants** (a blessedly shorter version of the name Crassulacean Acid Metabolism). Unlike C4 species, CAM plants do not fix carbon in separate cells. They fix it at different times.

During murderously prolonged droughts, when most plants wither and die, some CAM plants survive by keeping their stomata closed even at night. They get carbon by repeatedly fixing the carbon dioxide that forms during aerobic respiration. Not that much forms, but it is enough to allow these plants to maintain very low rates of metabolism.

As you may have deduced, CAM plants grow slowly. Try growing a cactus in Seattle or some other place with a mild climate and it will compete poorly with C3 and C4 plants.

As is true of other plant characteristics, modifications in the basic photosynthetic machinery are adaptations to specific kinds of environments.

7.7 CHEMOSYNTHESIS

Photosynthesis is so pervasive, it sometimes is easy to overlook other, less common energy-acquiring routes. Bacteria that are classified as chemoautotrophs obtain energy not from sunlight but rather by pulling away hydrogen and electrons from ammonium ions, iron or sulfur compounds, and other inorganic substances. Many influence the global cycling of nitrogen and other vital elements through the biosphere. We will return later to their environmental effects. In this unit, we turn next to pathways by which energy is released from glucose and other biological molecules—the chemical legacy of autotrophs.

SUMMARY

1. Plants and other photoautotrophs use sunlight (as an energy source) and carbon dioxide (as the carbon source) for building organic compounds. Animals and other heterotrophs obtain carbon and energy from organic compounds already synthesized by plants and other autotrophs.

2. Photosynthesis is the main biosynthetic pathway by which carbon and energy enter the web of life. It consists of two sets of reactions that start with the trapping of sunlight energy ("photo") and proceed through the assembly reactions ("synthesis").

   a. In chloroplasts, the light-dependent reactions take place at the thylakoid membrane system. These reactions produce ATP and NADPH.

   b. The light-independent reactions take place in the stroma around the membrane system. They produce sugar phosphates that are used in building sucrose, starch, and other end products of photosynthesis.

3. These are the key points concerning the light-dependent reactions:

   a. Photons are absorbed by photosynthetic pigments of photosystems (light-absorbing clusters of molecules embedded in the thylakoid membrane). Photon energy drives the transfer of electrons from a specific chlorophyll to an acceptor molecule, which donates them to a transport system in the membrane.

   b. In the cyclic pathway of ATP formation, excited electrons leave the P700 chlorophyll of photosystem I, give up energy in the transport system, and return to that photosystem.

   c. In the noncyclic pathway of ATP formation, electrons from the P680 chlorophyll of photosystem II pass through a transport system, enter photosystem I, then pass through another transport system, then end up in NADPH. Also, water molecules split into hydrogen ions, oxygen, and electrons. These electrons replace the ones that P680 gives up.

   d. Operation of electron transport systems moves many hydrogen ions into the inner compartment of the thylakoid membrane. Hydrogen ions split from water molecules accumulate here also. The accumulation sets up concentration and electric gradients that drive ATP formation.

4. These are the key points concerning the light-independent reactions:

   a. ATP delivers energy and NADPH delivers hydrogen and electrons to the stroma of chloroplasts, where sugar phosphates form and are combined into starch, cellulose, and other end products.

   b. Sugar phosphates form during the Calvin-Benson cycle. This cyclic pathway begins when carbon dioxide from the air is affixed to RuBP, making an unstable intermediate that splits into two PGA. ATP donates a phosphate group to each PGA. The resulting molecule receives H^+ and electrons from NADPH to form PGAL.

   c. For every six CO_2 molecules that enter the cycle, twelve PGAL are produced. Two of those are used to produce a six-carbon sugar phosphate. The remainder are used to regenerate RuBP for the cycle.

5. Plants differ in their carbon-fixing machinery. C3 plants can fix carbon by the Calvin-Benson cycle only. C4 plants can fix carbon twice, in two cell types. CAM plants fix and store carbon at night, then use it by day.

Review Questions

1. A caterpillar chewing on a weed is speared and eaten by a bird, which in turn is eaten by a cat. Which of these organisms are autotrophs? heterotrophs? *107*

2. Summarize the photosynthesis reactions as an equation. State the key events of the light-dependent reactions, then of the light-independent reactions. *107–108*

3. Which of the substances listed accumulates in the thylakoid compartment of chloroplasts: glucose, photosynthetic pigments, hydrogen ions, fatty acids? *109*

4. Which substance is *not* required for the light-independent reactions: ATP, NADPH, RuBP, carotenoids, free oxygen, carbon dioxide, enzymes? *115*

5. Suppose a plant busily photosynthesizing is exposed to CO_2 molecules that contain radioactively labeled carbon atoms ($^{14}CO_2$). Identify the compound in which the labeled carbon will first appear: NADPH, PGAL, pyruvate, or PGA. *115*

6. How many CO_2 molecules must enter the Calvin-Benson cycle to produce one sugar phosphate? Why? *115*

7. Fill in the blanks (red lines) in the diagram on the next page. *109*

Self-Quiz *(Answers in Appendix IV)*

1. Molecules with a backbone of _Carbon_ serve as the main building blocks of all organisms.

2. Photosynthetic autotrophs use _CO_2_ from the air as their carbon source and _sunlight_ as their energy source.

3. In plants, light-*dependent* reactions occur at the _____ .
   a. cytoplasm   c. stroma
   b. plasma membrane   d. thylakoid membrane

4. The light-*independent* reactions occur in the _____ .
   a. cytoplasm   c. stroma
   b. plasma membrane   d. grana

5. In the light-dependent reactions, _____ .
   a. carbon dioxide is fixed
   b. ATP and NADPH form
   c. carbon dioxide accepts electrons
   d. sugar phosphates form

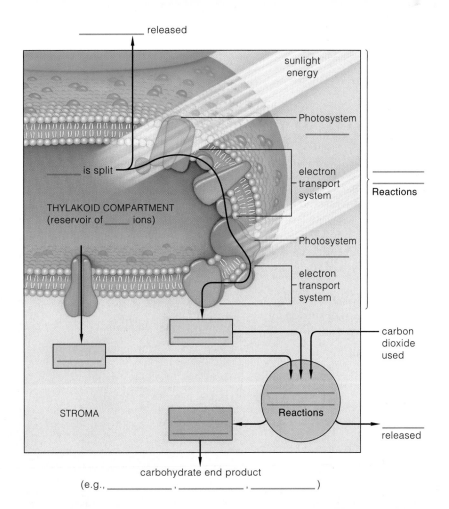

released

sunlight energy

Photosystem
_____

is split

electron transport system

Reactions

THYLAKOID COMPARTMENT
(reservoir of _____ ions)

Photosystem
_____

electron transport system

carbon dioxide used

STROMA

Reactions

released

carbohydrate end product
(e.g., _____ , _____ , _____)

6. When a photosystem absorbs light, _____ .
   a. sugar phosphates are produced
   b. electrons are transferred to an acceptor molecule
   c. RuBP accepts electrons
   d. light-dependent reactions begin
   e. both b and d are correct

7. The Calvin-Benson cycle starts when _____ .
   a. light is available
   b. light is not available
   c. carbon dioxide is attached to RuBP
   d. electrons leave a photosystem

8. In the light-independent reactions, ATP furnishes phosphate groups to _____ .
   a. RuBP   c. PGA
   b. NADP⁺   d. PGAL

9. Match each event in photosynthesis with its correct description.
   _c_ RuBP used; PGA formed   a. cyclic pathway
   _d_ ATP and NADPH used   b. noncyclic pathway
   _e_ NADPH formed   c. carbon dioxide fixation
   _b_ ATP and NADPH formed   d. PGAL formation
   _a_ only ATP formed   e. transfer of H^+ and electrons to NADP⁺

Selected Key Terms

autotroph *107*
C3 plant *118*
C4 plant *118*
Calvin-Benson cycle *115*
CAM plant *119*
carbon dioxide fixation *115*
carotenoid *110*
chlorophyll *110*
chloroplast *109*
cyclic pathway of ATP formation *112*
electron transport system *112*
heterotroph *107*
light-dependent reaction *112*
light-independent reaction *115*
noncyclic pathway of ATP formation *112*

PGA (phosphoglycerate) *115*
PGAL (phosphoglyceraldehyde) *115*
photolysis *112*
photon *110*
photosystem *112*
phycobilin *110*
pigment *110*
RuBP (ribulose bisphosphate) *115*
stroma *109*
succulent *119*
thylakoid membrane system *109*

Readings

Daviss, B. February 1992. "Going for the Green." *Discover* 13:20. Artificial systems for photosynthesis.

Hendry, George. May 1990. "Making, Breaking, and Remaking Chlorophyll." *Natural History*, pp. 36–41.

Youvan, D., and B. Marrs. 1987. "Molecular Mechanisms of Photosynthesis." *Scientific American* 256:42–50.

8 ENERGY-RELEASING PATHWAYS

The Killers Are Coming!

In 1990, "killer" bees from South America buzzed across the border between Mexico and Texas. In less than four years, they spread west to the border between Arizona and California. The bees are descended from African queen bees. When they are provoked, they can be terrifying.

In 1994, for instance, thousands of bees flew into action simply because a construction worker started up a tractor a few hundred yards away from their hive. The agitated bees entered a nearby subway station and started stinging passengers on the platform and in the trains. One person died, and a hundred others were injured.

Where did these bees come from? In the 1950s, some queen bees had been shipped from Africa to Brazil for selective breeding experiments. Why? Honeybees happen to be big business. Besides producing honey, they are rented out to pollinate commercial orchards—and they make a big difference. Put a screened cage around an orchard tree in bloom and less than 1 percent of the flowers will set fruit. Put a hive of honeybees in the same cage and 40 percent of the flowers will set fruit.

Compared with their African relatives, the bees in Brazil are sluggish pollinators and honey producers. By cross-breeding the two varieties, researchers hoped to

Figure 8.1 One of the mild-mannered honeybees buzzing in for a landing on a flower, wings beating with energy provided by ATP. If this were one of its Africanized relatives protecting a hive, possibly you would not stay around to watch the landing. Both kinds of bees look alike. How can we tell them apart? From our own biased perspective, the Africanized bees are the ones with an attitude problem.

produce a strain of mild-mannered but zippier bees. They put the local and imported bees together in artificial hives that were enclosed in nets; then they let nature take its course.

Twenty-six African queen bees escaped. That was bad enough. Then beekeepers got wind of preliminary experimental results. After learning that the first few generations of offspring were more energetic but not overly aggressive, they imported hundreds of African queens and encouraged them to mate with locals. And they set off a genetic time bomb.

Before long, African bees became established in commercial hives—and in wild bee populations. And their traits became dominant. The "Africanized" bees do everything other bees do, but they do more of it faster. Their eggs develop into adults more quickly. Adults fly more rapidly, outcompete other bees for nectar, and even die sooner.

When something disturbs their hives or swarms, Africanized bees become extremely agitated. Whereas a mild-mannered honeybee might chase an intruding animal 50 yards or so, a squadron of Africanized bees will chase it a quarter of a mile. If they catch up to it, they can sting it to death.

Doing things faster means having a nonstop supply of energy. The stomach of an Africanized bee can hold 30 milligrams of sugar-rich nectar—enough fuel to fly 60 kilometers (more than 35 miles). Compared to other kinds of bees, its flight muscle cells have larger mitochondria. And mitochondria are the organelles that specialize in releasing a great deal of energy from sugars and other organic compounds, then converting it to the energy of ATP.

In their ability to tap into the stored energy of organic compounds, Africanized bees are like all other organisms on earth. To be sure, the energy-releasing pathways differ in some details from one organism to the next. But they all require characteristic starting materials, then yield predictable products and by-products. And *all* of the energy-releasing pathways yield ATP.

In fact, throughout the biosphere, there is startling similarity in the uses to which energy and raw materials are put. *At the biochemical level, there is undeniable unity among all forms of life.* We will return to this idea in the *Commentary* that concludes this chapter.

KEY CONCEPTS

1. All organisms produce ATP by releasing energy stored in glucose and other organic compounds. Energy-releasing pathways differ from one another. But they all begin with the initial breakup of glucose by a set of reactions called glycolysis.

2. Three kinds of energy-releasing pathways dominate. Fermentation routes and anaerobic electron transport proceed in the cytoplasm. They yield a small amount of ATP for each glucose molecule metabolized. An oxygen-requiring pathway, aerobic respiration, releases far more energy from glucose. It alone proceeds in mitochondria.

3. Aerobic respiration has three stages. First (in glycolysis), glucose is broken down to pyruvate. Second, the pyruvate is broken down to carbon dioxide and water, and coenzymes pick up electrons liberated by these reactions. Third, coenzymes carry unbound hydrogen (H$^+$ ions) and electrons to a transport system. That system gives up electrons to oxygen, and its operation leads to ATP formation.

4. Over evolutionary time, photosynthesis and aerobic respiration have become linked on a global scale. Oxygen released during photosynthesis is required for the aerobic pathway. And the carbon dioxide and water released during aerobic respiration are raw materials used in building organic compounds during photosynthesis:

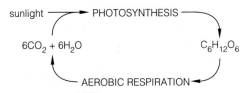

Organisms stay alive by taking in energy. Plants get energy from the sun. Animals get energy secondhand, thirdhand, and so on, by eating plants and one another. Regardless of its source, energy must be put in a form that can drive the life-sustaining metabolic reactions. Energy carried by adenosine triphosphate, ATP, serves that function.

Plants make ATP during photosynthesis. They and all other organisms also make ATP by breaking covalent bonds of carbohydrates (glucose especially), lipids, and proteins. When they do, energy stored in those bonds is released, and some of it is used to drive the formation of ATP. Electron transfers of the sort described on page 102 are at the heart of the energy-releasing pathways.

Comparison of Three Types of Energy-Releasing Pathways

For most types of cells, **aerobic respiration** is the main energy-releasing pathway leading to ATP formation. The "aerobic" part of the name means that the pathway cannot be completed without oxygen. With every breath, you provide your busily respiring cells with oxygen.

Other energy-releasing pathways are "anaerobic," in that they can be completed without using oxygen. The most common are the **fermentation pathways** and **anaerobic electron transport**. Many bacteria and many protistans depend exclusively on anaerobic pathways to make ATP. The cells of your own body run on ATP from aerobic respiration, but they also can use a fermentation pathway for short periods when they are not getting enough oxygen.

All three types of energy-releasing pathways start with the same reactions, called **glycolysis**. These reactions split and rearrange each glucose molecule into two pyruvate molecules. Glycolysis proceeds in the cytoplasm, and oxygen has no role in it.

Once this first stage is over, the energy-releasing pathways differ. Most important, the aerobic pathway continues inside a mitochondrion (Figure 8.2). There, oxygen serves as the final acceptor of electrons that are released during the reactions. Anaerobic pathways end in the cytoplasm, and a substance other than oxygen is the final electron acceptor.

As we examine the three types of pathways, keep in mind that the reaction steps do not proceed all by themselves. Enzymes catalyze each reaction step, and the intermediate produced at a given step serves as a substrate for the next enzyme in the pathway.

Overview of Aerobic Respiration

Of all energy-releasing pathways, aerobic respiration produces the most ATP for each glucose molecule. Whereas fermentation has a net yield of two ATP, the aerobic route may yield thirty-six or more. If you were a bacterium, you would not require much ATP. Being large, complex, and highly active, you depend on the high yield of the aerobic route.

When glucose is the starting material, aerobic respiration can be summarized this way:

$$C_6H_{12}O_6 + 6O_2 \longrightarrow 6CO_2 + 6H_2O$$

glucose   carbon dioxide

The summary equation tells us only what the substances are at the start and finish of the pathway. In between are three stages of reactions.

Figure 8.3 outlines the reactions. The first stage, again, is glycolysis. In the second stage, which includes the **Krebs cycle**, pyruvate is broken down completely to carbon dioxide and water. Neither stage produces much ATP. But hydrogen and electrons are stripped from intermediates during both stages—and coenzymes deliver these to an electron transport system. Such systems, along with neighboring enzymes, serve as the machinery for **electron transport phosphorylation**. This third and final stage of reactions yields many ATP molecules. As it draws to a close, oxygen accepts the "spent" electrons from the transport system.

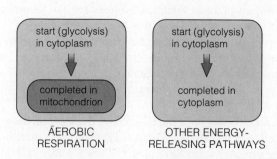

Figure 8.2 Where the main energy-releasing pathways proceed. All three start with glycolysis in the cytoplasm. Aerobic respiration alone is completed in mitochondria.

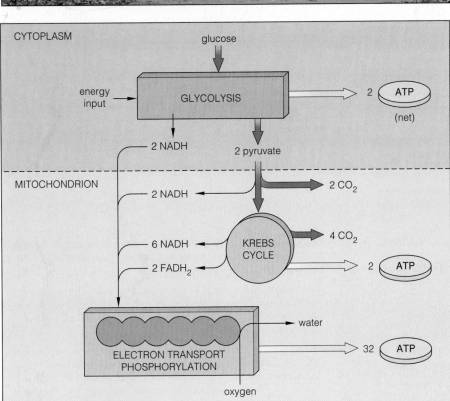

CYTOPLASM

glucose

energy input →

GLYCOLYSIS → 2 ATP (net)

2 NADH

2 pyruvate

MITOCHONDRION

2 NADH ← 2 CO_2

6 NADH ← KREBS CYCLE → 4 CO_2

2 $FADH_2$ ← 2 ATP

ELECTRON TRANSPORT PHOSPHORYLATION → water

→ 32 ATP

oxygen

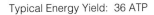

Typical Energy Yield: 36 ATP

Figure 8.3 Overview of the three stages of aerobic respiration, the main energy-releasing pathway. Only this pathway delivers enough ATP to build and maintain giant redwoods and other large, multicelled organisms. Only ATP delivers enough energy to sustain birds, humans, and other highly active animals.

During the first stage (glycolysis), glucose is partially broken down to pyruvate. During the second stage, which includes the Krebs cycle, pyruvate is broken down to carbon dioxide. In these two stages, hydrogen and electrons stripped from intermediates are loaded onto coenzymes (NAD^+ and FAD).

During the final stage, electron transport phosphorylation, the loaded coenzymes (NADH and $FADH_2$) give up hydrogen and electrons. As the electrons pass through transport systems, energy is released that indirectly drives ATP formation. Oxygen accepts the electrons at the end of the transport system.

From start (glycolysis) to finish, the aerobic pathway typically has a net energy yield of thirty-six ATP for every glucose molecule.

8.2 GLYCOLYSIS: FIRST STAGE OF THE ENERGY-RELEASING PATHWAYS

Recall, from Figure 3.4, that glucose is a simple sugar with a backbone of six carbon atoms. In glycolysis, glucose (or some other carbohydrate) present in the cytoplasm is partially broken down to **pyruvate**, a molecule with a backbone of three carbon atoms.

We can think about the glucose backbone in this simplified way:

As Figure 8.4 indicates, the first steps of glycolysis are energy-requiring. They proceed only when two ATP molecules each donate energy to the glucose backbone by transferring a phosphate group to it. Such transfers are called phosphorylations. The attachments cause the backbone to split apart into two molecules of PGAL (phosphoglyceraldehyde).

PGAL formation marks the start of the energy-releasing steps of glycolysis. Each PGAL is converted to an unstable intermediate that gives up a phosphate group to ADP, forming ATP. The next intermediate in the sequence does the same thing. Thus, a total of four ATP molecules have been formed by **substrate-level phosphorylation**—the direct transfer of a phosphate group from a substrate of the reactions to ADP. Remember, though, two ATP were invested to start the reactions. So the net energy yield is only two ATP.

Meanwhile, hydrogen atoms and electrons that were released from each PGAL are transferred to an enzyme helper—the coenzyme NAD^+. Like other coenzymes, NAD^+ is reusable. As you read on page 101, it picks up hydrogen and electrons stripped from a substrate and so becomes NADH. When the coenzyme gives them up at a different site, it becomes NAD^+ again.

In sum, glycolysis converts a bit of the energy stored in glucose to ATP energy. Hydrogen and electrons stripped from glucose have been loaded onto a coenzyme, and these have roles in the next stage of reactions. So do the end products of glycolysis—two molecules of pyruvate.

---

1. Aerobic respiration and other energy-releasing pathways start with glycolysis, a series of reactions that partially break down a glucose molecule.

2. Two NADH and four ATP form during glycolysis, which ends with the formation of two pyruvate molecules. After subtracting the two ATP required to start the reactions, the net energy yield of glycolysis is two ATP.

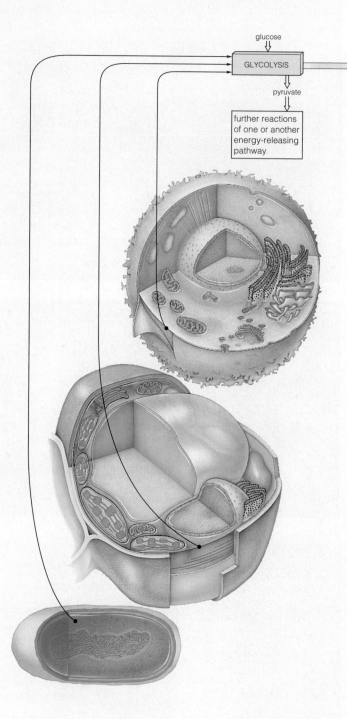

Figure 8.4 Glycolysis, first stage of the main energy-releasing pathways. The reactions proceed in the cytoplasm of all prokaryotic and eukaryotic cells.

In this example, glucose is the starting material. The reactions produce two pyruvate, two NADH, and four ATP molecules. Two ATP are invested to start glycolysis, so the net energy yield of glycolysis is two ATP.

Depending on the cell type and on conditions in its environment, the pyruvate molecules may be used in the second set of reactions of the aerobic pathway, which includes the Krebs cycle. Or they may be used in different reactions, such as those of fermentation pathways.

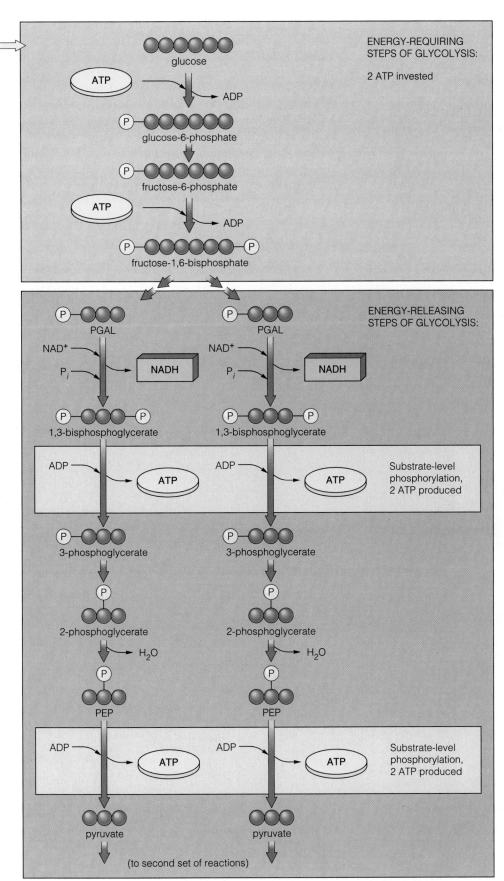

ENERGY-REQUIRING
STEPS OF GLYCOLYSIS:

2 ATP invested

glucose

ATP → ADP

glucose-6-phosphate

fructose-6-phosphate

ATP → ADP

fructose-1,6-bisphosphate

ENERGY-RELEASING
STEPS OF GLYCOLYSIS:

PGAL

NAD⁺
P_i → NADH

PGAL

NAD⁺
P_i → NADH

1,3-bisphosphoglycerate

1,3-bisphosphoglycerate

ADP → ATP

ADP → ATP

Substrate-level
phosphorylation,
2 ATP produced

3-phosphoglycerate

3-phosphoglycerate

2-phosphoglycerate

2-phosphoglycerate

H_2O

H_2O

PEP

PEP

ADP → ATP

ADP → ATP

Substrate-level
phosphorylation,
2 ATP produced

pyruvate

pyruvate

(to second set of reactions)

NET ENERGY YIELD: 2 ATP

a *Glycolysis starts with an energy investment of two ATP.* First, enzyme action promotes the transfer of a phosphate group from ATP to glucose, which has a backbone of six carbon atoms. With this transfer, the glucose molecule becomes slightly rearranged.

b Enzyme action promotes the transfer of a phosphate group from another ATP to the rearranged molecule.

c The resulting fructose-1,6-bisphosphate molecule splits at once into two molecules, each with a three-carbon backbone. We can call these two PGAL.

d During enzyme-mediated reactions, two NADH form after each PGAL gives up two electrons and a hydrogen atom to two NAD⁺. Each PGAL also combines with inorganic phosphate (P_i) present in the cytoplasm, then donates a phosphate group to ADP.

e *Thus two ATP have formed by the direct transfer of phosphate from two intermediate molecules that serve as substrates in the reactions.* With this formation of two ATP, the original energy investment of two ATP is paid off.

f In the next two enzyme-mediated reactions, each of the two intermediate molecules releases a hydrogen atom and an —OH group, which combine to form water.

g The resulting intermediates (two molecules of 3–phosphoenolpyruvate, or PEP) are rather unstable. Each gives up a phosphate group to ADP. *Once again, two ATP have formed by substrate-level phosphorylation.*

h Thus the net energy yield from glycolysis is two ATP for each glucose molecule entering the reactions. The end products of glycolysis are two molecules of pyruvate, each with a three-carbon backbone.

8.3 COMPLETING THE AEROBIC PATHWAY

Preparatory Steps and the Krebs Cycle

Suppose pyruvate, the product of glycolysis, leaves the cytoplasm and enters a mitochondrion, as shown in Figure 8.5. Only the aerobic pathway continues inside this organelle, and two more stages are required to complete it. During the *second* stage, a bit more ATP forms. Carbon and oxygen atoms depart from pyruvate, in carbon dioxide and water. And coenzymes accept hydrogen and electrons released during the reactions.

Take a look at Figure 8.6. During a few preparatory steps, a carbon atom is stripped from each pyruvate molecule, leaving an acetyl group that gets picked up by a coenzyme (forming acetyl-CoA). The acetyl group becomes attached to oxaloacetate, the point of entry into the Krebs cycle. The name of this cyclic pathway honors Hans Krebs, who began working out its details in the 1930s. Notice that three carbon atoms enter the second stage of reactions (as a pyruvate backbone) and three leave (in three carbon dioxide molecules) during the preparatory reactions and the cycle proper.

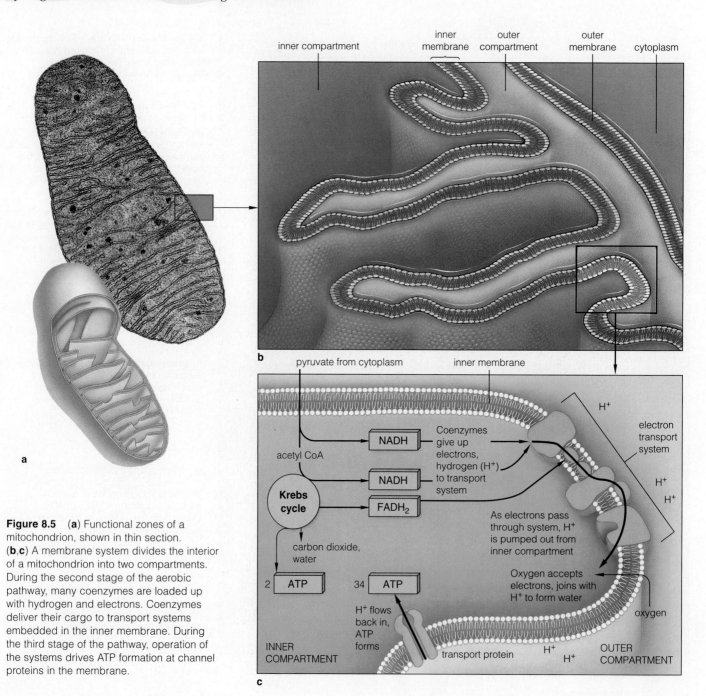

Figure 8.5 (a) Functional zones of a mitochondrion, shown in thin section. (b,c) A membrane system divides the interior of a mitochondrion into two compartments. During the second stage of the aerobic pathway, many coenzymes are loaded up with hydrogen and electrons. Coenzymes deliver their cargo to transport systems embedded in the inner membrane. During the third stage of the pathway, operation of the systems drives ATP formation at channel proteins in the membrane.

Functions of the Second Stage

The second-stage reactions serve three functions. First, hydrogen and electrons are transferred to NAD$^+$ and FAD, forming NADH and FADH$_2$. Second, substrate-level phosphorylations produce two ATP. Third, intermediates are rearranged into oxaloacetate. (Cells have only so much oxaloacetate, and it must be regenerated to keep the cyclic reactions going.)

The two ATP do not add much to the small yield from glycolysis. But the reactions also load up *many* coenzymes with hydrogen and electrons that can be used during the third stage of the aerobic pathway:

Glycolysis:	2 NADH
Pyruvate conversion that precedes Krebs cycle:	2 NADH
Krebs cycle:	2 FADH$_2$ + 6 NADH
Total coenzymes sent to third stage:	2 FADH$_2$ + 10 NADH

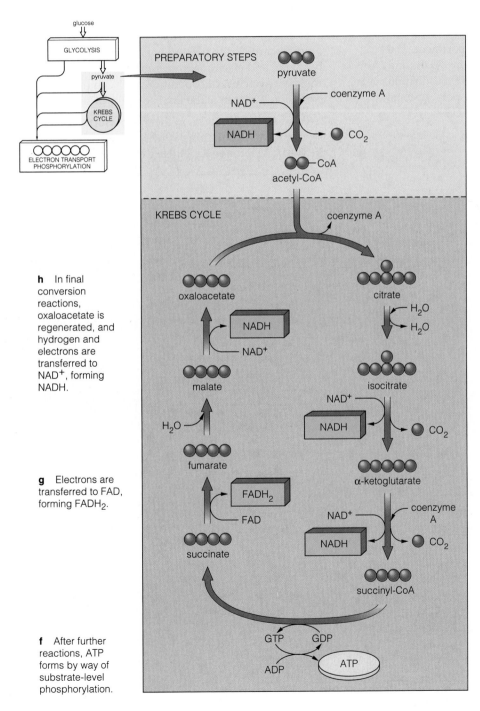

a A pyruvate molecule enters a mitochondrion. It undergoes preparatory conversions before entering cyclic reactions (Krebs cycle).

b First, the pyruvate is stripped of a functional group (COO$^-$), which departs as CO$_2$. Next, it gives up hydrogen and electrons to NAD$^+$, forming NADH. A coenzyme joins with the two-carbon fragment, forming acetyl-CoA.

c The acetyl-CoA is transferred to oxaloacetate, a four-carbon compound that is the point of entry into the Krebs cycle. The result is citrate, with a six-carbon backbone.

d Citrate enters conversion reactions in which a COO$^-$ group departs (as CO$_2$). Also, hydrogen and electrons are transferred to NAD$^+$, forming NADH.

e Another COO$^-$ group departs (as CO$_2$) and another NADH forms. *At this point, three carbon atoms have been released, balancing out the three that entered the mitochondrion (in pyruvate).*

h In final conversion reactions, oxaloacetate is regenerated, and hydrogen and electrons are transferred to NAD$^+$, forming NADH.

g Electrons are transferred to FAD, forming FADH$_2$.

f After further reactions, ATP forms by way of substrate-level phosphorylation.

Figure 8.6 Second stage of aerobic respiration: the Krebs cycle *and* a few reactions that immediately precede it. For each three-carbon pyruvate molecule that enters the cycle, three CO$_2$, one ATP, four NADH, and one FADH$_2$ are formed. The steps shown proceed twice (remember, the glucose molecule was broken down initially to two pyruvate molecules).

Third Stage of the Aerobic Pathway— Electron Transport Phosphorylation

ATP production goes into high gear in the third stage of the aerobic pathway. The production machinery runs on electrons and unbound hydrogen (that is, H^+ ions) delivered by coenzymes. Electron transport systems and neighboring channel proteins serve as the machinery. These are embedded in the inner membrane that divides the mitochondrion into two compartments (Figure 8.7).

As electrons are transferred through the transport systems, H^+ ions are tossed into the outer compartment. This sets up H^+ concentration and electric gradients across the membrane. Channel proteins allow H^+ to follow the gradients, back into the inner compartment. The flow drives the formation of ATP from ADP and unbound phosphate. Free oxygen keeps ATP production going. It withdraws electrons from transport systems and combines with H^+ to form water molecules.

Summary of the Energy Harvest

In many cells, the third stage of the aerobic pathway produces thirty-two ATP. Add these to the net yield from the preceding stages, and we have a total energy harvest of thirty-six ATP from one glucose molecule. Think of this as a typical yield only. The amount depends on cellular conditions, as when molecules of some intermediate are required elsewhere and pulled out of a reaction sequence. As Figure 8.8 indicates, the amount also depends on how a given cell uses the electrons from NADH that formed in the cytoplasm.

---

1. In aerobic respiration, glucose is completely broken down to carbon dioxide and water. Coenzymes transfer hydrogen and electrons from substrates to electron transport systems, the operation of which drives ATP formation. Oxygen is the final acceptor of the electrons.

2. From start (glycolysis in the cytoplasm) to finish (in the mitochondrion), this pathway commonly yields thirty-six ATP for every glucose molecule.

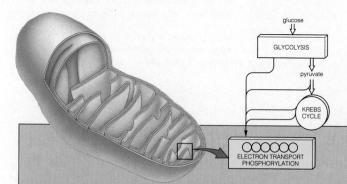

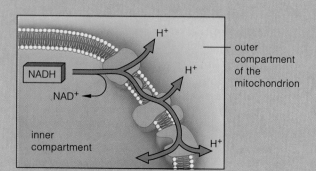

Figure 8.7 Electron transport phosphorylation, the third and final stage of aerobic respiration. The reactions proceed at transport systems and at channel proteins (ATP synthases) that are embedded in the inner mitochondrial membrane. Each transport system consists of enzymes, cytochromes, and other proteins that act in sequence.

The inner membrane divides the mitochondrion into two compartments. The third-stage reactions start in the inner compartment, when NADH and $FADH_2$ give up hydrogen (as H^+ ions) and electrons to the transport system. The electrons are transferred through the system, but the ions are left behind—in the outer compartment:

Soon there is a higher concentration of H^+ in the outer compartment than in the inner one. Concentration and electric gradients now exist across the membrane. The ions follow the gradients and flow across the membrane, through the interior of the channel proteins. Energy associated with the flow drives the formation of ATP from ADP and unbound phosphate (hence the name, electron transport *phosphorylation*):

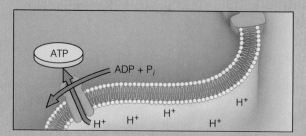

Do these events sound familiar? They should. ATP forms in much the same way in chloroplasts (Figure 7.8). The *chemiosmotic theory* that concentration and electric gradients across a membrane drive ATP formation applies also to mitochondria—even though ions flow in the opposite direction.

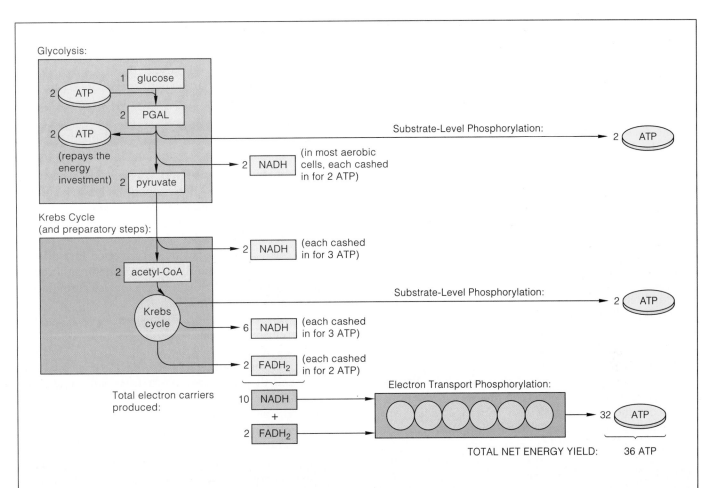

Figure 8.8 Summary of the harvest from one glucose molecule sent through the energy-releasing pathway of aerobic respiration. Commonly, 36 ATP form for each glucose molecule. By contrast, the anaerobic pathways would require 18 glucose molecules to produce as many ATP.

Remember, glucose has more energy (stored in more covalent bonds) than carbon dioxide or water. When it is broken down completely to those more stable end products, about 686 kilocalories of energy are released. Most escapes (as heat energy), but about 7.5 kilocalories are conserved in each ATP molecule. Thus, when 36 ATP form during the breakdown of one glucose molecule, the energy-conserving efficiency of this pathway is (36)(7.5)/(686), or 39 percent.

Bear in mind, the net energy yield is not *always* 36 ATP. The outcome depends not only on cellular conditions, but also on the type of cell and its electron-shuttling mechanisms. Two examples will make the point.

First, think about what happens when NADH forms *inside* a mitochondrion (during the second stage of reactions). It delivers electrons to the highest possible point of entry into a transport system. When it does, enough H^+ can be pumped across the inner mitochondrial membrane to produce *three* ATP. When $FADH_2$ forms inside the mitochondrion, it makes its delivery at a lower point of entry into the transport system. Fewer H^+ can be pumped, and only *two* ATP can be produced.

Now think about NADH formed in the cytoplasm. It can only deliver electrons *to* the mitochondrion, not *into* it. In liver, heart, and kidney cells, a protein shuttle built into the outer membrane accepts the electrons and donates them to NAD^+ in the mitochondrion. NADH formed this way puts electrons at the top of a transport system, so three ATP form, and the overall energy harvest is 38 ATP.

More commonly, as in skeletal muscle and brain cells, *a different shuttle accepts electrons from cytoplasmic NADH*. It donates these to FAD inside the mitochondrion—so only two ATP form. In such cells, the overall energy harvest is 36 ATP.

8.4 ANAEROBIC ROUTES

So far, we have tracked the fate of a glucose molecule through the pathway of aerobic respiration. We turn now to its use as a substrate for fermentation pathways. These are anaerobic pathways; they do *not* use oxygen as the final acceptor of the electrons that ultimately drive the ATP-forming machinery.

Fermentation Pathways

Which kinds of organisms use fermentation pathways? Many are bacteria and protistans that make their homes in marshes, bogs, mud, the animal gut, canned foods, sewage treatment ponds, and other oxygen-free settings. Some kinds actually die if exposed to oxygen. Bacteria responsible for many diseases, including botulism and tetanus, are like this. Other kinds, such as the bacterial "employees" of yogurt manufacturers, are indifferent to the presence of oxygen. Still others can use oxygen, but they switch to fermentation when oxygen levels decline. Even your muscle cells do this.

As is true of aerobic respiration, glycolysis also is the first stage of the fermentation pathways. Here also, a glucose molecule is split and rearranged into two pyruvate molecules, two NADH form, and the net energy yield is two ATP. However, as Figure 8.9 shows, the reactions do not break down glucose completely to carbon dioxide and water, and they produce no more ATP beyond the yield from glycolysis. *The final steps serve only to regenerate NAD$^+$—the coenzyme that is central to the pathway's operation.*

The energy yield from fermentation is enough for many single-celled anaerobic organisms. It even helps carry some aerobic cells through times of stress. It is not enough to sustain large, multicelled organisms (this being one reason why you never will come across an anaerobic elephant).

Lactate Fermentation Figure 8.10 shows the main reaction steps of the energy-releasing pathway called **lactate fermentation**. As you can see, pyruvate itself, formed during the first stage of reactions (glycolysis), accepts hydrogen and electrons from NADH. With this transfer, pyruvate is converted to a different three-carbon compound, lactate. Sometimes lactate is called "lactic acid." However, the ionized form (lactate) is far more common in cells.

A few bacteria, including species of *Lactobacillus*, rely exclusively on this anaerobic pathway. Left to their own devices, their fermentation activities can spoil food. Yet some fermenters also have commercial uses, as when they produce yogurt from milk, and sauerkraut from raw cabbage.

Some types of animal cells also can switch to lactate fermentation for a quick ATP fix. When your demands for energy are intense but brief—say, during a short race—your muscle cells use this anaerobic pathway. They cannot do this for long—too much of glucose's stored energy would be thrown away for too little ATP. When glucose stores are depleted, muscles fatigue and lose their ability to contract.

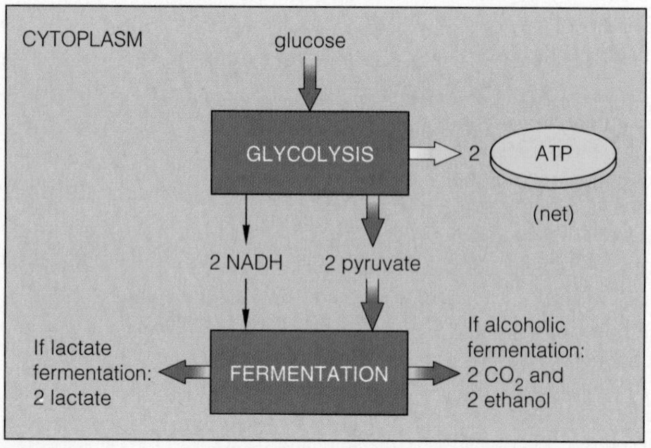

Figure 8.9 Overview of two fermentation routes. In this type of energy-releasing pathway, the net ATP yield is from the initial reactions (glycolysis).

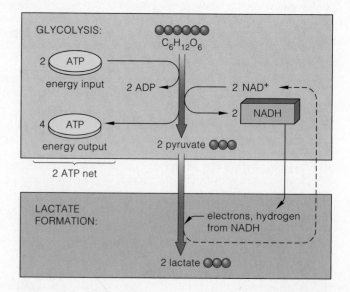

Figure 8.10 Lactate fermentation. In this anaerobic pathway, electrons end up in the reaction product (lactate).

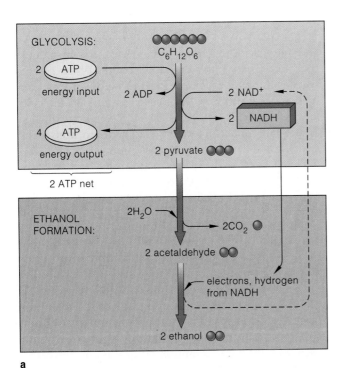

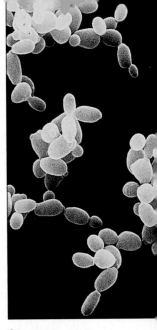

b   c

Figure 8.11 (**a**) Alcoholic fermentation. In this anaerobic pathway, an intermediate of the reactions (acetaldehyde) is the final electron acceptor, and ethanol is the end product. Yeasts, single-celled organisms, use this pathway. (**b**) One species of *Saccharomyces* lives on sugar-rich, ripened grapes. Another (**c**) makes bread dough rise.

Alcoholic Fermentation In this anaerobic pathway, each pyruvate molecule from the first stage of reactions (glycolysis) is rearranged into an intermediate form called acetaldehyde. When this intermediate accepts hydrogen and electrons from NADH, it is converted to ethanol, the end product of this pathway (Figure 8.11).

The single-celled fungi called yeasts use this pathway. And commercial bakers use certain yeasts. *Saccharomyces cerevisiae* can make bread dough rise. Bakers mix this yeast with sugar and blend the mixture into dough. As the yeast cells use the sugar, they release carbon dioxide, a gas that expands the dough (makes it "rise"). Oven heat forces the gas out, leaving a porous product.

Beer and wine manufacturers use yeasts on a large scale. Vintners use wild yeasts that live on grapes. They also use cultivated strains of *S. ellipsoideus*, which remain active until the alcohol concentration in the vats exceeds 14 percent. Wild yeasts die when it exceeds 4 percent—but 4 percent still packs a punch. Robins get drunk on the naturally fermenting berries of pyracantha shrubs. So do wild turkeys when they gobble up fermenting apples in untended orchards.

In fermentation pathways, the net energy yield of two ATP is from glycolysis. The remaining reactions simply regenerate the NAD+ required for glycolysis.

Anaerobic Electron Transport

Although aerobic respiration and fermentation are the most common, there are other kinds of energy-releasing pathways, especially among the bacteria. As described in Chapter 48, some of these less common pathways influence the global cycling of sulfur, nitrogen, and other vital elements—and so influence the availability of nutrients for organisms everywhere.

Consider anaerobic electron transport, as employed by a variety of bacteria. Electrons stripped from an organic compound are sent through transport systems bound in the plasma membrane. The energy yield varies. An inorganic compound in the environment often serves as the final electron acceptor. While you are reading this, for example, certain anaerobic bacteria living in waterlogged soil are stripping electrons from a variety of compounds. Then they dump the electrons on sulfate (SO_4^-), leaving hydrogen sulfide (H_2S), a very foul-smelling gas. Sulfate-reducing bacteria also live in aquatic habitats that are rich in decomposed organic material. They even live deep in the ocean, around hydrothermal vents. As described in Chapter 49, these bacteria form the food production base for unique communities.

In anaerobic electron transport, an inorganic substance (but not oxygen) usually serves as the final electron acceptor.

8.5 ALTERNATIVE ENERGY SOURCES IN THE HUMAN BODY

Carbohydrate Breakdown in Perspective

So far, you've looked at what happens after a lone glucose molecule enters an energy-releasing pathway. Now you can start thinking about what cells do when they have too many or too few of these molecules.

Consider what happens after you or any other mammal finishes a meal. Your body absorbs a great deal of glucose and other small organic molecules, which your bloodstream delivers to tissues throughout your body. There, cells rapidly take up excess glucose, then "trap" it by converting it to glucose-6-phosphate (which cannot be transported across the plasma membrane). Glucose-6-phosphate, recall, is the first intermediate of glycolysis.

When your food intake exceeds cellular demands for energy, ATP-producing machinery goes into high gear. In time, the increased ATP concentration inhibits glycolysis. Now glucose-6-phosphate is diverted into a biosynthesis pathway that ends with glycogen—a storage polysaccharide composed of glucose units.

Between meals, when the input of free glucose dwindles, cells break down the glycogen to glucose-6-phosphate, and they use this in glycolysis. Liver cells do more. They convert glucose-6-phosphate back to free glucose and release it. The bloodstream delivers this glucose to energy-demanding cells of muscles and other organs that have depleted their own glycogen stores.

Don't let the preceding paragraphs lead you to believe that *all* cells squirrel away large amounts of glycogen. Liver and muscle cells maintain the largest stores. Even then, glycogen represents a mere 1 percent or so of the total stored energy in the body of an adult human. On the average, 78 percent is stored in fats and another 21 percent in proteins.

Energy from Fats

Maybe you avoid butter, ice cream, and other fatty foods, thinking it is better to fill up on carbohydrates and proteins. This is a good idea, as long as you don't stuff yourself with these organic compounds—because your body converts the excess to fats. Fat molecules, recall, have a glycerol head and one, two, or three fatty acid tails. Most of the fats stored in your body are in the form of triglycerides (with three tails). These accumulate inside the fat cells of adipose tissues, at strategic locations beneath the skin.

Between meals or during exercise, triglycerides are tapped as alternatives to glucose. At such times, enzymes in fat cells cleave the bonds holding glycerol and fatty acids together, then release the breakdown products. These enter the bloodstream. When glycerol reaches the liver, it is converted to PGAL—a key intermediate of glycolysis. Most cells can take up the circulating fatty acids. Enzymes cleave the carbon backbone of the tails and convert the fragments to acetyl-CoA, which can enter the Krebs cycle (Figures 8.6 and 8.12).

Each fatty acid tail has many more carbon-bound hydrogen atoms than glucose—so its breakdown yields much more ATP. Between meals or during sustained exercise, fatty acid conversions supply about half of the ATP that your muscle, liver, and kidney cells require.

Energy from Proteins

Eat more protein molecules than your body requires to grow and maintain itself, and your cells won't store them. Enzymes split the proteins into amino acid units, and then they remove the amino group (NH_3) from each unit. Depending on cellular conditions, the carbon backbones that remain may be converted to fats or carbohydrates. Or they may enter the Krebs cycle. There, hydrogen and electrons stripped from the carbon atoms are transferred to the cycle's coenzymes. The amino groups undergo conversions that produce urea, a nitrogen-containing waste product that is excreted from the body in urine.

In humans and other mammals, the entrance of glucose or some other organic compound into an energy-releasing pathway depends on its concentrations inside and outside the cell, and on the type of cell.

This concludes our look at aerobic respiration and other energy-releasing pathways. To gain insight into how these pathways fit into the greater picture of life's evolution and interconnectedness, take a moment to read the *Commentary* that concludes this unit.

Figure 8.12 Points of entry into the aerobic pathway for complex carbohydrates, fats, and proteins after they have been reduced to their simpler components in the human digestive system.

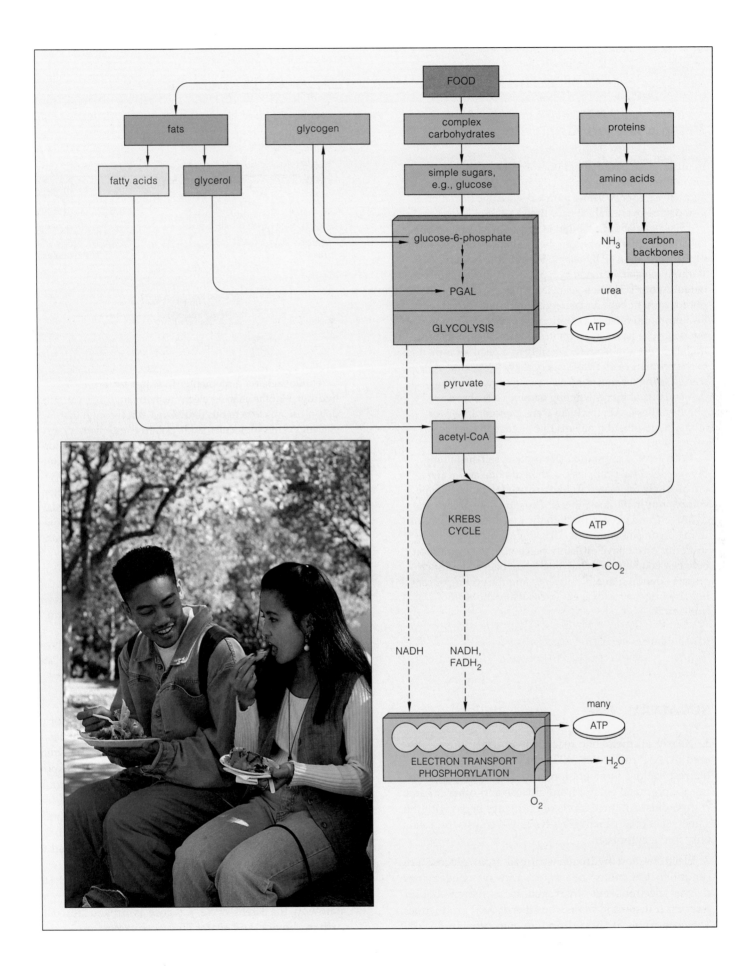

FOOD

fats   glycogen   complex carbohydrates   proteins

fatty acids   glycerol

simple sugars, e.g., glucose

amino acids

glucose-6-phosphate

NH$_3$ carbon backbones

urea

PGAL

GLYCOLYSIS   →   ATP

pyruvate

acetyl-CoA

KREBS CYCLE   →   ATP

CO$_2$

NADH   NADH, FADH$_2$

many ATP

ELECTRON TRANSPORT PHOSPHORYLATION

→ H$_2$O

O$_2$

Commentary

Perspective on Life

In this unit, you read about photosynthesis and aerobic respiration—the main pathways by which cells trap, store, and release energy. Those pathways became linked on a grand scale over evolutionary time.

Life began about 3.8 billion years ago, when the atmosphere had little free oxygen. Early single-celled organisms probably made ATP by reactions similar to glycolysis. Without oxygen, fermentation pathways must have dominated. About 1.5 billion years later, oxygen-producing photosynthetic cells had emerged, and they turned out to be a profound force in evolution. Oxygen, a by-product of the noncyclic pathway of photosynthesis, began to accumulate in the atmosphere. Possibly as a result of mutations in the proteins of electron transport systems, some cells started using oxygen as an electron acceptor. In time, descendants of those fledgling aerobic cells abandoned photosynthesis. Among them were the forerunners of animals and other organisms that engage in aerobic respiration.

With aerobic respiration, life became self-sustaining, because the final products—carbon dioxide and water—are precisely the materials that are used to build organic compounds in photosynthesis! Thus the flow of carbon, hydrogen, and oxygen through the metabolic pathways of living organisms came full circle (Figure *a*).

Perhaps you have difficulty perceiving the connection between yourself—a living, intelligent being—and such remote-sounding things as energy and the cycling of carbon, hydrogen, and oxygen. Is this really the stuff of humanity?

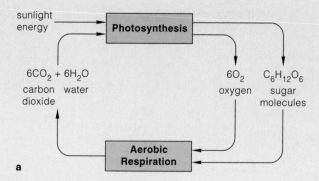

Think back, for a moment, to a water molecule. Two hydrogen atoms sharing electrons with an oxygen atom doesn't seem close to our daily lives. But through that sharing, water molecules show polarity, and they hydrogen-bond with one another. That is a beginning for the organization of lifeless matter that leads to the organization of all living things.

For now you can imagine different molecules in water. The nonpolar ones resist interaction with water; the polar ones dissolve in it. The phospholipids among them spontaneously assemble into a two-layered film. Such lipid bilayers are the basis for all cell membranes, hence all cells. From the beginning, the cell has been the fundamental *living* unit.

The essence of life is not some mysterious force. It is metabolic control. With a cell membrane to contain them,

SUMMARY

1. Nearly all metabolic reactions run on energy delivered by ATP molecules. ATP can be produced by aerobic respiration, fermentation, and other pathways that release chemical energy from glucose and other organic compounds. All three kinds of pathways begin with the same reactions, called glycolysis. Glycolysis proceeds only in the cytoplasm.

2. Electrons and hydrogen stripped from glucose are central to the energy-releasing pathways. Coenzymes deliver electrons and hydrogen to membrane-bound electron transport systems, the sites of ATP production. The coenzymes are NAD^+ and, in one pathway, FAD.

3. In aerobic respiration, oxygen is the final acceptor of electrons stripped from glucose. The pathway proceeds through three stages of reactions: glycolysis, the Krebs cycle (and preparatory steps), and electron transport phosphorylation. Its net energy yield is commonly thirty-six or more ATP molecules.

   a. Glycolysis partly breaks down a glucose molecule. Two pyruvate, two NADH, and four ATP form. The net energy yield is only two ATP (two ATP had to be invested up front to get the reactions going).

   b. The next stage proceeds in mitochondria. Pyruvate is converted to an intermediate that enters a cyclic pathway, the Krebs cycle. Glucose is broken down to carbon dioxide and water. Ten coenyzmes are loaded

metabolic reactions *can* be controlled. Cells can respond to energy changes and to the kinds of molecules in their environment. Their response mechanisms operate by "telling" protein molecules—enzymes—when and what to build or tear down.

And it is not some mysterious force that creates the proteins themselves. DNA, the slender double strand of heredity, has the chemical structure—*the chemical message*—that allows molecule to reproduce molecule, one generation after the next. Those DNA strands tell trillions of cells in your body how countless molecules must be built and torn apart for their stored energy.

So yes, carbon, hydrogen, oxygen, and other organic molecules represent the stuff of you, and us, and all of life. But it takes more than molecules to complete the picture. Life exists only as long as a constant flow of energy maintains its organization. Molecules are assembled into cells, cells into organisms, organisms into communities, and so on up through the biosphere. It takes energy, primarily from the sun, to maintain all these levels of organization. And energy flows through time in one direction—from organized to less organized forms. Only as long as sunlight flows into the web of life can life continue in all its rich diversity.

In short, life is no more *and no less* than a marvelously complex system of prolonging order. Sustained by energy transfusions, life continues because of a capacity for self-reproduction—a handing down of hereditary instructions in DNA. With DNA, energy and materials can be organized, generation after generation. Even with the death of

b

the individual, life is prolonged. With death, molecules are released and recycled once more, providing raw materials for new generations. In this flow of energy and cycling of material through time, each birth is affirmation of our ongoing capacity for organization, each death a renewal.

with hydrogen and electrons (eight NADH and two $FADH_2$), and two ATP form.

c. The third stage also proceeds in mitochondria. An inner membrane divides the mitochondrion into two compartments. Electron transport systems and channel proteins are embedded in this inner membrane.

d. Coenzymes deliver electrons to a transport system. Operation of the system sets up H^+ concentration and electric gradients across the membrane. H^+ follows the gradients and leaves the inner compartment, through channel proteins. Energy associated with the flow drives the formation of ATP from ADP and unbound phosphate. Oxygen withdraws electrons from the transport system and combines with H^+ to form water.

4. Alcoholic fermentation, lactate fermentation, and anaerobic electron transport use an organic or inorganic compound (not oxygen) as the final acceptor of electrons from the reactions.

a. The fermentation pathways start with glycolysis. Two ATP are invested, glucose is partially broken down to pyruvate, and two NADH and four ATP form.

b. In lactate fermentation, pyruvate accepts hydrogen and electrons from coenzymes and is itself rearranged into the end product, lactate.

c. In alcoholic fermentation, pyruvate is converted to an intermediate (acetaldehyde), and carbon dioxide is released. That intermediate accepts hydrogen and electrons from coenzymes and so becomes ethanol.

Review Questions

1. Which energy-releasing pathway yields the most ATP for each glucose molecule? *124*

2. Think of the various energy-releasing pathways. Which of their reactions occur only in cytoplasm? Which occur only in mitochondria? *124*

3. State the function of coenzymes in the energy-releasing pathways. *124*

4. Briefly describe glycolysis, the first stage of the energy-releasing pathways. *126–127*

5. Describe the two stages of aerobic respiration that follow glycolysis:
   a. the Krebs cycle (and preparatory conversions). *128–129*
   b. electron transport phosphorylation. *130*

6. Describe the functional zones of a mitochondrion. Explain where the electron transport systems and transport proteins required for ATP formation are located. *128*

7. Is the following statement true? Your muscle cells cannot function at all without oxygen. *132*

8. In fermentation, conversions following the net production of two ATP do not yield more energy. What do they accomplish? *132*

9. Cells do not use nucleic acids as an alternative energy source. Speculate why.

10. Summarize the net energy yield from one glucose molecule for aerobic respiration. *131*

Self-Quiz *(Answers in Appendix IV)*

1. Stored energy can be released from glucose and used to produce ~~ATP~~ , an energy carrier.

2. In the first stage of the main energy-releasing pathways, glucose is partly broken down to pyruvate

3. ATP is _____ .
   a. a phosphate compound
   b. an energy carrier
   c. produced by all organisms
   d. all of the above

4. Which of the following is *not* produced during glycolysis?
   a. NADH
   b. pyruvate
   c. FAD
   d. ATP

5. Glycolysis occurs in the _____ .
   a. nucleus
   b. mitochondrion
   c. plasma membrane
   d. cytoplasm

6. The final acceptor of electrons stripped from glucose during aerobic respiration is _____ .
   a. water
   b. hydrogen
   c. oxygen
   d. NADH

7. For the aerobic pathway, electron transport systems are located in the _____ .
   a. cytoplasm
   b. inner mitochondrial membrane
   c. outer mitochondrial compartment
   d. stroma

8. The flow of _____ through channel proteins drives the formation of ATP from ADP and phosphate.
   a. electrons
   b. hydrogen ions
   c. NADH
   d. FADH$_2$

9. Match the events with the metabolic reaction.
   b glycolysis
   c fermentation
   a Krebs cycle
   d electron transport phosphorylation
   a. ATP, NADH, and CO$_2$ form
   b. glucose to two pyruvate
   c. NAD$^+$ regenerated, a net yield of two ATP
   d. H$^+$ flows through channel proteins, ATP forms

Selected Key Terms

aerobic respiration *124*
anaerobic electron transport *124*
electron transport phosphorylation *124*
fermentation pathway *124*
glycolysis *124*
Krebs cycle *124*
lactate fermentation *132*
pyruvate *126*
substrate-level phosphorylation *126*

Readings

Levi, P. October 1984. "Travels with C." *The Sciences.* Journey of a carbon atom through the world of life.

Roberts, L. August 28, 1987. "Discovering Microbes with a Taste for PCBs." *Science* 237:975–977. Bacteria make ATP by breaking down pollutants.

Wolfe, S., 1993. *Molecular and Cellular Biology.* Belmont, California: Wadsworth.

FACING PAGE: *Human sperm, one of which will penetrate this mature egg and so set the stage for the development of a new individual in the image of its parents.*

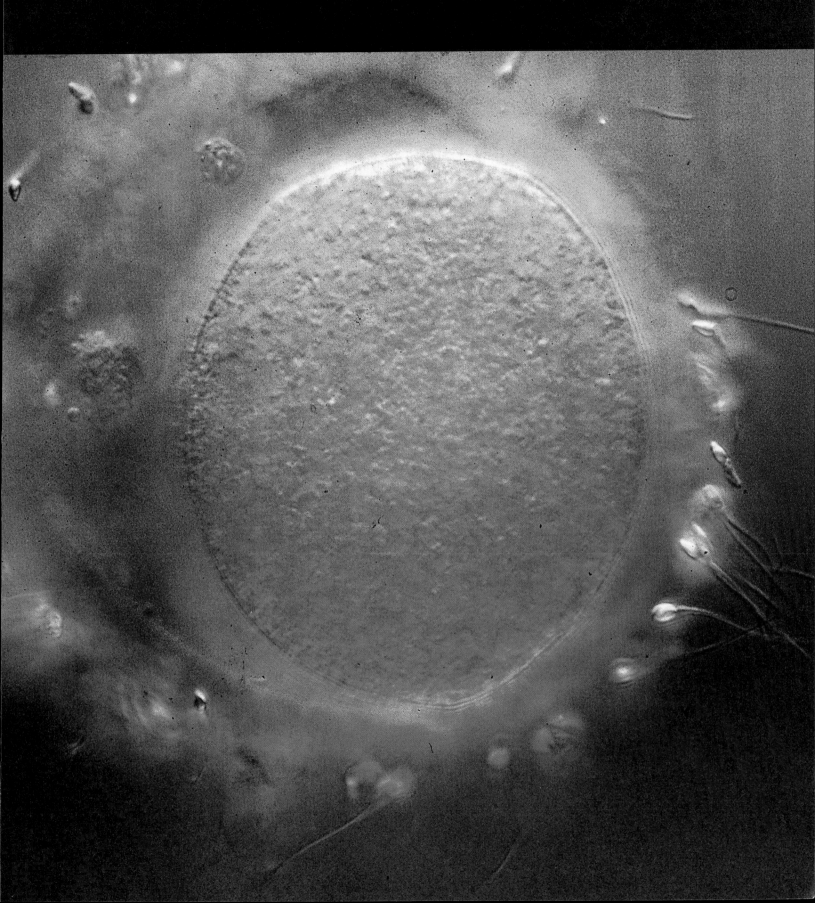

9 CELL DIVISION AND MITOSIS

Silver in the Stream of Time

Five o'clock, and the first rays of the sun dance over the wild Alagnak River of the Alaskan tundra. It is September, and life is both ending and beginning in the clear, frigid waters. By the thousands, mature silver salmon have returned from the open ocean to spawn in their shallow native home. The females are tinged with red, the color of spawners, and they are dying.

On this morning, a female salmon releases translucent pink eggs into a "nest," hollowed out by her fins in the gravel riverbed (Figure 9.1). Within moments, a male salmon sheds a cloud of sperm, and fertilization follows. Trout and other predators eat most of the eggs, but some survive and give rise to a new generation.

Within three years, the pea-size eggs have become streamlined salmon, fashioned from billions of cells. A few of those cells will develop into eggs or sperm. In time, on some September morning, they will take part in an ongoing story of birth, growth, death, and rebirth.

For you, as for salmon and all other multicelled organisms, growth as well as reproduction depends on *cell division*. In your mother's body, a single fertilized egg divided in two, then the two into four, and so on until billions of cells were growing, developing in specialized ways, and dividing at different times to produce your genetically prescribed body parts. Your body now has roughly 65 trillion cells—and many of the cells are still dividing. Every five days, for instance, cell divisions replace the lining of your small intestine.

Understanding cell division—and, ultimately, how new individuals are put together in the image of their parents—begins with answers to three questions. *First*, what instructions are necessary for inheritance? *Second*, how are those instructions duplicated for distribution into daughter cells? *Third*, by what mechanisms are those instructions divided into daughter cells? We will require more than one chapter to consider cell reproduction and other mechanisms of inheritance. However, the points made in the first part of this chapter can help you keep the overall picture in focus.

Figure 9.1 The last of one generation and the first of the next in the Alagnak River of Alaska.

KEY CONCEPTS

1. When a cell divides, its two daughter cells must each receive a required number of DNA molecules as well as cytoplasm. In eukaryotes, mitosis sorts out the DNA into two new nuclei. A separate mechanism divides the cytoplasm in two.

2. Mitotic cell division is the basis of growth and tissue repair in multicelled eukaryotes. It also is the means by which single-celled eukaryotes and many multicelled eukaryotes reproduce asexually.

3. DNA molecules are also called chromosomes. Members of the same species normally have the same total number of chromosomes in their body cells. The chromosomes have different lengths and shapes, and they carry different portions of the hereditary instructions.

4. When cells prepare for division, every chromosome is duplicated. Each now consists of two DNA molecules. Mitosis separates the two for distribution to daughter cells. In this way, each cell gets the same total number and the same types of chromosomes as the parent cell.

5. For many species, body cells have two of each type of chromosome, inherited from two parents. "Chromosome number" is the number of *each type* of chromosome for the species. Mitosis keeps the chromosome number constant from one cell generation to the next.

DIVIDING CELLS: THE BRIDGE BETWEEN GENERATIONS

Overview of Division Mechanisms

In biology, the word **reproduction** means producing a new generation of cells or multicelled individuals. Reproduction begins with the division of single cells. And the ground rule for cell division is this:

Parent cells must provide their daughter cells with hereditary instructions (encoded in DNA) and enough cytoplasmic machinery to start up their own operation.

DNA, recall, contains instructions for synthesizing proteins. Some proteins are structural materials. Many are enzymes that put together carbohydrates, lipids, and other building blocks of cells. Unless a daughter cell receives the necessary instructions for making proteins, it cannot grow or function properly.

Also, the parent cell's cytoplasm already has operating machinery—enzymes, organelles, and so on. When a daughter cell inherits what looks like a blob of cytoplasm, it really is getting start-up machinery for its operation, until it has time to use its inherited DNA for growing and developing on its own.

The cells of plants, animals, and other eukaryotic organisms divide the DNA by **mitosis** or **meiosis**. Both mechanisms sort out and package DNA molecules into new nuclei for forthcoming daughter cells. In other words, mitosis and meiosis only divide the nuclear material. Different mechanisms divide the cytoplasm and so split a parent cell into daughter cells.

Multicelled organisms grow by mitosis and cytoplasmic division of body cells, which are called **somatic cells**. They also repair tissues that way. Nick yourself peeling a potato, and mitotic cell divisions will replace the cells the knife sliced away. Besides this, many protistans, fungi, plants, and even some animals reproduce asexually by mitotic cell division.

By contrast, meiosis occurs only in **germ cells**, a cell lineage set aside for sexual reproduction. Meiosis must precede the formation of gametes, such as sperm and eggs. As you will see in the next chapter, it has much in common with mitosis, but the end result is different.

What about prokaryotic cells—the bacteria? They reproduce asexually by way of a different mechanism. We will consider the bacteria later, in Chapter 22. For now, take a moment to study Table 9.1, which lists the major division mechanisms.

Some Key Points About Chromosomes

Before a cell starts preparing for mitosis, its DNA molecules are stretched out like threads, with many proteins attached to them. Each DNA molecule, along with its attached proteins, is a **chromosome**.

Chromosomes undergo duplication while they are in this threadlike form. Once duplicated, each consists of *two* DNA molecules, which will stay together until late in mitosis. For as long as they remain attached, the two are called **sister chromatids** of the chromosome:

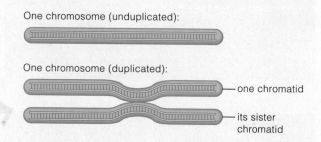

One chromosome (unduplicated):

One chromosome (duplicated):
— one chromatid
— its sister chromatid

"Sister chromatids" is a name that seems to confuse almost everybody. It might help to stretch their dictionary meaning a bit and think of them as the forthcoming daughters of a parent chromosome.

Notice how the duplicated chromosome narrows down in a small region. This is the **centromere**, a region having attachment sites for microtubules that will help move the chromosome during nuclear division:

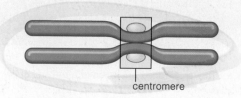

centromere

Table 9.1	Cell Division Mechanisms
Mechanisms	Used by
Mitosis, cytoplasmic division	*Single-celled eukaryotes (for asexual reproduction)*
	Multicelled eukaryotes (for bodily growth; also for asexual reproduction in many species)
Meiosis, cytoplasmic division	*Eukaryotes (basis of gamete formation and sexual reproduction)*
Prokaryotic fission	*Bacterial cells (for asexual reproduction)*

Keep in mind, the preceding sketches are simplified. As Figure 9.2 suggests, the centromere location is not the same for all chromosomes. Also, a DNA molecule's two parallel strands don't look like a ladder—they are twisted together repeatedly like a spiral staircase and are much longer than can be shown here.

Mitosis and the Chromosome Number

All normal members of the same species have the same total number of chromosomes in their somatic cells. Humans have forty-six, gorillas have forty-eight, and pea plants have fourteen.

Figure 9.3 shows all forty-six human chromosomes, lined up as twenty-three pairs. Think of them as two

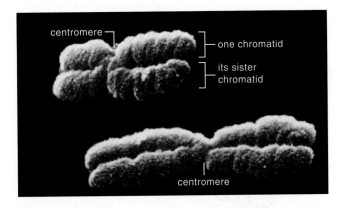

Figure 9.2 Scanning electron micrograph of two human chromosomes, each in the duplicated state.

sets of books on how to build a house. Your father gave you one set. Your mother had her own ideas about plumbing, storage, and so forth, so she gave you a revised edition. Her set covers the exact same topics but has different things to say about many of them.

Your chromosomes are like the volumes of two sets of books. Each set is numbered 1 to 23. Thus, for example, you have two "volumes" of chromosome 22—that is, a pair of them. Generally, both members of each pair have the same length and shape, and they carry instructions for the same traits. In these respects, they are not like any of the other pairs of chromosomes.

Any cell having two of each type of chromosome is a **diploid cell**. Such cells exist in gorillas, pea plants, and a great many other organisms besides humans.

With mitosis, a diploid parent cell produces two diploid daughter cells. This doesn't mean each merely gets forty-six or forty-eight or fourteen chromosomes. If only the total mattered, one cell might get, say, two pairs of chromosome 22 and no pairs of chromosome 9. However, neither cell would function properly without *one pair of each type of chromosome*.

Chromosome number tells you how many of each type of chromosome are present in a cell. Diploid cells, with two of each type, are $2n$. The n stands for chromosome number.

With mitosis, the chromosome number remains constant, division after division, from one cell generation to the next. Thus, if a parent cell is diploid, its two daughter cells will be diploid also.

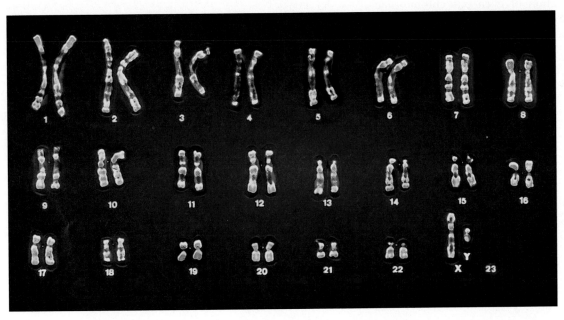

Figure 9.3 Forty-six chromosomes from a human male. Each chromosome is in the duplicated state.

9.2 MITOSIS AND THE CELL CYCLE

Mitosis is only one phase of the **cell cycle**. Such cycles start at the time new cells are produced, and they end when those cells complete their own division. The cycle starts again for each new daughter cell. **Interphase** usually is the longest part of a cell cycle. At this time, a cell increases its mass, roughly doubles the number of its cytoplasmic components, and finally duplicates its chromosomes. The following abbreviations designate the phases of the cell cycle (see also Figure 9.4):

M   *Mitosis;* nuclear division, commonly followed by cytoplasmic division

G_1 Of interphase, a *"Gap"* (interval) before the onset of DNA replication

S   Of interphase, the time of *"Synthesis"* (replication) of DNA and associated proteins

G_2 Of interphase, a second *"Gap"* between the completion of DNA replication and the onset of mitosis

The cell cycle lasts about the same length of time for cells of a given type. Its duration differs among cells of different types. For example, all of the neurons in your brain are arrested at interphase and typically will not divide again. Cells of a new sea urchin may double in number every two hours.

Adverse conditions may disrupt a cell cycle. When deprived of a vital nutrient, for instance, the free-living cells called amoebas do not leave interphase. Even so, when a cell proceeds past a certain point in interphase, the cycle normally continues regardless of outside conditions, owing to built-in controls over its duration.

Good health depends on the proper timing and completion of events in the cell cycle. Mistakes in the duplication or distribution of even one chromosome during interphase may lead to a genetic disorder. As described on page 234, a mature tissue will be destroyed

if controls are lost that otherwise keep its cells from dividing, and cancer may follow. The *Focus* essay describes a landmark case of unchecked cell divisions.

We turn now to mitosis and how it can maintain the chromosome number through turn after turn of the cell cycle. Figure 9.5 only hints at the precise division mechanisms that take over as a cell leaves interphase.

A cell cycle begins at interphase, when a new cell (formed by mitosis and cytoplasmic division) increases its mass, doubles the number of its cytoplasmic components, then duplicates its chromosomes. The cycle ends when the cell divides.

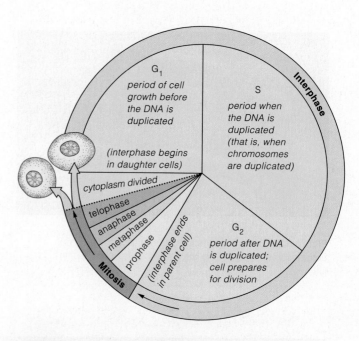

Figure 9.4 Generalized eukaryotic cell cycle. The length of different stages differs among cells.

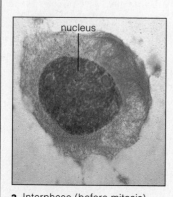

a Interphase (before mitosis)

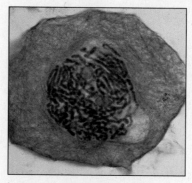

b Early prophase

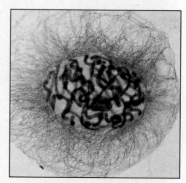

c Prophase

Henrietta's Immortal Cells

Each human starts out as a single fertilized egg. At birth a human body has about a trillion cells. Even in an adult, trillions of cells are still dividing. Cells in the stomach's lining divide every day. Liver cells usually don't divide—but if part of the liver becomes injured or diseased, they will divide repeatedly and produce new cells until the damaged part is replaced.

In 1951, George and Margaret Gey of Johns Hopkins University were trying to develop a way to keep human cells dividing outside the body. (Researchers could study basic life processes with such cells. They also could study cancer and other diseases, without having to experiment directly on humans.) Local physicians had provided them with normal or diseased human cells from patients. But the Geys just couldn't stop the cell lines from dying out within a few weeks.

Mary Kubicek, one of their assistants, was about to give up after dozens of failed attempts. Still, in 1951, she prepared another sample of cancer cells for culture. The sample was code-named HeLa, for the first two letters of the patient's first and last names.

The cells began to divide. And divide. And divide again. By the fourth day there were so many cells that they had to be subdivided into more tubes. As months passed, the culture continued to thrive. Unfortunately, the tumor cells inside the patient's body were just as vigorous. Six months after the patient was first diagnosed as having cancer, tumor cells had spread through her body. Two months later, Henrietta Lacks, a young woman from Baltimore, was dead.

Although Henrietta was gone, some of her cells lived on in the Geys' laboratory as the first successful human cell culture. HeLa cells were soon being shipped to other researchers, who passed cells on to others, and so HeLa cells came to live in laboratories all over the world. Some even traveled into space aboard the *Discoverer XVII* satellite. Every year, research that is described in hundreds of scientific papers is based on work with HeLa cells.

Henrietta was only thirty-one years old when runaway cell divisions killed her. Now, more than forty years later, her legacy is still benefiting humans everywhere, in cells that are still alive and dividing, day after day after day.

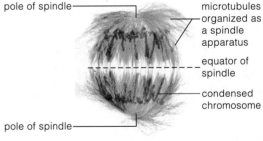

pole of spindle

microtubules organized as a spindle apparatus

equator of spindle

condensed chromosome

pole of spindle

Figure 9.5 (*Below*) Light micrographs showing the progress of mitosis in a cell from the African blood lily (*Haemanthus*). As indicated by the icon to the right, the chromosomes are stained blue, and microtubules that are moving them about are stained red.

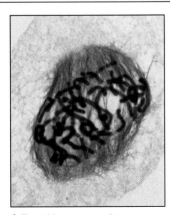

d Transition to metaphase

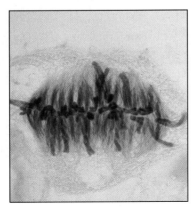

e Metaphase

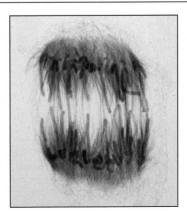

f Anaphase

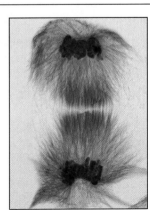

g Telophase

9.3 STAGES OF MITOSIS: AN OVERVIEW

When a cell makes the transition from interphase to mitosis, it has stopped constructing new cell parts, and its DNA has been replicated. Within that cell, profound changes now proceed smoothly, one after the other, through four stages. The sequential stages of mitosis are **prophase**, **metaphase**, **anaphase**, and **telophase**.

Figure 9.6 shows these stages for a dividing animal cell. Compare this illustration with the preceding one of a plant cell. When you look closely, it becomes quite clear that the chromosomes within both cells move about dramatically during mitosis. They do not move about on their own. A **spindle apparatus** moves them.

In all eukaryotic cells, a fully formed spindle consists of two organized sets of microtubules. The microtubules extend from the spindle's two end points (poles) and overlap at its equator, which lies midway

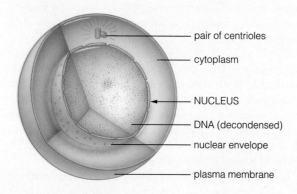

Cell at Interphase
The DNA is duplicated, then the cell prepares for division.

- pair of centrioles
- cytoplasm
- NUCLEUS
- DNA (decondensed)
- nuclear envelope
- plasma membrane

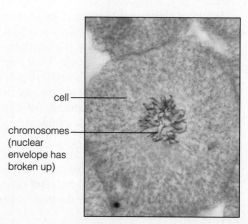

- cell
- chromosomes (nuclear envelope has broken up)

MITOSIS

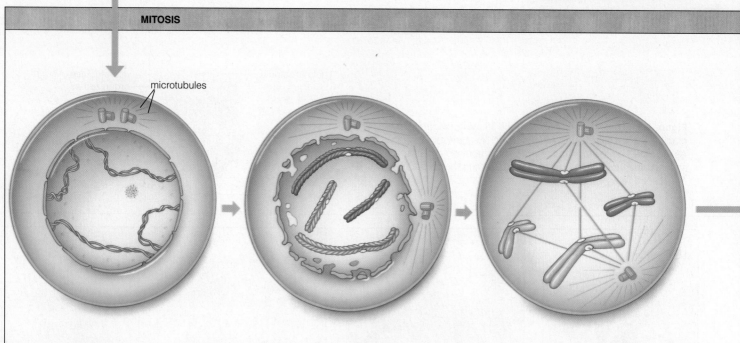

microtubules

Early Prophase
The DNA and its associated proteins start to condense. The two chromosomes shaded purple were inherited from the male parent. The other two (blue) are their counterparts, inherited from the female parent.

Late Prophase
Chromosomes continue to condense. New microtubules are assembled, and they move one of two centriole pairs toward the opposite end of the cell. The nuclear envelope starts to break up.

Transition to Metaphase
Microtubules penetrate the nuclear region. Together they form a spindle apparatus. They become attached to the sister chromatids of each chromosome.

between the poles. The spindle poles establish the destinations of chromosomes during mitosis.

Mitosis proceeds through four stages, called prophase, metaphase, anaphase, and telophase.

A bipolar spindle, composed of two sets of microtubules, positions the chromosomes and moves them to specific locations during these stages.

Figure 9.6 Mitosis. This nuclear division mechanism assures that daughter cells will have the same chromosome number as the parent cell. For clarity, the diagram shows only two pairs of chromosomes from a diploid (2*n*) animal cell. With rare exceptions, the picture is more involved than this, as indicated by the micrographs of mitosis in a whitefish cell.

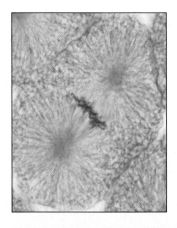 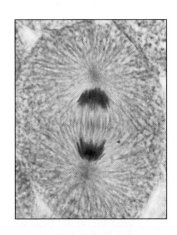 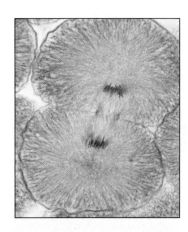

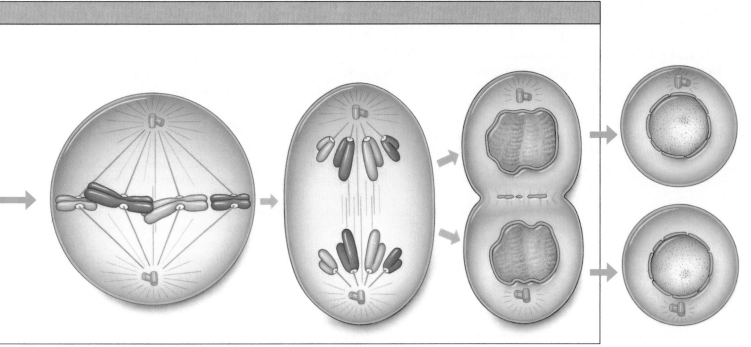

Metaphase

All chromosomes are lined up at the equator of the spindle. They are now in their most condensed form.

Anaphase

The attachment between the two sister chromatids of each chromosome breaks; the two are now chromosomes in their own right. They move to opposite spindle poles.

Telophase

Chromosomes decondense. New patches of membrane join to form nuclear envelopes around them. Most often, cytoplasmic division occurs before telophase is over.

Interphase

Two daughter cells have formed. Each is diploid, with two of each type of chromosome—just like the parent cell.

A CLOSER LOOK AT MITOSIS

Prophase: Mitosis Begins

Chromosomes Start Condensing Prophase, the first stage of mitosis, is evident when chromosomes become visible in the light microscope as threadlike forms. ("Mitosis" comes from the Greek *mitos*, meaning thread.) Every chromosome was duplicated earlier, during interphase. In other words, each already consists of two sister chromatids joined at the centromere. Early in prophase, both chromatids are twisting and folding into a more compact form. By late prophase, all of the chromosomes will be condensed to a thicker, rod-shaped form.

Figure 9.7 shows what one type of duplicated chromosome will look like when it reaches its most condensed form. Notice the centromere. At this constricted region, each sister chromatid of the chromosome has a disk-shaped structure on its surface. This structure, a **kinetochore**, will serve as an attachment site for microtubules of the spindle.

The Spindle Starts Forming Meanwhile, in the cytoplasm near the nucleus, microtubules are forming. Microtubules, recall, are tube-shaped proteins of the cytoskeleton, the internal scaffold of eukaryotic cells. They are composed of numerous subunits called tubulins (Figure 4.19). When a nucleus is about to divide, most of the cytoskeletal microtubules disassemble, then the unbound subunits reassemble as *new* microtubules.

Depending on the species, a completed spindle may consist of ten to many thousands of microtubules. (The plant cell shown in Figure 9.5 probably has 10,000 of them.) Some microtubules will extend continuously between a kinetochore and one spindle pole or the other. The rest will not interact at all with the chromosomes. Rather, they will extend from both poles and overlap each other.

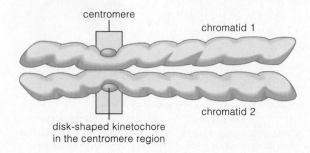

centromere

chromatid 1

disk-shaped kinetochore
in the centromere region

chromatid 2

Figure 9.7 Kinetochores of a duplicated chromosome. Spindle microtubules become attached to the kinetochore of each sister chromatid.

In many cells, an MTOC near the nucleus organizes the assembly of microtubules so that they lengthen in specific directions (hence "MTOC," for microtubule organizing center). While the new microtubules are lengthening, the nuclear envelope prevents them from interacting with the chromosomes inside the nucleus. However, as prophase draws to a close, the nuclear envelope starts to break up.

The Spindle Separates the Centrioles In many cells, an MTOC near the nucleus includes two barrel-shaped **centrioles**. Each centriole started duplicating itself during interphase, and by the time prophase is under way, the cell has two pairs of them. Now microtubules start moving one pair to the opposite pole of the newly forming spindle. You can see this carefully orchestrated movement in Figure 9.6.

Centrioles, recall, give rise to flagella or cilia. If you observe them in the cells of a particular organism, you can safely bet that flagellated or ciliated cells develop during at least some part of its life cycle. For example, the protistan responsible for *trichomoniasis*, a sexually transmitted disease, swims through vaginal fluids by beating its flagella. Even "stationary" plants and fungi commonly produce flagellated, motile gametes. And ciliated cells line airways to your lungs. Ultimately, these and other functions depend on centrioles, so it is no wonder that they are parceled out carefully for the forthcoming daughter cells.

Transition to Metaphase

So much happens during the transition from prophase to metaphase that researchers give this transitional period its own name, "prometaphase."

The nuclear envelope breaks up completely. Its two membrane layers have fused together at scattered sites, producing numerous tiny, flattened vesicles. Now the chromosomes are free to interact with microtubules that extend toward them, from both poles of the developing spindle. At first the chromosomes appear to go into a frenzy. This happens when kinetochores are making their first random contacts with the microtubules.

Once a chromosome has both of its kinetochores harnessed, microtubules from *both* poles start pulling on it. The two-way pulling orients the two sister chromatids of that chromosome toward opposite poles. Meanwhile, the overlapping spindle microtubules are ratcheting past each other in such a way that the two poles are pushed farther apart. All of the push–pull forces are balanced when the chromosomes reach the midpoint of the spindle. The formation of the bipolar spindle is now complete.

When all of the duplicated chromosomes are aligned midway between the poles of a completed spindle, we call this metaphase (*meta-* means midway between). The alignment is crucial for the next stage of mitosis.

Anaphase

During anaphase, the two sister chromatids of each chromosome are separated from each other and moved to opposite poles. They do this at about the same time, and they move at the same rate. Once they do separate, they are no longer referred to as chromatids. Each is now a chromosome in its own right:

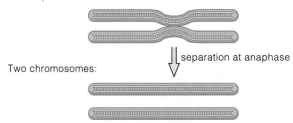

One duplicated chromosome:

separation at anaphase

Two chromosomes:

Said another way, every chromosome that was present in the parent cell now has a daughter chromosome at both spindle poles.

Two mechanisms account for the anaphase movements. First, microtubules attached to the kinetochores shorten, pulling the chromosomes toward the spindle poles (Figure 9.8*a*). Second, the spindle elongates when overlapping microtubules ratchet past each other and push the two spindle poles farther apart (Figure 9.8*b*).

Telophase

Telophase begins once the two daughter chromosomes arrive at opposite spindle poles. The chromosomes are no longer harnessed to microtubules, and they return to threadlike form. Vesicles of the old nuclear envelope fuse together to form patches of membrane around the chromosomes. Patch joins with patch, and soon a new nuclear envelope separates each cluster of chromosomes from the cytoplasm. If the parent cell was diploid, each cluster will contain a pair of each type of chromosome. With mitosis, each new nucleus has the same chromosome number as the parent nucleus.

Once the two nuclei form, telophase is over—and so is mitosis.

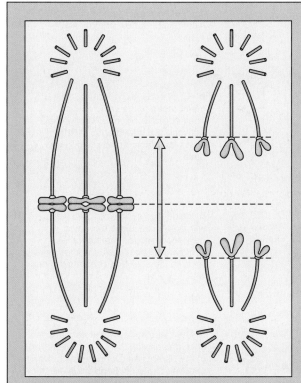

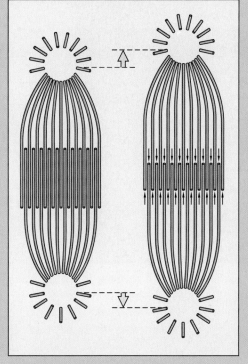

Figure 9.8 Two of the mechanisms responsible for separating the sister chromatids of a chromosome at anaphase.

a Microtubules attached to the chromatids shorten, decreasing the distance between the kinetochore and the spindle poles.

b Overlapping microtubules ratchet past each other, moving the poles farther apart and so increasing the distance between sister chromatids.

9.5 DIVISION OF THE CYTOPLASM

The cytoplasm usually divides at some time between late anaphase and the end of telophase. Cytoplasmic division (or cytokinesis, as it is often called) proceeds by different mechanisms in plant and animal cells.

The cells of most land plants have fairly rigid walls, and they cannot be merely pinched in two. Their cytoplasm divides by a mechanism that involves **cell plate formation**. As Figure 9.9 shows, vesicles filled with wall-building material fuse with remnants from the spindle, forming a disklike structure (the "cell plate"). At this location, cellulose deposits accumulate and form a crosswall that divides the parent cell into two daughter cells.

By contrast, animal cells do not have a cell wall, and most of them can divide by "pinching in two." By a mechanism called **cleavage**, a layer of deposits forms around microtubules at the midsection of most animal cells. A shallow, ringlike depression appears above the layer, at the cell surface (Figure 9.10). At this depression, the cleavage furrow, microfilaments are at work.

Microfilaments, recall, are components of the cytoskeleton. These particular ones are arranged to slide past one another. As they do, they also pull the plasma membrane inward and cut the cell in two.

Most plant cells divide by the formation of a cell plate and then a crosswall between the developing plasma membranes of daughter cells.

Most animal cells divide by cleavage. Microfilaments, arranged in a ring around the cell's midsection, slide past one another in such a way that the cell is cut in two.

This concludes our picture of mitotic cell division. Look now at your hands—and think of all the cells in your palms, thumbs, and fingers. Find a microscope and look at cells in a sample from your skin. Imagine all the divisions of all the cells that preceded them when you were developing early on, inside your mother (Figure 9.11). And be grateful for the astonishing precision that led to their formation. Even from the *Focus* essay presented earlier in this chapter, you have an inkling that the alternatives can be terrible indeed.

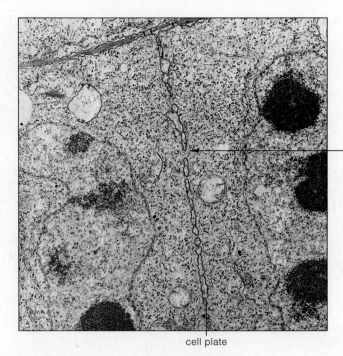

cell plate

Figure 9.9 Cytoplasmic division of a plant cell, as brought about by cell plate formation.

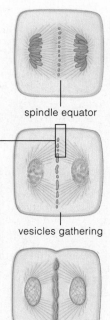

spindle equator

vesicles gathering

cell plate growing

two new primary walls

a As mitosis draws to a close, vesicles gather at the equator of the microtubular spindle. They contain cementing materials and starting materials for a new primary wall.

b A cell plate starts forming as the membranes of the vesicles fuse. Their contents become sandwiched between two membranes that form along the plane of the growing cell plate.

c Inside the "sandwich," two walls will form as cellulose becomes deposited over both membranes. Other substances inside will form the "middle lamella" that cements the new walls together.

d The cell plate grows at its margins until it fuses with the plasma membrane of the parent cell. During growth, when cells are expanding and walls are thin, new material also is deposited over the old primary wall.

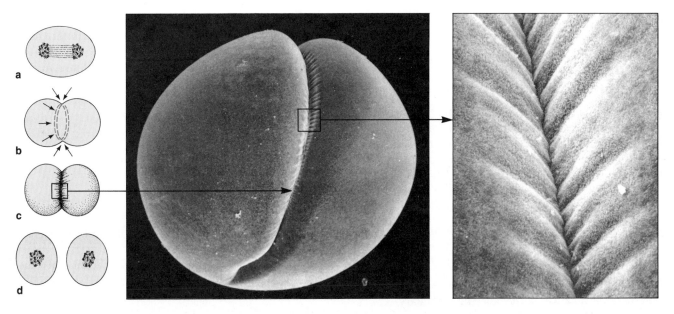

Figure 9.10 Cytoplasmic division of an animal cell. (**a**) Mitosis is completed; the microtubular spindle is disassembling. (**b**) Just beneath the plasma membrane of the parent cell, microfilament rings at the former spindle equator contract, like a purse string. (**c,d**) Contractions continue and in time will divide the cell in two. The micrographs show how the plasma membrane sinks inward, defining the cleavage plane. Cleavages helped form the human hand shown below in Figure 9.11*d*.

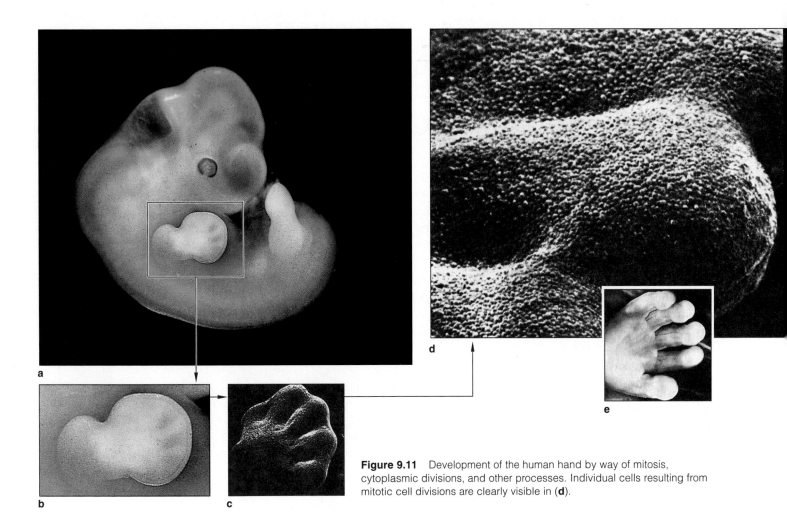

Figure 9.11 Development of the human hand by way of mitosis, cytoplasmic divisions, and other processes. Individual cells resulting from mitotic cell divisions are clearly visible in (**d**).

SUMMARY

1. Parent cells provide each daughter cell with the hereditary instructions (DNA) and cytoplasmic machinery necessary to start up its own operation.

2. Mitosis divides the hereditary instructions in the nucleus of a parent cell into two equivalent parcels. It is the basis for growth and tissue repair in multicelled eukaryotes. It also is the basis of asexual reproduction among many single-celled and multicelled eukaryotes. (Meiosis, another nuclear division mechanism, occurs only in germ cells set aside for sexual reproduction.)

3. The cytoplasm divides by different mechanisms near the end of nuclear division or at some point thereafter. (Animal cells are cleaved in two. A cell plate forms and divides plant cells.)

4. A chromosome is a single DNA molecule. In eukaryotes, many proteins are attached to it. A cell's chromosomes differ from one another in length, shape, and which portion of the hereditary instructions they carry. Members of the same species have the same total number of chromosomes in their somatic cells.

5. "Chromosome number" is not the same thing as the total number of chromosomes. It is the number of *each type* of chromosome characteristic of a species. Mitosis keeps the chromosome number constant from one cell generation to the next. Thus, if a parent cell is diploid (with two chromosomes of each type), the daughter cells resulting from mitosis will be diploid also.

6. The cell cycle starts when a new cell forms, proceeds through interphase, and ends when the cell reproduces by mitosis and cytoplasmic division. At interphase, a cell increases in mass, doubles its number of cytoplasmic components, and duplicates all chromosomes.

7. After duplication, each chromosome consists of two DNA molecules attached at the centromere. For as long as the two stay attached, they are called sister chromatids of the chromosome.

8. Mitosis proceeds through four continuous stages:
   a. Prophase. Duplicated, threadlike chromosomes start to condense. New microtubules assemble in organized arrays near the nucleus; they will form a spindle. The nuclear envelope starts to break up.
   b. Metaphase. During the *transition* to metaphase, the nuclear envelope breaks up completely. Microtubules of a forming spindle attach to the sister chromatids of each chromosome and orient the two toward opposite spindle poles. *At* metaphase, all chromosomes are aligned at the spindle equator.
   c. Anaphase. Microtubules pull the sister chromatids of each chromosome away from each other, to opposite spindle poles. Every chromosome that was present in the parent cell is now represented by a daughter chromosome at both poles.
   d. Telophase. The chromosomes decondense to the threadlike form. A new nuclear envelope forms around them. Each nucleus has the same chromosome number as the parent cell. Mitosis is completed.

Review Questions

1. Define the two types of nuclear division mechanisms that occur in eukaryotes. Does either one divide the cytoplasm? *142*

2. Define somatic cell and germ cell. *142*

3. What is a chromosome? What is a chromosome called in its unduplicated state? In its duplicated state (that is, with two sister chromatids)? *142–143*

4. Describe the spindle apparatus and its general function in nuclear division. *148–149*

5. Name and describe the main features of the four stages of mitosis. At what stages were the following micrographs of a plant cell taken? *148–150*

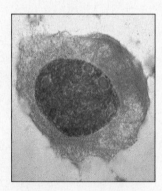

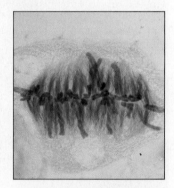

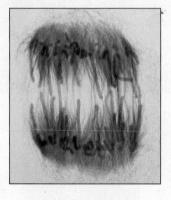

6. Correctly label the component parts of this human chromosome. *148*

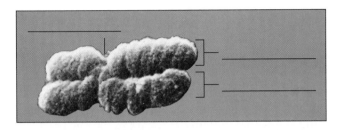

7. Describe cytoplasmic division as it occurs in plant cells and animal cells. *150*

8. The function of mitosis and cytoplasmic division is _____ .
   a. asexual reproduction of single-celled eukaryotes
   b. growth, repair, and asexual reproduction in many eukaryotes
   c. gamete formation in eukaryotes
   d. both a and b are correct

9. Only _____ is not a stage of mitosis.
   a. prophase   c. metaphase
   b. interphase   d. anaphase

10. Match each stage with its key events.
   __d__ metaphase   a. sister chromatids of each
   __b__ prophase   chromosome separate
   __c__ telophase   b. threadlike chromosomes start
   __a__ anaphase   to condense
                          c. chromosomes decondense,
                             daughter nuclei form
                          d. all chromosomes are aligned
                             at the equator

Self-Quiz *(Answers in Appendix IV)*

1. A eukaryotic chromosome consists of _____ .
   a. DNA only   c. DNA plus membrane
   b. DNA plus proteins   d. DNA plus lipids

2. A cell with pairs of the chromosomes characteristic of the species is a(an) _____ cell.
   a. diploid   c. abnormal
   b. mitotic   d. a and c

3. A duplicated chromosome has _____ chromatid(s).
   a. one   c. three
   b. two   d. four

4. The _____ is the constricted region of a duplicated chromosome, with attachment sites for microtubules.
   a. chromatid   c. cell plate
   b. centromere   d. cleavage furrow

5. In the cell cycle, the interphase stage immediately following DNA replication is _____ .
   a. G$_2$ c. S
   b. G$_1$ d. none of the above

6. Interphase is the stage when _____ .
   a. nothing occurs
   b. a germ cell forms its spindle apparatus
   c. a cell grows and duplicates its DNA
   d. mitosis occurs

7. After mitosis, the chromosome number of a daughter cell is _____ the parent cell's.
   a. the same as   c. rearranged compared to
   b. one-half   d. doubled compared to

Selected Key Terms

anaphase *146*
cell cycle *144*
cell plate formation *150*
centriole *148*
centromere *142*
chromosome *142*
chromosome number *143*
cleavage *150*
diploid cell *143*
germ cell *142*
interphase *144*

kinetochore *148*
meiosis *142*
metaphase *146*
mitosis *142*
prophase *146*
reproduction *142*
sister chromatid *142*
somatic cell *142*
spindle apparatus *146*
telophase *146*

Readings

Gallagher, G. March 1990. "Evolution: The Mitotic Spindle." *The Journal of NIH Research.*

Murray, A., and M. Kirschner. March 1991. "What Controls the Cell Cycle?" *Scientific American* 264(3): 56–63.

Science. November 3, 1989. "Frontiers in Biology: The Cell Cycle."

Wolfe, S. 1993. *Molecular and Cellular Biology.* Belmont, California: Wadsworth.

10 MEIOSIS

Octopus Sex and Other Stories

The couple clearly are interested in each other. He caresses her with one tentacle, then another—then another and another. She reciprocates with a hug here, a squeeze there. This goes on for hours. Finally the male reaches under his mantle, a tissue fold that drapes around most of his body. He removes a packet of sperm and inserts it into an egg chamber under the female's mantle. For every sperm that fertilizes an egg, a new octopus may develop.

Unlike the coupling between a male and female octopus, sex for the slipper limpet is a group activity. Slipper limpets are marine animals, relatives of land snails. Before one of them becomes transformed into a sexually mature adult, it passes through a free-living larval stage. When the time comes for the larva to undergo transformation, it settles down on sand or sediments. If it settles down alone, it will become a female. If a second larva settles down on the first one, *it* will become a male—and so will any subsequent larvae. Adult slipper limpets almost always live in such piles, with the bottom one always being female (Figure 10.1*a*). The males release sperm on a perpetual basis. When the female finally dies, the male at the bottom of the pile becomes transformed into a female, and so it goes.

For many sexually reproducing organisms, the life cycle also has asexual episodes, based on mitotic cell divisions. Many kinds of plants, including orchids, can reproduce without sex. Flatworms can split lengthwise into two roughly equivalent parts that each grow into a new flatworm. In summer, nearly all aphids are females, which busily produce more females from unfertilized egg cells (Figure 10.1*b*). When autumn approaches, male aphids finally develop and do their part in the sexual phase of the life cycle. Still, females

Figure 10.1 Variations in reproductive modes. (**a**) Slipper limpets busily perpetuating the species by group participation in sexual reproduction. (**b**) Live birth of an aphid, a type of insect that reproduces sexually in autumn but switches to an asexual mode in summer.

a

that survive the winter can do without the males, and come summer they begin another round of producing offspring all by themselves.

These examples only hint at the immense variation in reproductive modes. Yet despite the variation, sexual reproduction dominates eukaryotic life cycles, and it always involves certain events. Chromosomes are duplicated in germ cells. The germ cells undergo meiosis and cytoplasmic division, and then gametes develop. When gametes join at fertilization, they form the first cell of a new individual. *Meiosis, gamete formation, and fertilization are the hallmarks of sexual reproduction.* As you will see in this chapter, these processes contribute to life's splendid diversity.

b

KEY CONCEPTS

1. Sexual reproduction proceeds through three events: meiosis, formation of gametes (such as sperm and eggs), and fertilization.

2. Meiosis is a nuclear division mechanism that is the first step in sexual reproduction. It sorts out the chromosomes of a germ cell into four new nuclei. In most plants, spore formation and other events intervene between meiosis and gamete formation.

3. The body cells of many species are diploid, with two of each type of chromosome. The two function as a pair during meiosis. Most often, one chromosome of the pair is maternal (with hereditary instructions from a female parent), and the other is paternal (with comparable instructions from a male parent).

4. Meiosis divides the chromosome number in half for each forthcoming gamete. Thus, if both parents are diploid ($2n$), the union of two gametes at fertilization will restore the diploid number in the new individual ($n + n = 2n$).

5. During meiosis, each pair of chromosomes may swap segments and thereby swap hereditary instructions about particular traits. Also, meiosis randomly assigns either the maternal or paternal chromosome to a gamete—and it does this for each pair. Hereditary instructions are further shuffled at fertilization. All three of these reproductive events lead to variations in traits among offspring.

10.1 ON ASEXUAL AND SEXUAL REPRODUCTION

What kind of offspring do orchids, flatworms, and aphids get with **asexual reproduction**? By this process, one parent alone passes on a duplicate of all its genes to each new individual. **Genes** are specific stretches of chromosomes—that is, DNA molecules. Genes are all the heritable bits of information that are required to produce new individuals. Rare mutations aside, this means that asexually produced offspring can only be clones—genetically identical copies of the parent.

Inheritance is much more interesting with **sexual reproduction**. In humans and many other species, *two* parents—each with pairs of genes—engage in this process. Both parents pass on one of each gene to offspring by way of meiosis, gamete formation, and fertilization (the union of two gametes). Thus the first cell of a new individual also ends up with pairs of genes.

If instructions in every pair of genes were identical down to the last detail, then sexual reproduction would produce clones, also. Just imagine—you, everyone you know, and the entire human population would be clones and might end up looking exactly alike.

But a pair of genes may *not* be identical. Why not? The molecular structure of genes can change; this is what we mean by mutation. Depending on its structure, one gene of a pair may "say" slightly different things about a trait. Each unique molecular form of the same gene is called an **allele**.

Such tiny differences affect thousands of traits. Figure 10.2 is an example. One allele that may occur at a certain location in a human chromosome says "put a dimple in the chin" and another says "no dimple."

This brings us to a key reason why members of any sexually reproducing species don't all look alike. *Through sexual reproduction, offspring end up with new combinations of alleles, and these lead to variations in their physical and behavioral traits.*

This chapter describes the cellular basis of sexual reproduction. More importantly, it starts us thinking about the far-reaching effects of gene shufflings at different stages of the process. The resulting variation among offspring may be acted upon by agents of natural selection. Thus, *variation in traits is a foundation for evolutionary change.*

Asexual reproduction produces genetically identical copies of the parent. Sexual reproduction introduces variations in the details of traits among offspring.

10.2 OVERVIEW OF MEIOSIS

Think "Homologues"

Think back on the preceding chapter, which focused on mitotic cell division. Unlike mitosis, **meiosis** divides the chromosomes in a nucleus not once but twice prior to cell division. Unlike mitosis, it is the first step leading to gamete formation. **Gametes** are sex cells, such as sperm and eggs. In multicelled organisms, gametes arise only from germ cells, which are produced in specific reproductive structures and organs (Figure 10.3).

Figure 10.2 (a) The chin fissure, a heritable trait arising from an uncommon form of a gene. Actor Kirk Douglas received a gene that influences this trait from each of his parents. One gene called for a chin fissure and the other didn't, but one is all it takes in this case. (b) This photograph shows what his chin might have looked like if he had inherited two ordinary forms of the gene instead.

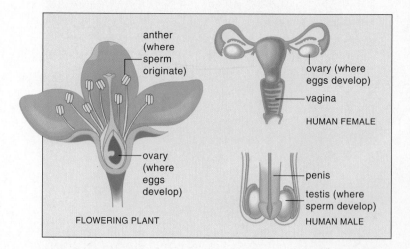

Figure 10.3 Examples of gamete-producing structures.

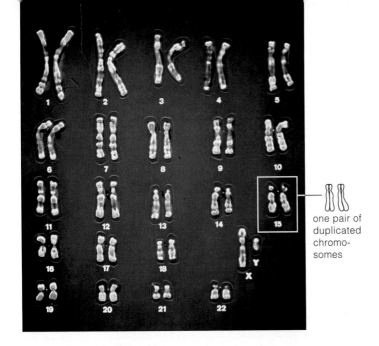

Figure 10.4 From a diploid cell of a human male, twenty-three pairs of homologous chromosomes. The sketch corresponding to the two boxed chromosomes indicates that the chromosomes are in the duplicated state (each consists of two sister chromatids).

Germ cells have the same chromosome number as the rest of the body's cells. The ones that are **diploid** (2n) have *two* of each type of chromosome, often from two parents. The two chromosomes have the same length and shape. Their genes deal with the same traits. And they line up with each other at meiosis. Think of them as **homologous chromosomes** (*hom-* means alike).

As you can deduce from Figure 10.4, there are 23 + 23 homologous chromosomes in your germ cells. After meiosis, 23 chromosomes—one of each type—end up in gametes. Said another way, meiosis halves the chromosome number, so the gametes are **haploid** (n).

Overview of the Two Divisions

Meiosis resembles mitosis in some respects, even though the outcome is different. Before interphase gives way to meiosis, a germ cell duplicates its DNA. Now each duplicated chromosome consists of two DNA molecules. These remain attached at a narrowed-down region (the centromere). For as long as the two remain attached, they are called **sister chromatids** of the chromosome:

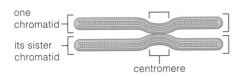

As in mitosis, microtubules of a spindle apparatus move the chromosomes in prescribed directions.

With meiosis alone, however, *chromosomes proceed through two consecutive divisions, which end with the formation of four haploid nuclei.* The two divisions are called meiosis I and II:

	MEIOSIS I		MEIOSIS II
DNA duplication during interphase	Prophase I Metaphase I Anaphase I Telophase I	No DNA duplication between divisions	Prophase II Metaphase II Anaphase II Telophase II

During meiosis I, each duplicated chromosome lines up with its partner, *homologue to homologue*; then the partners are separated from each other:

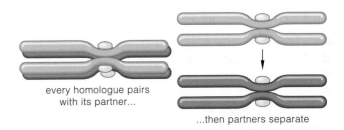

every homologue pairs with its partner...

...then partners separate

The cytoplasm typically divides after the separation of homologues. The two daughter cells are haploid, with only one of each type of chromosome. But remember, those chromosomes are still duplicated.

During meiosis II, *the two sister chromatids of each chromosome are separated from each other:*

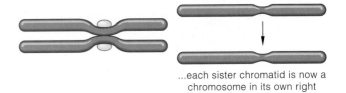

...each sister chromatid is now a chromosome in its own right

After four nuclei form, the cytoplasm often divides once more, the outcome being four haploid cells.

On the next two pages, Figure 10.5 illustrates the key events of meiosis I and II.

Meiosis is a type of nuclear division that reduces the parental chromosome number by half—to the haploid number (n).

Meiosis is the first step leading to the formation of gametes, which are required for sexual reproduction.

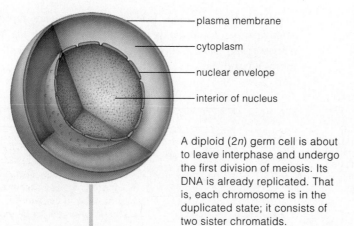

plasma membrane

cytoplasm

nuclear envelope

interior of nucleus

Figure 10.5 Meiosis: the nuclear division mechanism by which the parental number of chromosomes is reduced by half (to the haploid number) for forthcoming gametes. Only two pairs of homologous chromosomes are shown. Maternal chromosomes are shaded *purple,* and paternal ones *blue.*

A diploid (*2n*) germ cell is about to leave interphase and undergo the first division of meiosis. Its DNA is already replicated. That is, each chromosome is in the duplicated state; it consists of two sister chromatids.

Meiosis I

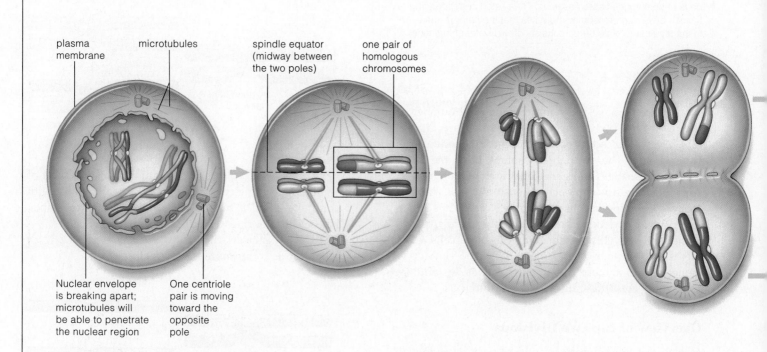

plasma membrane

microtubules

spindle equator (midway between the two poles)

one pair of homologous chromosomes

Nuclear envelope is breaking apart; microtubules will be able to penetrate the nuclear region

One centriole pair is moving toward the opposite pole

Prophase I

Each chromosome starts to twist and fold into more condensed form. It pairs up with its homologue, and the two typically swap segments. This swapping, called crossing over, is indicated by the break in color on the pair of larger chromosomes. As in mitosis, chromosomes become attached to a newly forming microtubular spindle.

Metaphase I

Microtubules have pushed and pulled all chromosomes into position, midway between the two spindle poles. The spindle is now completely formed, owing to the chromosome-microtubule interactions.

Anaphase I

Each chromosome is separated from its homologue. The two are moved to opposite poles of the spindle.

Telophase I

When the cytoplasm divides, two haploid (*2n*) cells are the result. Each cell has one chromosome of each type, although these are still in the duplicated state.

Of the four haploid cells that may form by way of meiosis, one or all may develop into gametes. (In plants, the cells that form by way of meiosis may develop into spores, which precede the formation of gametes.)

Meiosis II

There is no DNA replication between the two divisions

Prophase II

During the transition to prophase II, the two centrioles were moved apart and a new spindle was assembled in each newly formed cell. Now, at prophase II, chromosomes become attached to microtubules and start moving toward the spindle equator.

Metaphase II

All of the duplicated chromosomes are now positioned midway between the two poles of the spindle.

Anaphase II

The attachment between the two chromatids of each chromosome breaks. Now the former "sister chromatids" are chromosomes in their own right and are moved to opposite poles of the spindle.

Telophase II

Four daughter nuclei form. When the cytoplasm divides, each new cell has a haploid number of chromosomes, all in the unduplicated state.

10.3 KEY EVENTS DURING MEIOSIS I

The preceding overview is enough to convey the over-riding function of meiosis—that is, *reduction of the chromosome number by half for forthcoming gametes.* However, as you will now see, two other events that unfold during prophase I and metaphase I contribute enormously to the key adaptive advantage of sexual reproduction—*the production of offspring with new combinations of alleles, hence variations in traits.*

Prophase I Activities

Prophase I of meiosis is a time of major gene shufflings. Consider Figure 10.6a, which shows two chromosomes that have condensed to threadlike form. All chromosomes in a germ cell condense this way. When they do, each is drawn close to its homologue by a process called synapsis. It is as if they become stitched together point by point along their length, with little space between. The intimate, parallel array favors **crossing over**, a type of molecular interaction between the chromatids of a pair of homologous chromosomes. These *nonsister* chromatids break at the same places along their length and exchange corresponding segments—that is, genes—at the break points. Figure 10.6c is a simplified picture of this interaction.

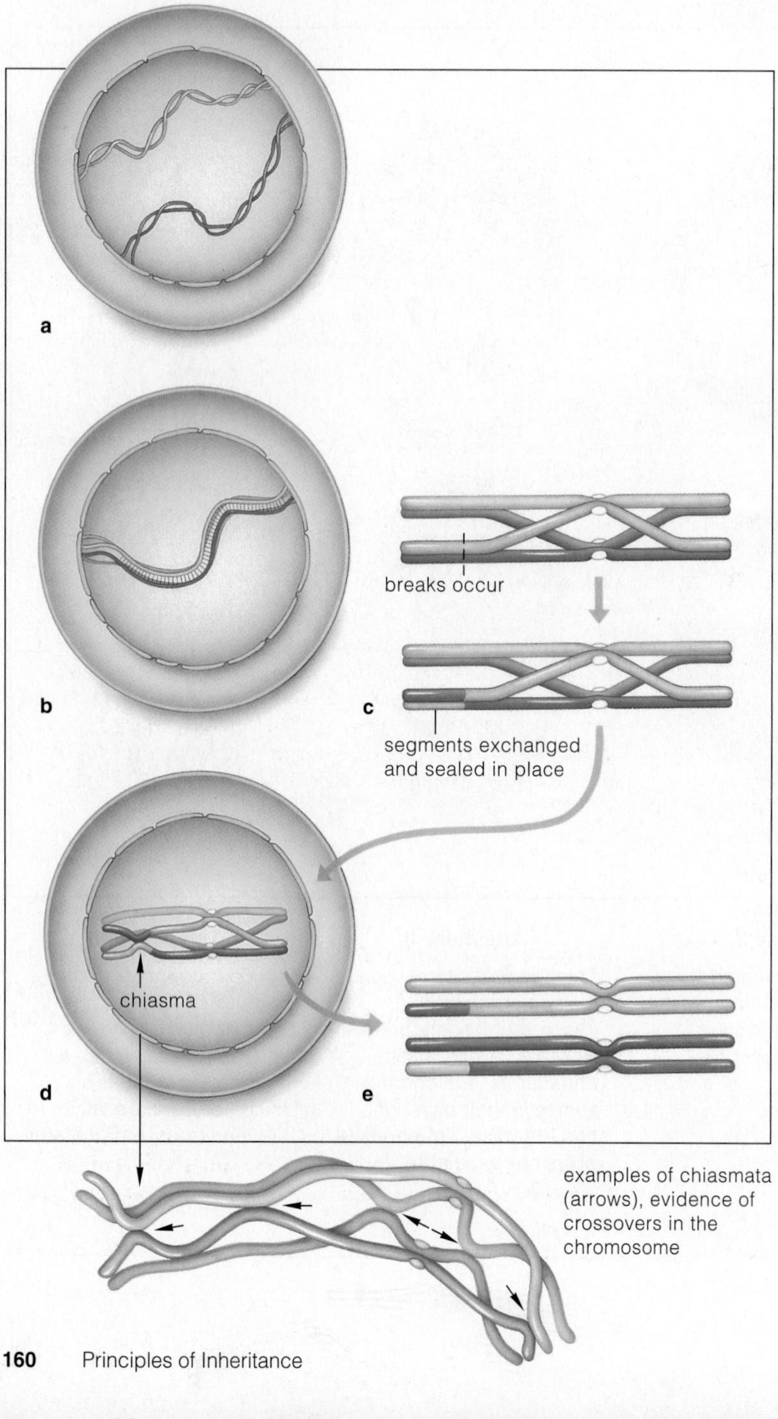

breaks occur

segments exchanged and sealed in place

chiasma

examples of chiasmata (arrows), evidence of crossovers in the chromosome

Figure 10.6 Key events during prophase I, the first stage of meiosis. For clarity, only a single pair of homologous chromosomes and only one crossover event are shown. *Blue* signifies the paternal chromosome, and *purple* signifies its maternal homologue.

a Chromosomes become duplicated before meiosis begins. Early in prophase I, each duplicated chromosome is in threadlike form, attached at both ends to the nuclear envelope. Its two sister chromatids are so close together, they look like a single thread.

b The two chromosomes become zippered together, so that all four chromatids are intimately aligned. (X and Y chromosomes, which are two forms of sex chromosomes, also become zippered together for a small part of their length.)

c One or more crossovers occur at intervals along the chromosomes. In each crossover, two nonsister chromatids break at identical sites. They swap segments at the breaks, then enzymes seal the broken ends. (For clarity, the two chromosomes are shown in condensed form and pulled apart. Crossing over may seem more plausible when you realize that it occurs while chromosomes are extended like threads and tightly aligned.)

d As prophase I ends, the chromosomes continue to condense, becoming thicker, rodlike forms. They detach from the nuclear envelope and from each other—except at "chiasmata." Each chiasma is indirect evidence that a crossover occurred at some point in the chromosomes.

e Crossing over breaks up old combinations of alleles and puts new ones together in pairs of homologous chromosomes.

Gene swapping would be rather pointless if each type of gene never varied from one chromosome to the next. But remember, a gene can have slightly different forms—alleles. You can safely bet that some alleles on one chromosome will *not* be identical to those on the homologue. So each crossover represents a chance to swap a *slightly different version* of the hereditary instructions for a particular trait.

We will look at the mechanism of crossing over in later chapters. For now, it is enough to know that crossing over leads to genetic recombination, which in turn leads to variation in the traits of offspring.

Crossing over is an interaction between a pair of homologous chromosomes. It breaks up old combinations of alleles and puts new ones together.

Metaphase I Alignments

Major shufflings of whole chromosomes begin during the transition from prophase I to metaphase I, the second stage of meiosis. Suppose the shufflings are proceeding right now in one of your germ cells. By now, crossovers have made genetic mosaics out of the chromosomes, but let's put this aside to simplify tracking. Just call the twenty-three chromosomes inherited from your mother the *maternal* chromosomes and call their twenty-three homologues from your father, the *paternal* chromosomes.

Microtubules have harnessed and oriented one chromosome of each pair toward one spindle pole and its homologue toward the other. Now they are moving all the chromosomes, which soon will be positioned midway between the spindle poles.

Are all the maternal chromosomes attached to one pole and all the paternal ones attached to the other? Maybe, but probably not. Remember, the first contacts between spindle microtubules and chromosomes are random. Because of this random harnessing, *the eventual positioning of a maternal or paternal chromosome at the spindle equator at metaphase I follows no particular pattern.* Carrying this one step further, either chromosome of a pair may end up at either pole of the spindle when the two move apart at anaphase I.

Think of the possibilities when merely three pairs of homologues are involved. As Figure 10.7 shows, by metaphase I, these may be arranged in any one of four possible positions. In this case, 2^3 or 8 combinations of maternal and paternal chromosomes are possible for the forthcoming gametes.

Of course, a human germ cell has twenty-three pairs of homologous chromosomes, not just three. So 2^{23} or *8,388,608 combinations* of maternal and paternal chromo-

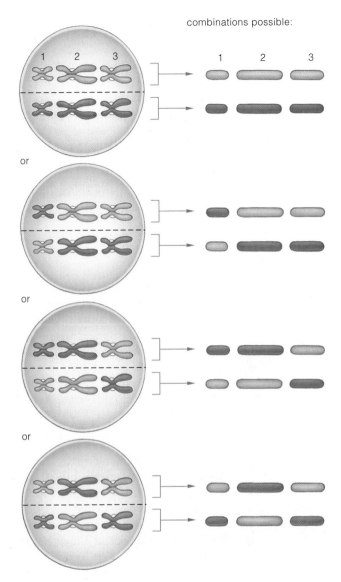

combinations possible:

Figure 10.7 The possible outcomes of the random alignment of three pairs of homologous chromosomes at metaphase I of meiosis. The three types of chromosomes are labeled 1, 2, and 3. Maternal chromosomes are *purple;* paternal ones are *blue.* With merely four possible alignments, eight different combinations of maternal and paternal chromosomes are possible in forthcoming gametes.

somes are possible every time a germ cell gives rise to sperm or eggs! And in each sperm or egg, many hundreds of alleles inherited from the mother might differ from their counterparts from the father. Are you beginning to get an idea of why such splendid mixes of traits show up even in the same family?

The random attachment and subsequent positioning of each pair of maternal and paternal chromosomes at metaphase I lead to different combinations of maternal and paternal traits in offspring.

10.4 FORMATION OF GAMETES

The gametes that form after meiosis are not all the same. Human sperm have one tail, opossum sperm have two, and roundworm sperm have none. Crayfish sperm look like pinwheels. Most eggs are microscopic, yet ostrich eggs tucked inside a shell are as large as a softball. From appearance alone, you might not believe a plant gamete is even remotely like an animal's.

Later chapters contain details of gamete formation in the life cycles of representative organisms, including humans. Figure 10.8 and the following points may help you keep the details in perspective.

Gamete Formation in Plants

For pine trees, roses, and other familiar plants, certain events intervene between the time of meiosis and gamete formation. Among other things, spores form. **Spores** are haploid cells, often with walls that allow them to resist dry periods or other adverse environmental conditions. Under favorable conditions, spores germinate and develop into a haploid body or structure that will produce gametes. Thus gamete-producing bodies and spore-producing bodies develop during the life cycle (Figure 10.8a).

Gamete Formation in Animals

In male animals, gametes form by a process called spermatogenesis. Inside a male reproductive system, a diploid germ cell increases in size. It becomes a large, immature cell (the primary spermatocyte) that undergoes meiosis and cytoplasmic divisions. Its four haploid daughter cells develop into immature cells called spermatids (Figure 10.9). Spermatids change in form and develop a tail. In this way, each becomes a **sperm**, a male gamete.

In female animals, gametes form by a process called oogenesis. Each diploid germ cell develops into an immature egg (oocyte). Compared to a sperm, an oocyte accumulates more cytoplasmic components. Also, its daughter cells differ in size and function (Figure 10.10). When the oocyte divides after meiosis I, one cell (the secondary oocyte) gets nearly all the cytoplasm. So the other cell (the first polar body) is quite small. Both may undergo meiosis II and cytoplasmic division. Here again, one cell gets most of the cytoplasm. It develops into a gamete. A mature female gamete is called an ovum or, more commonly, an **egg**.

Division of the smaller cell means there are now three polar bodies. These do not function as gametes. In effect, they are dumping grounds for chromosomes, so the egg ends up with the required haploid number. Because polar bodies do not have much cytoplasm, they do not have much in the way of nutrients and metabolic machinery. In time they degenerate.

10.5 MORE GENE SHUFFLINGS AT FERTILIZATION

The parental chromosome number is restored at **fertilization**, when the nuclei of two haploid gametes fuse. Unless meiosis precedes it, fertilization would result in a doubling of the chromosome number in every new generation.

Fertilization contributes to variation in offspring. Reflect on the possibilities for humans alone. During prophase I of meiosis, an average of two or three crossovers takes place in each human chromosome. Even without crossovers, the random positioning of pairs of paternal and maternal chromosomes at metaphase I results in one of 8,388,608 possible chromosome combinations in each gamete. And of all the male and female gametes that are produced, *which two* actually get together is a matter of chance. Thus the sheer number of combinations that can exist at fertilization is staggering!

Cumulatively, crossing over, the distribution of random mixes of homologous chromosomes into gametes, and fertilization contribute to variation in the traits of offspring.

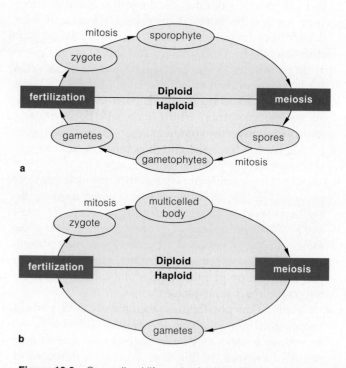

Figure 10.8 Generalized life cycles for (**a**) multicelled plants and (**b**) animals. A zygote is the first cell formed after two gametes fuse at fertilization. For plants, a sporophyte (spore-producing body) forms from the zygote, and gametophytes (gamete-producing bodies) form after meiosis. A pine tree is a sporophyte. Gametophytes develop in its cones (page 401).

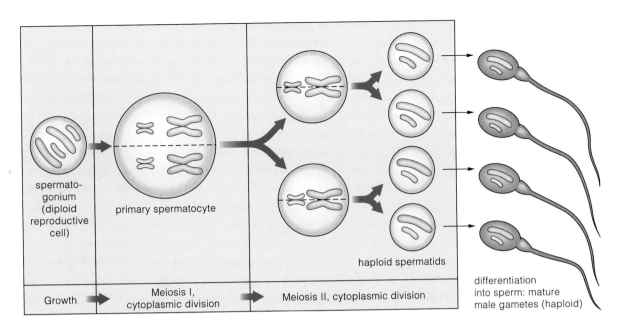

Figure 10.9 Generalized picture of sperm formation in male animals.

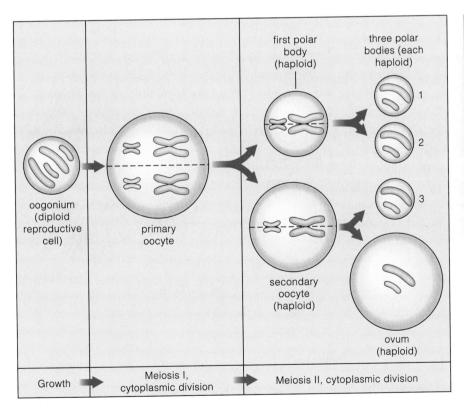

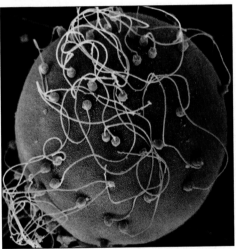

Figure 10.10 Generalized picture of egg formation in female animals. The diagram is not at the same scale as Figure 10.9. An egg is *much* larger than a sperm, as indicated by the micrograph of sea urchin gametes. Also, the three polar bodies are extremely small compared to an egg.

10.6 MEIOSIS AND MITOSIS COMPARED

In this unit, our focus has been on two different mechanisms that divide the nuclear DNA of eukaryotic cells. Single-celled species use mitosis in asexual reproduction; multicelled species use mitosis during growth and tissue repair. Meiosis is the basis of gamete formation and sexual reproduction. Figure 10.11 summarizes the similarities and differences between the two mechanisms.

The two mechanisms differ in a crucial way. *Mitotic cell division produces clones* (genetically identical copies of a parent cell). *Meiosis, together with fertilization, promotes variation in traits among offspring.* First, crossing over at prophase I of meiosis puts new combinations of alleles in chromosomes. Second, the movement of either member of each pair of homologous chromosomes to either spindle pole after metaphase I puts different mixes of maternal and paternal alleles into gametes. Third, different combinations of alleles are brought together by chance at fertilization. Later chapters describe how meiosis and fertilization contribute to the evolution of sexually reproducing organisms.

Figure 10.11 Comparison of mitosis and meiosis, using a diploid (2n) animal cell as the example. This diagram is arranged to help you compare the similarities and differences between the two division mechanisms. Maternal chromosomes are shaded *purple*, and paternal chromosomes are *blue*.

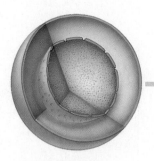

A diploid (2n) *somatic* cell is at interphase. DNA is replicated (all chromosomes are duplicated) before nuclear division begins.

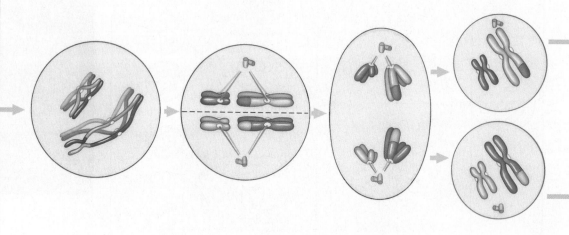

Meiosis I

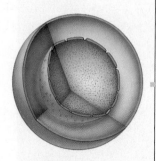

A diploid (2n) *reproductive* cell is at interphase. DNA is replicated (all chromosomes are duplicated) before nuclear division begins.

Prophase I

Each duplicated chromosome (consisting of two sister chromatids) condenses to threadlike form, then rodlike form. *Crossing over* occurs. Each chromosome unzips from its homologue. Each gets attached to the spindle in transition to metaphase.

Metaphase I

All chromosomes are now positioned at the spindle's equator.

Anaphase I

Each chromosome is separated from its homologue. They are moved to opposite poles of the spindle.

Telophase I

When the cytoplasm divides, there are two cells. Each has a haploid (n) number of chromosomes, but these are still in the duplicated state.

Mitosis

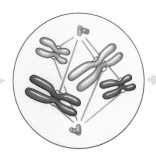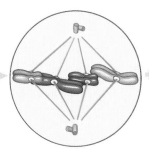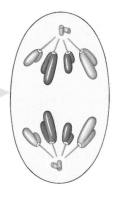

Prophase

Each duplicated chromosome (consisting of two sister chromatids) condenses from threadlike form to rodlike form. Each gets attached to the spindle during the transition to metaphase.

Metaphase

All chromosomes are now positioned at the spindle's equator.

Anaphase

Sister chromatids of each chromosome are separated from each other. These new, daughter chromosomes are moved to opposite poles of the spindle.

Telophase

When the cytoplasm divides, there are two cells. Each is diploid (2*n*)—*it has the same chromosome number as the parent cell.*

Meiosis II

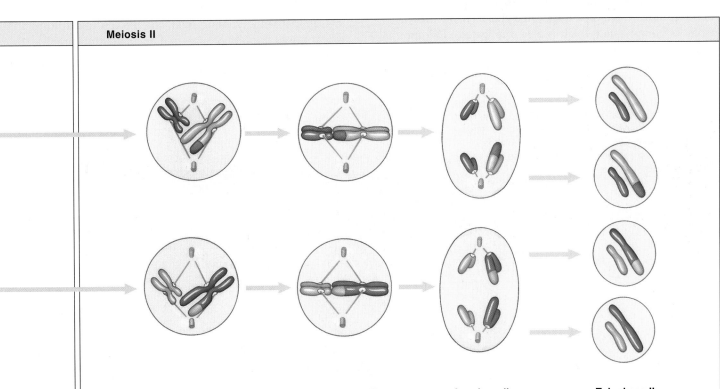

Prophase II

Before prophase II, the two centrioles in each new cell were moved apart and a new spindle formed. Now, each chromosome becomes attached to the spindle and starts moving toward its equator.

Metaphase II

All chromosomes are now positioned at the spindle's equator.

Anaphase II

Sister chromatids of each chromosome are separated from each other. These new, daughter chromosomes are moved to opposite poles of the spindle.

Telophase II

Four daughter nuclei form. When the cytoplasm divides, each new cell is haploid (*n*). *The original chromosome number has been reduced by half.* One or all of these cells may become gametes.

SUMMARY

1. Life cycles of sexually reproducing species proceed through meiosis, gamete formation, and fertilization. Meiosis reduces the chromosome number of a parent cell by half. It precedes the formation of haploid gametes (typically, sperm in males, eggs in females). Fusion of a sperm nucleus and an egg nucleus at fertilization restores the chromosome number (Figure 10.12).

2. If cells of sexually reproducing organisms are diploid (2n), they have two of each type of chromosome characteristic of the species. Commonly, one of the two is maternal (inherited from a female parent), and the other is paternal (from a male parent).

3. Each pair of maternal and paternal chromosomes shows homology (the two are alike). Generally, the two have the same length, same shape, and same sequence of genes. They interact during meiosis.

4. Chromosomes are duplicated during interphase. So before meiosis, each consists of two DNA molecules that remain attached (as sister chromatids).

5. Meiosis consists of two consecutive divisions that require a spindle apparatus. In meiosis I, microtubules of the spindle move each duplicated chromosome away from its homologue (which is also duplicated). During meiosis II, microtubules separate the sister chromatids of each chromosome.

6. The following activities proceed during meiosis I:

   a. At prophase I, nonsister chromatids of homologous chromosomes break at corresponding sites and exchange segments. Crossing over puts new combinations of alleles together. Alleles (slightly different molecular forms of the same gene) code for different forms of the same trait. So new allelic combinations lead to variation in the details of a given trait among offspring.

   b. At metaphase I, all pairs of homologous chromosomes are positioned at the spindle equator. In each case, either the maternal chromosome or its homologue can be oriented toward either pole.

   c. At anaphase I, each chromosome separates from its homologue, and the two move to opposite poles.

7. The following activities proceed during meiosis II:

   a. At metaphase II, chromosomes are still duplicated, and they are positioned at the spindle equator.

   b. During anaphase II, the sister chromatids of each chromosome are moved apart. Each is now a separate, unduplicated chromosome.

   c. During telophase II, four haploid nuclei form.

8. When the cytoplasm divides, there are four haploid cells. One or all of those cells may function as gametes (or as spores, in the case of flowering plants).

9. Cumulatively, crossing over, the distribution of different mixes of pairs of maternal and paternal chromosomes into gametes, and fertilization contribute to immense variation in the traits of offspring.

Figure 10.12 Summary of changes in the chromosome number at different stages of sexual reproduction, using diploid (2n) germ cells as the example. Meiosis reduces the chromosome number by half (n). The union of haploid nuclei of two gametes at fertilization restores the diploid number.

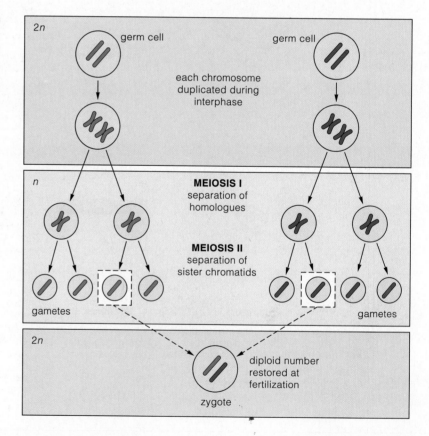

1. Name the three key events of sexual reproduction. *156*

2. What is the key difference between sexual and asexual reproduction? *156*

3. Refer to the following numbers of chromosomes in the diploid body cells of a few organisms. In each case, how many chromosomes would end up in gametes? *157*

Fruit fly, *Drosophila melanogaster*	8
Garden pea, *Pisum sativum*	14
Corn, *Zea mays*	20
Frog, *Rana pipiens*	26
Earthworm, *Lumbricus terrestris*	36
Human, *Homo sapiens*	46
Chimpanzee, *Pan troglodytes*	48
Amoeba, *Amoeba*	50
Horsetail, *Equisetum*	216

4. Suppose a diploid germ cell has four pairs of homologous chromosomes, designated AA, BB, CC, and DD. How would the chromosomes of the gametes be designated? *156–157*

5. Define meiosis and characterize its main stages. In what respects is meiosis *not* like mitosis? *156–159, 164–165*

6. Is this cell at anaphase I or anaphase II?

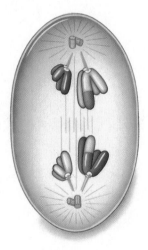

7. Outline the steps involved in the formation of sperm and eggs in animals. *162–163*

1. Sexual reproduction requires _____ .
   a. meiosis
   b. gamete formation
   c. fertilization
   d. all of the above

2. Meiosis is a division mechanism that produces _____ .
   a. two cells
   b. two nuclei
   c. four cells
   d. four nuclei

3. An animal cell with two of each type of chromosome characteristic of the species is _____ .
   a. diploid
   b. haploid
   c. probably not a normal gamete
   d. both b and c

4. Meiosis _____ the parental chromosome number.
   a. doubles
   b. reduces
   c. maintains
   d. corrupts

5. Generally, a pair of homologous chromosomes _____ .
   a. carry the same genes
   b. are the same length and shape
   c. interact at meiosis
   d. all of the above

6. Before the onset of meiosis, all chromosomes are _____ .
   a. condensed
   b. released from protein
   c. duplicated
   d. b and c

7. Each chromosome moves away from its homologue and ends up at the opposite spindle pole during _____ .
   a. prophase I
   b. prophase II
   c. anaphase I
   d. anaphase II

8. Sister chromatids of each chromosome move apart and end up at opposite spindle poles during _____ .
   a. prophase I
   b. prophase II
   c. anaphase I
   d. anaphase II

9. Match each term and its description.
   ___ chromosome number
   ___ alleles
   ___ metaphase I
   ___ interphase

   a. different molecular forms of the same gene
   b. none between meiosis I, II
   c. pairs of homologues aligned at spindle equator
   d. the number of each type of chromosome present in cell

allele *156*
asexual reproduction *156*
crossing over *160*
diploid *157*
egg *162*
fertilization *162*
gamete *156*
gene *156*

haploid *157*
homologous chromosome *157*
meiosis *156*
sexual reproduction *156*
sister chromatid *157*
sperm *162*
spore *162*

Klug, W., and M. Cummings. 1994. *Concepts of Genetics*. Fourth edition. New York: Macmillan.

Wolfe, S. 1993. *Molecular and Cellular Biology*. Belmont, California: Wadsworth.

11 OBSERVABLE PATTERNS OF INHERITANCE

A Smorgasbord of Ears and Other Traits

Basketball ace Charles Barkley has them. So does actor Tom Cruise. Actress Joan Chen doesn't, and neither did a monk named Gregor Mendel. To see how *you* fit in with these folks, use a mirror to check your ears. Is the fleshy lobe at the base of each ear attached to the side of your head? If so, you and Barkley and Cruise have something in common. Or is the fleshy lobe unattached, so that you can flap it back and forth? If so, you are like Chen and Mendel (Figure 11.1).

Whether a person is born with detached or attached earlobes depends on a single kind of gene. That gene comes in slightly different molecular forms—alleles. Only one form has information about detached lobes. The information is put to use while a human body is developing inside the mother. It calls for a death signal, which is sent to all the cells positioned between the newly forming lobes and the head. Without the signal, the cells don't die, and earlobes don't detach.

We all have genes for thousands of traits, including earlobes, cheeks, lashes, and eyeballs. Most of the traits vary in their details from one person to the next. Remember, humans inherit pairs of genes, on pairs of chromosomes. In some pairings, one allele has powerful effects and overwhelms the other's contribution to a trait. The outgunned allele is said to be recessive to the dominant one. If you have *detached* earlobes, *dimpled* cheeks, *long* lashes, or *large* eyeballs, you carry at least one and possibly two dominant alleles that influence the trait in a particular way.

When both alleles of a pair are recessive, nothing masks their effect on a trait. You get *attached* earlobes with one pair of recessive alleles (and *flat* feet with another, a *straight* nose with another, and so on).

How did we discover such remarkable things about our genes? It all started with Gregor Mendel. By analyzing pea plants generation after generation, Mendel found indirect but *observable* evidence of how parents transmit units of hereditary information—genes—to offspring. This chapter focuses on Mendel's experimental methods and results. They remain a classic example of how a scientific approach can pry open important secrets about the natural world. And to this day, they serve as the foundation for modern genetics.

a Tom Cruise

b Charles Barkley

c Joan Chen

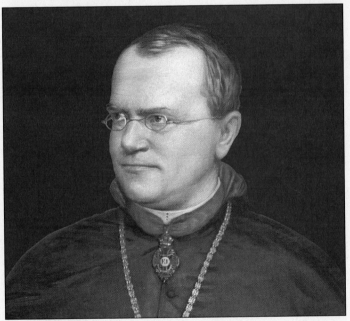

d Gregor Mendel

Figure 11.1 The attached and detached earlobes of a few representative humans. This sampling provides observable evidence of a trait that is governed by a single gene. Do you have one or the other version of this trait? It depends on whether you inherited a specific molecular form of that gene from your mother, your father, or both of your parents.

Earlobe attachment, chin dimpling, cheek dimpling, and many other single-gene traits vary from one individual to the next. As Gregor Mendel perceived, such easily observable traits can be used to identify patterns of inheritance that exist from one generation to the next.

KEY CONCEPTS

1. Genes are units of information about heritable traits. Each gene has a specific location in the chromosomes of a species. But its molecular form may differ slightly from one chromosome to the next. Different molecular forms of a gene (alleles) specify different versions of the same trait.

2. The two genes of a pair are segregated from each other during meiosis and end up in different gametes. Gregor Mendel found indirect evidence of this when he crossbred plants having observable differences in the same trait.

3. The sorting of each gene pair into different gametes tends to be independent of how the other gene pairs are sorted out. Mendel found evidence of this when he tracked plants having observable differences in *two* traits, such as flower color and height.

4. If the two genes of a pair specify different versions of a trait (if they are nonidentical alleles), one may have more pronounced effects on the trait. Besides this, two or more gene pairs often influence the same trait, and some single genes influence many traits. Finally, environmental conditions may alter gene expression.

More than a century ago, Charles Darwin explained how natural selection might bring about evolutionary change. According to his key premise, individuals vary in heritable traits. Variations that improve chances of surviving and reproducing are favored more often in each generation. Those that don't become less frequent. Thus the population changes over time—it evolves.

Darwin's theory did not fit with a prevailing idea about inheritance. It was common knowledge that sperm and eggs both transmit information about traits to offspring. But almost no one suspected the information is organized in *units* (genes). Instead, the idea was that a father's blob of information "blended" with a mother's blob at fertilization, like cream into coffee.

Carried to its logical conclusion, blending would eventually dilute hereditary information until there was only one version left of each trait. Yet why did *freckled* children keep turning up among nonfreckled generations? Why weren't all the descendants of a herd of white stallions and black mares *gray*? The blending theory scarcely explained what people could see with their own eyes. But few disputed it. Blending happened, so populations "had to be" uniform—and with uniformity in traits, evolution could not occur.

Even before Darwin presented his theory, however, someone was gathering evidence that eventually would support his key premise. A monk, Gregor Mendel, was about to prove that sperm and eggs do indeed carry "units" that deal with separate heritable traits.

MENDEL'S INSIGHT INTO PATTERNS OF INHERITANCE

Mendel's Experimental Approach

Mendel spent most of his life in a monastery in Brünn, an Austrian village that has since become part of the Czech Republic. The monastery of St. Thomas was somewhat removed from the European capitals, which were then the centers of scientific inquiry. Yet Mendel was not a man of narrow interests who simply stumbled by chance onto principles of great import. Having been raised on a farm, he was well aware of agricultural principles and their application. He kept abreast of breeding experiments and developments described in the available books. Mendel was a member of the regional agricultural society. He won several awards for developing improved varieties of vegetables and fruits. After entering the monastery, he spent two years studying mathematics at the University of Vienna. Few schol-

ars of his time had similarly combined talents in plant breeding and mathematics.

Shortly after his university training, Mendel began experiments with the garden pea plant, *Pisum sativum* (Figure 11.2). This plant is self-fertilizing. Its flowers produce sperm and eggs, which meet up in the same flower. Some of the plants are **true-breeding**. In other words, successive generations are just like the parents in one or more traits, as when all plants grown from seeds of white-flowered plants have white flowers.

As Mendel knew, pea plants also will cross-fertilize when sperm and eggs from different plants are brought together under controlled conditions. For his experiments, he could open flower buds of a plant that bred true for a trait (say, white flowers) and snip out the stamens. (Stamens bear pollen grains in which sperm develop.) Then he could brush the "castrated" buds with pollen from a plant that bred true for a *different* version of the same trait (purple flowers). Mendel hypothesized that such clearly observable differences

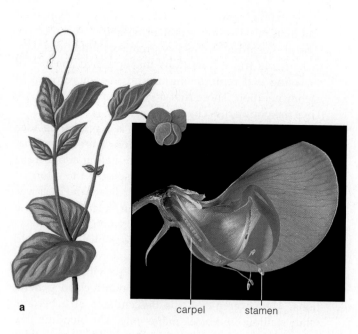

a

carpel   stamen

Figure 11.2 The garden pea plant (*Pisum sativum*), the focus of Mendel's experiments. A flower has been sectioned to show the location of its stamens and carpel. Sperm-producing pollen grains form in stamens. Eggs develop, fertilization takes place, and seeds mature inside the carpel.

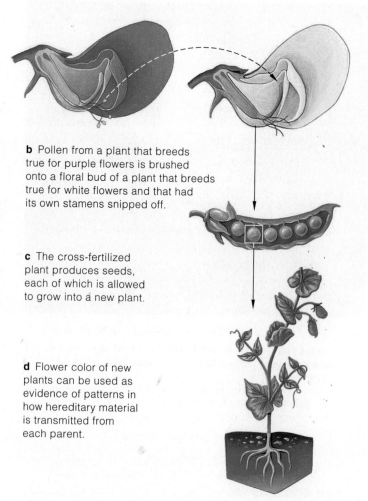

b Pollen from a plant that breeds true for purple flowers is brushed onto a floral bud of a plant that breeds true for white flowers and that had its own stamens snipped off.

c The cross-fertilized plant produces seeds, each of which is allowed to grow into a new plant.

d Flower color of new plants can be used as evidence of patterns in how hereditary material is transmitted from each parent.

could be used to track a trait through many generations. If there were patterns to the trait's inheritance, *those patterns might tell him something about the hereditary material itself.*

Some Terms Used in Genetics

Having read the chapter on meiosis, you already have insight into the mechanisms of sexual reproduction—which is more than Mendel had. He did not know about chromosomes. So he could not have known that the chromosome number is reduced by half in gametes, then restored at fertilization. Yet Mendel sensed what was going on. As we follow his thinking, let's simplify things by substituting a few modern terms used in studies of inheritance (see also Figure 11.3):

1. **Genes** are units of information about specific traits, and they are passed from parents to offspring. Each gene has a specific location (locus) on a chromosome.

2. Diploid cells have a pair of genes for each trait, on a pair of homologous chromosomes.

3. Although both genes of a pair deal with the same trait, they may vary in their information about it. This happens when they have slight molecular differences, as when one gene for flower color specifies purple and another specifies white. The different molecular forms of a gene are called **alleles** of that gene.

4. If it turns out that the two alleles of a pair are the same, this is a *homozygous* condition. If different, this is a *heterozygous* condition.

5. An allele is *dominant* when its effect on a trait masks that of any *recessive* allele paired with it. We use capital letters for dominant alleles and lowercase letters for recessive ones (for instance, *A* and *a*).

6. Putting this together, a **homozygous dominant** individual has a pair of dominant alleles (*AA*) for the trait that is being studied. A **homozygous recessive** individual has a pair of recessive alleles (*aa*) for the trait. A **heterozygous** individual has a pair of nonidentical alleles (*Aa*) for the trait.

7. Two terms help keep the distinction clear between genes and the traits they specify. **Genotype** refers to the genes present in an individual. **Phenotype** refers to an individual's observable traits.

8. When tracking the inheritance of traits through generations of offspring, these abbreviations apply:

P parental generation
F_1 first-generation offspring
F_2 second-generation offspring

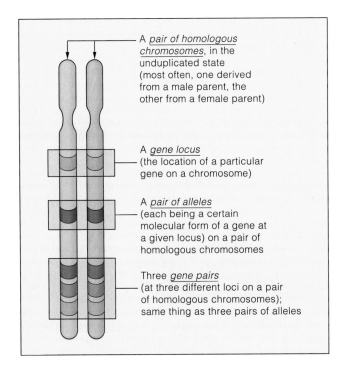

Figure 11.3 A few genetic terms illustrated. Diploid organisms have pairs of genes, on pairs of homologous chromosomes. (In your case, one chromosome of each pair was inherited from your mother, and the other homologous chromosome was inherited from your father.)

The genes themselves may have different molecular forms, called alleles. Different alleles specify slightly different versions of the same trait. An allele at one location on a chromosome may or may not be identical to its partner on the homologous chromosome.

When the offspring of genetic crosses inherit identical alleles for a given trait, generation after generation, they are said to be a true-breeding lineage. When offspring of a genetic cross inherit nonidentical alleles for a trait under study, they are said to be hybrid offspring.

11.2 MENDEL'S THEORY OF SEGREGATION

Predicting the Outcome of Monohybrid Crosses

Mendel had an idea that in every generation, a plant inherits two "units" (genes) of information for a trait, one from each parent. To test his idea, he performed what we now call **monohybrid crosses**. Offspring of such crosses are heterozygous for the one trait being studied (which is what monohybrid means). Their parents breed true for different versions of the trait, so the offspring inherit a pair of nonidentical alleles.

Mendel tracked many single traits through two generations. For instance, in one series of experiments, he crossed true-breeding purple-flowered plants and true-breeding white-flowered ones. All plants grown from the seeds that resulted from this cross had purple flowers. Mendel let these plants self-fertilize. Some plants grown from those seeds had white flowers!

If Mendel's hypothesis were correct—if each plant had inherited two units of information about flower color—then the "purple" unit would have to be dominant, for it masked the "white" unit in F_1 plants.

Let's rephrase Mendel's thinking. Germ cells of garden pea plants are diploid, with pairs of homologous chromosomes. Assume one parent plant is homozygous dominant (*AA*) for flower color and the other is homozygous recessive (*aa*). After meiosis, their sperm or eggs will carry only one allele for flower color (Figure 11.4). Thus, when a sperm fertilizes an egg, only one outcome is possible: $A + a = Aa$.

Before continuing, you should know that Mendel crossed hundreds of plants and tracked thousands of offspring. He also counted and recorded the plants showing dominance or recessiveness. As Figure 11.5 indicates, an intriguing ratio emerged. On the average, three of every four F_2 plants had the dominant phenotype and one had the recessive phenotype.

To Mendel, this ratio suggested that fertilization is a chance event, with a number of possible outcomes. And

homozygous dominant parent
homozygous recessive parent
(chromosomes duplicated before meiosis)
meiosis I
meiosis II
(gametes)
(gametes)
fertilization produces heterozygous offspring

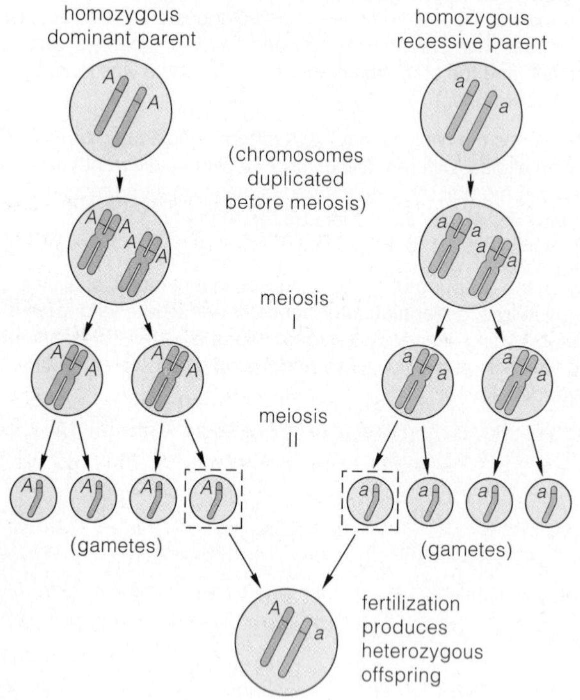

Figure 11.4 Example of a monohybrid cross, showing how one gene of a pair segregates from the other. Two parents that breed true for two different versions of a trait produce heterozygous offspring only.

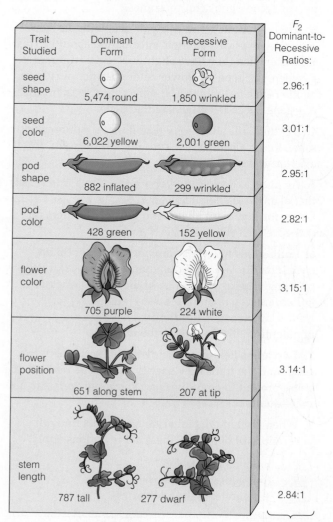

Trait Studied
Dominant Form
Recessive Form
F2 Dominant-to-Recessive Ratios:
seed shape
5,474 round
1,850 wrinkled
2.96:1
seed color
6,022 yellow
2,001 green
3.01:1
pod shape
882 inflated
299 wrinkled
2.95:1
pod color
428 green
152 yellow
2.82:1
flower color
705 purple
224 white
3.15:1
flower position
651 along stem
207 at tip
3.14:1
stem length
787 tall
277 dwarf
2.84:1
Average ratio for all traits studied:
3:1

Figure 11.5 (*Right*) Results from Mendel's monohybrid cross experiments with the garden pea. The numbers are his counts of F_2 plants that he assumed were carrying dominant or recessive hereditary "units" (alleles) for the trait. On the average, the dominant-to-recessive ratio was 3:1.

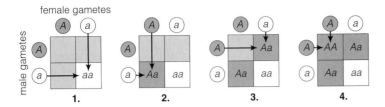

female gametes

male gametes

1. 2. 3. 4.

Figure 11.6 Punnett-square method of predicting the probable outcome of a genetic cross. Circles represent gametes. Letters on gametes represent dominant or recessive alleles. The different squares depict the different genotypes possible among offspring. In this example, gametes are from a self-fertilizing plant that is heterozygous (*Aa*) for a trait.

Figure 11.7 Results from one of Mendel's monohybrid crosses. On the average, the dominant-to-recessive ratio among F_2 plants was 3:1.

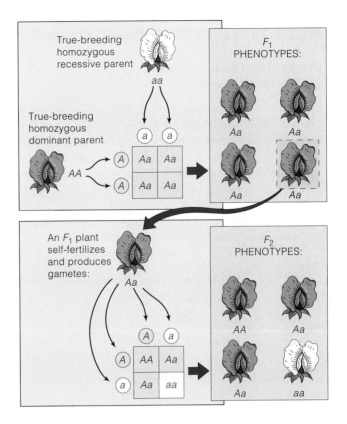

he had an understanding of **probability**, which applies to chance events *and so could help him predict the possible outcomes of crosses.* (Probability simply means that the chance of each outcome occurring is proportional to the number of ways it can be reached.) The **Punnett-square method**, explained in Figure 11.6 and applied in Figure 11.7, can help you visualize the possibilities. As you can see, if half of a plant's sperm (or eggs) were *a* and half were *A*, four outcomes were possible every time a sperm fertilized an egg:

Possible event:	Probable outcome:
sperm *A* meets egg *A*	1/4 *AA* offspring
sperm *A* meets egg *a*	1/4 *Aa*
sperm *a* meets egg *A*	1/4 *Aa*
sperm *a* meets egg *a*	1/4 *aa*

By this prediction, an F_2 plant had three chances in four of getting at least one dominant allele (purple flowers). It had one chance in four of getting two recessive alleles (white flowers). That is a probable phenotypic ratio of three purple to one white, or 3:1.

Mendel's observed ratios weren't *exactly* 3:1. You can see this for yourself by studying the numerical results in Figure 11.5. Why did Mendel put aside the deviations? To understand why, flip a coin a couple of times. As we all know, a coin is just as likely to end up heads as tails. But often it ends up heads, or tails, several times in a row. So if you flip the coin only a few times, the observed ratio may differ greatly from the predicted ratio of 1:1. Only by flipping the coin many, many times will you come close to the predicted ratio. Mendel knew the rules of probability—and he performed a large number of crosses. Almost certainly, this kept him from being confused by minor deviations from the predicted results of the experimental crosses.

Testcrosses

Mendel gained support for his prediction with the **testcross**. In this type of experimental cross, an organism shows dominance for a trait but its genotype is unknown, so it is crossed to a known homozygous recessive individual. Results may reveal whether the organism is homozygous dominant or heterozygous.

With respect to the monohybrid crosses described above, Mendel tested his prediction that the purple-flowered F_1 offspring were heterozygous. He crossed F_1 plants with true-breeding, white-flowered plants. If the F_1 plants were homozygous dominant, all of their offspring would show dominance. If heterozygous, there would be about as many recessive as dominant offspring. Sure enough, about half the offspring had purple flowers (*Aa*) and half had white (*aa*). Can you construct two Punnett squares that show the two possible outcomes of this testcross?

On the basis of results from Mendel's monohybrid crosses and testcrosses, it is possible to formulate a theory, which we state here in modern terms:

MENDEL'S THEORY OF SEGREGATION. **Diploid cells have pairs of genes (on pairs of homologous chromosomes). During meiosis, the two genes of each pair segregate from each other. As a result, they end up in different gametes.**

11.3 INDEPENDENT ASSORTMENT

Predicting the Outcome of Dihybrid Crosses

By another series of experiments, Mendel attempted to explain how *two* pairs of genes are assorted into gametes. He selected true-breeding pea plants that differed in two traits, such as flower color and height. In such **dihybrid crosses**, the F_1 offspring inherit two gene pairs, each consisting of two nonidentical alleles.

Let's diagram one of his dihybrid crosses, using dominant alleles *A* for flower color and *B* for height, and *a* and *b* as their recessive counterparts:

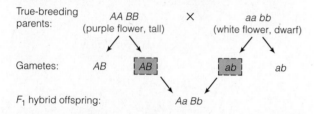

As Mendel would have predicted, all the F_1 offspring are purple-flowered and tall (*Aa Bb*). How will the two

gene pairs be assorted into gametes when those plants mature and reproduce?

The answer partly depends on the chromosomal locations of the gene pairs. Assume the *Aa* alleles and the *Bb* alleles are located on two different pairs of homologous chromosomes. Next, think about how all chromosomes become positioned at the spindle equator during metaphase I of meiosis (page 161 and Figure 11.8). The chromosome with the *A* allele may be positioned to move to either spindle pole (and on into one of four gametes). The same is true for its homologue. And the same is true for the chromosomes with the *B* and *b* alleles. Therefore, after meiosis, four combinations of alleles are possible in sperm or eggs: 1/4 *AB*, 1/4 *Ab*, 1/4 *aB*, and 1/4 *ab*.

This permits several different combinations at fertilization. Simple multiplication (four kinds of sperm times four kinds of eggs) tells us sixteen combinations of gametes are possible in the F_2 offspring of a dihybrid cross. Use the Punnett-square method to diagram the probabilities (Figure 11.9). Add up the possible phenotypes and you get 9/16 tall purple-flowered, 3/16 dwarf purple-flowered, 3/16 tall white-flowered, and 1/16 dwarf white-flowered plants. That is a probable phenotypic ratio of 9:3:3:1. Results from all of Mendel's dihybrid F_2 crosses were close to a 9:3:3:1 ratio.

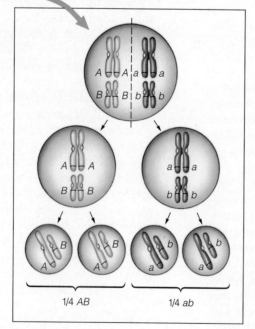
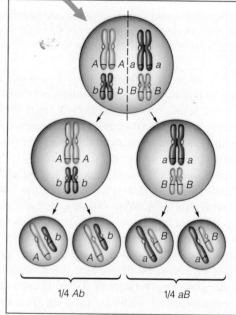
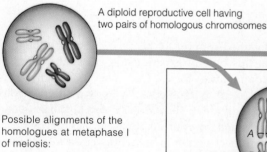

Figure 11.8 Example of independent assortment using just two pairs of homologous chromosomes. An allele at one chromosome locus may or may not be identical to its partner allele on the homologous chromosome. During meiosis, either chromosome of a pair may get attached to either spindle pole. So two different lineups are possible at metaphase I.

Figure 11.9 Results from Mendel's dihybrid cross between parent plants that bred true for different versions of two traits (flower color and height). *A* and *a* represent dominant and recessive alleles for flower color. *B* and *b* represent dominant and recessive alleles for height. The probabilities of certain combinations of phenotypes among F_2 offspring occur in a 9:3:3:1 ratio, on the average.

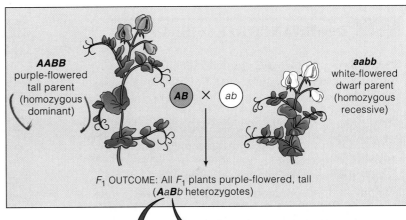

AABB
purple-flowered
tall parent
(homozygous
dominant)

×

aabb
white-flowered
dwarf parent
(homozygous
recessive)

F_1 OUTCOME: All F_1 plants purple-flowered, tall
(**Aa**B*b* heterozygotes)

The Theory in Modern Form

Mendel didn't know that a pea plant's "units" of inheritance are divided up among a number of different chromosomes, so he could do no more than analyze the numbers from his dihybrid crosses. It seemed to him that the units for different traits were assorting independently into gametes, and so he formulated a theory of independent assortment. As you will read in the next chapter, we now know there are exceptions to this. So we state Mendel's theory in updated form:

MENDEL'S THEORY OF INDEPENDENT ASSORTMENT. **At meiosis, the gene pairs on homologous chromosomes tend to be sorted out for distribution into one gamete or another independently of how gene pairs on other chromosomes are sorted out.**

Independent assortment and hybrid crossing can lead to staggering variety among offspring. In a monohybrid cross (involving a single gene pair), three genotypes are possible (*AA*, *Aa*, and *aa*). We can represent this as 3^n, where *n* is the number of gene pairs. With more gene pairs, the number of possible combinations increases dramatically. Even if parents differ in only ten gene pairs, almost 60,000 different genotypes are possible among their offspring. If they differ in twenty gene pairs, the number approaches 3.5 billion!

On the basis of his experimental results, Mendel was convinced that the hereditary material comes in units that retain their physical identity from one generation to the next, despite being segregated and assorted before gametes form. In 1865 he reported this idea before the Brünn Society for the Study of Natural Science. His report made no impact whatsoever. The following year his paper was published. Apparently it was read by few, and its near-universal applicability was understood by no one.

In 1871 Mendel became an abbot of the monastery, and his experiments gave way to administrative tasks. He died in 1884, never to know that his work was the starting point for the development of modern genetics.

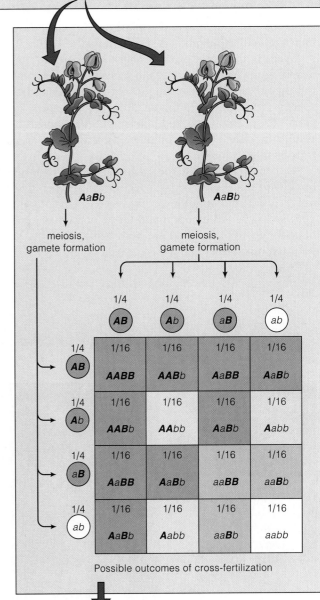

Possible outcomes of cross-fertilization

ADDING UP THE F_1 COMBINATIONS POSSIBLE:

■ 9/16 or 9 purple-flowered, tall

□ 3/16 or 3 purple-flowered, dwarf

■ 3/16 or 3 white-flowered, tall

□ 1/16 or 1 white-flowered, dwarf

11.4 DOMINANCE RELATIONS

For the most part, Mendel studied traits having clearly dominant or recessive forms. As the following examples will make clear, the expression of other traits is not as straightforward.

Sometimes one allele is incompletely dominant over another. This condition, called **incomplete dominance**, results in a heterozygous phenotype that is *somewhere between* the two homozygous phenotypes. Cross true-breeding red snapdragons with true-breeding white ones, and all F_1 offspring will have pink flowers. Cross two F_1 plants, and you get red, pink, or white snapdragons in a predictable ratio, as shown in Figure 11.10.

What is the basis of this inheritance pattern? Red snapdragons have two alleles of a gene that specifies a red pigment. White snapdragons have two alleles that specify "no pigment." Pink snapdragons are heterozygous. They have one red allele, but its expression results in only enough pigment molecules to make flowers pink, not red.

Or consider **codominance**, a condition in which two nonidentical alleles of a pair specify two different phenotypes, yet one cannot mask expression of the other. If you have type AB blood, for instance, you have a pair of codominant alleles that are both being expressed in your red blood cells (see the *Focus* essay).

11.5 MULTIPLE EFFECTS OF SINGLE GENES

Expression of a single gene can influence two or more traits, an effect called **pleiotropy**. Consider the genetic disorder *sickle-cell anemia*. It arises from a mutated gene for hemoglobin (Hb^S instead of Hb^A). Hemoglobin is an oxygen-transporting protein in red blood cells. Heterozygotes (Hb^A/Hb^S) typically show few symptoms; they produce enough normal hemoglobin molecules to cover for the abnormal ones. Homozygotes (Hb^S/Hb^S) can only produce abnormal hemoglobin, and drastic changes in phenotype may follow. The changes start with disruptions to the oxygen concentration in blood.

Humans depend on oxygen for aerobic respiration. Oxygen enters the lungs, then is picked up by the bloodstream and distributed to tissues throughout the body. In those tissues, oxygen diffuses out of thin-walled blood vessels called capillaries, then diffuses into cells. The concentration gradient for oxygen is steep across capillary walls. Most of the oxygen dissolved in blood moves out, into the surrounding tissues. In response to the decreased oxygen concentration, abnormal hemoglobin molecules in the red blood cells stick together in long, rodlike arrangements. The rods distort the cells into a shape rather like a sickle, a short-handled farm tool with a crescent-shaped blade. The cells rupture easily, and they clog and rupture capillaries. Body tissues become oxygen-starved, and metabolic wastes build up. Figure 11.11 tracks the resulting phenotypic changes. Other aspects of this disorder are described in later chapters.

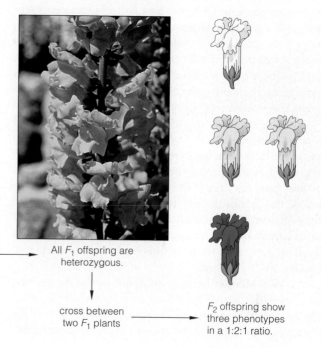

homozygous dominant parent ✕ homozygous recessive parent

All F_1 offspring are heterozygous.

cross between two F_1 plants

F_2 offspring show three phenotypes in a 1:2:1 ratio.

Figure 11.10 Example of incomplete dominance. Red-flowering and white-flowering homozygous snapdragons produce pink-flowering plants in the first generation. In heterozygotes, the red allele is only partly dominant over the white allele.

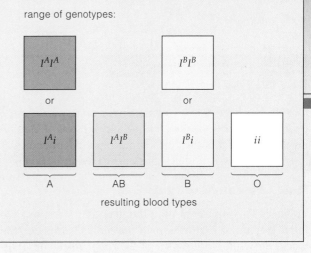

a Possible combinations of alleles that are associated with ABO blood typing.

Focus on Health

ABO Blood Typing

Various molecules at the surface of your cells are "self" markers—they identify the cells as being part of your own body. One kind of marker on red blood cells has different molecular forms. **ABO blood typing**, a method of analysis, can reveal which form a person has.

In the human population, there are three alleles of the gene that specifies this marker. They influence its form in different ways. Two alleles, I^A and I^B, are codominant when paired with each other. A third allele, i, is recessive. When paired with either I^A or I^B, its effect is masked. Whenever three or more alleles of a gene exist in a population, we call this a **multiple allele system**.

Think about a cell that is making new molecules of the marker. Before each molecule becomes positioned at the cell surface, it takes on final form in the cytomembrane system (page 64). A carbohydrate chain is attached to a lipid. Then an enzyme attaches a sugar unit to the chain. Alleles I^A and I^B code for two versions of that enzyme. The two attach different sugar units, and this gives the marker a special identity—either A or B.

Which two alleles for this enzyme do you have? With either I^AI^A or I^Ai, your blood is type A. With I^BI^B or I^Bi, it is type B. With codominant alleles I^AI^B, *both* versions of the enzyme are produced. And both attach sugar units to the marker molecules. In this case, your blood is type AB. If you are homozygous recessive (*ii*), the markers never did get a sugar unit attached to them. Then your blood type is neither A nor B (that's what type "O" means).

When blood of two people mixes during transfusions, self markers must be compatible. Without the proper markers, red blood cells from a donor will be recognized as foreign. The immune system, which defends the body against "nonself" will act against the foreign cells and may cause death (Chapter 40).

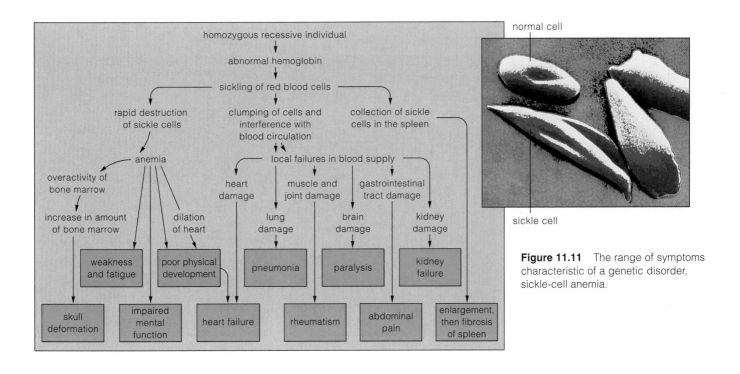

Figure 11.11 The range of symptoms characteristic of a genetic disorder, sickle-cell anemia.

11.6 INTERACTIONS BETWEEN GENE PAIRS

Often a trait results from interactions among two or more gene pairs. For example, two alleles of one gene can mask alleles of another, and some expected phenotypes may not occur. Such interactions are called **epistasis**.

Hair Color in Mammals

Epistasis is common among the gene pairs responsible for the color of fur or skin in mammals. Consider the black, brown, or yellow fur of Labrador retrievers (Figure 11.12). Variations in the amount and distribution of melanin, a brownish black pigment, produce the different colors. Many gene pairs affect different steps in melanin production and its deposition in certain body regions. Alleles of one gene specify an enzyme required to produce melanin. Allele B (black) has a more pronounced effect and is dominant to b (brown). Alleles of a different gene control whether melanin will be deposited in a retriever's hairs. Allele E permits deposition, but a pair of recessive alleles (ee) will block it, and the fur will be yellow.

The interactions between the two gene pairs just described may not even be possible if a certain allelic combination exists at still another gene locus. Here, a gene (C) codes for tyrosinase, the first of several enzymes in the metabolic pathway that produces melanin. An individual with one or two dominant alleles (CC or Cc) can produce the functional enzyme. An individual with two recessive alleles (cc) cannot. When the biosynthetic pathway leading to melanin production is blocked, the resulting condition is called *albinism*. Figure 11.13 shows an example.

Figure 11.12 Coat color of Labrador retrievers, as dictated by interactions among alleles of two gene pairs. Allele B (black) of one kind of gene involved in melanin production is dominant to allele b (brown). Allele E of a different gene allows melanin pigment to be deposited in individual hairs. Two recessive alleles (ee) of this gene block deposition, and a yellow coat results.

F_1 offspring of a dihybrid cross produce F_2 offspring in a 9:3:4 ratio, as the Punnett-square diagram shows. The yellow Labrador in photograph (**b**) probably has genotype BB ee, because it can produce melanin but can't deposit pigment in hairs. (Looking at that photograph, can you say why?)

P homozygotes: *BB EE* × *bb ee*

↓

All F_1:   *Bb Ee*

↓

F_2 combinations possible:

	BE	Be	bE	be
BE	BB EE	BB Ee	Bb EE	Bb Ee
Be	BB Ee	BB ee	Bb Ee	Bb ee
bE	Bb EE	Bb Ee	bb EE	bb Ee
be	Bb Ee	Bb ee	bb Ee	bb ee

RESULTING PHENOTYPES:

☐ 9/16 or 9 black

☐ 3/16 or 3 brown

☐ 4/16 or 4 yellow

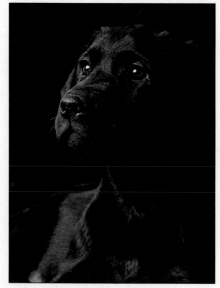

a Black Labrador

b Yellow Labrador

c Chocolate Labrador

Figure 11.13 A rare albino rattlesnake. Like other animals that cannot produce melanin, it has pink eyes and white surface coloration. (Eyes look pink because the absence of melanin from a tissue layer in the inner eyeball allows red light to be reflected from blood vessels in the eyes.) In birds and mammals, surface coloration is due to pigments in feathers, fur, or skin. In fishes, amphibians, and reptiles, color-bearing cells give skin its surface coloration. Some of the cells contain melanin pigments or yellow-to-red pigments. Others contain crystals that reflect light and alter the effect of other pigments present.

The mutation affecting melanin production in the snake shown here had no effect on the production of yellow-to-red pigments and light-reflecting crystals. So the snake's skin appears iridescent yellow as well as white.

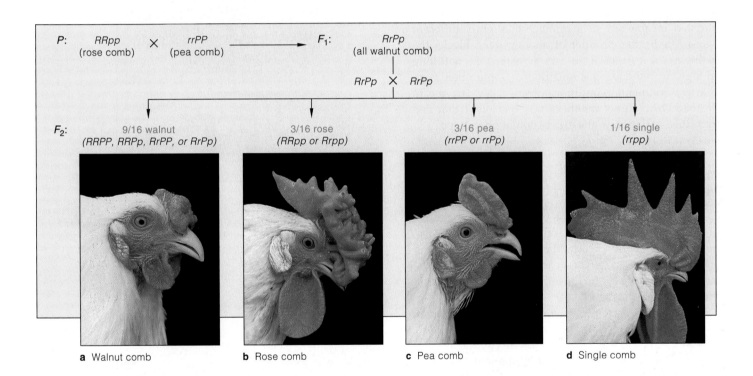

P: *RRpp*   ×   *rrPP*   →   F₁: *RrPp*
      (rose comb)   (pea comb)   (all walnut comb)

RrPp   ×   *RrPp*

F₂: 9/16 walnut   3/16 rose   3/16 pea   1/16 single
(RRPP, RRPp, RrPP, or RrPp)   (RRpp or Rrpp)   (rrPP or rrPp)   (rrpp)

a Walnut comb   **b** Rose comb   **c** Pea comb   **d** Single comb

Comb Shape in Poultry

In some cases of epistasis, interaction between two gene pairs results in a phenotype that neither pair can produce alone. The geneticists W. Bateson and R. Punnett identified two interacting gene pairs (*R* and *P*) that produce comb shape in chickens. The allelic combinations of *rr* at one chromosome locus and *pp* at the other locus result in the most common phenotype, the single comb. The presence of dominant alleles (*R*, *P*, or both) results in different phenotypes. Figure 11.14 shows the allelic combinations that lead to rose, pea, or walnut combs.

Figure 11.14 Interaction between two genes affecting the same trait in domestic breeds of chickens. The initial cross is between a Wyandotte (with a rose comb, **b**, on the crest of its head) and a brahma (with a pea comb, **c**). With complete dominance at the locus for pea comb and at the locus for rose comb, the products of these two gene pairs interact and give a walnut comb (**a**). With complete recessiveness at both loci, the products interact and give rise to a single comb (**d**).

11.7 LESS PREDICTABLE VARIATIONS IN TRAITS

As Mendel demonstrated, the phenotypic effects of one or two pairs of certain genes show up in predictable ratios from one generation to the next. Certain interactions among gene pairs also may produce phenotypes in predictable ratios, as the example of Labrador coat color demonstrated.

However, even when you are tracking a single gene through the generations, the resulting phenotypes may be not quite as expected. Consider *camptodactyly*. Some people who carry the gene that is responsible for this rare genetic disorder have immobile, bent fingers on both hands. Others have immobile, bent fingers on the left hand only—or on the right hand only. Still others who carry the mutated gene have normal fingers.

Why the confounding variation? Bear in mind, the path from most genes to their end products (proteins) is a series of small metabolic steps. Maybe one gene is mutated in any one of a number of slightly different ways. Maybe the product of another gene blocks the pathway, or causes it to run nonstop, or not long enough. Or maybe poor nutrition or some other variable environmental condition affects a key enzyme in the pathway. These are the kinds of factors that can introduce variation in the expression of phenotype.

Continuous Variation in Populations

Generally, the individuals of a population show a range of small differences in most traits. We call this **continuous variation**. The greater the number of genes and environmental factors that can influence a trait, the more continuous will be the expected distribution of all the versions of that trait.

Think about the color of your eyes. As is true of all humans, the color is the cumulative result of many genes that are involved in the stepwise production and distribution of a light-absorbing pigment, melanin. Black eyes have abundant melanin deposits in the iris and absorb most of the incoming light. Dark brown eyes have less melanin, so some light is not absorbed but rather is reflected from the iris. Pale brown or hazel eyes have even less (Figure 11.15). Green, gray, and blue eyes do not have green, gray, or blue pigments. In these cases, the iris contains different quantities of melanin, but not very much of it, and most of the blue wavelengths of light are reflected.

How can you describe the continuous variation of a trait within a group—say, the students in Figure 11.16? They range from very short to very tall, with average heights more common than either extreme. Start out by

Figure 11.15 Samples from the range of continuous variation in human eye color. Different gene pairs interact to produce and deposit melanin. Among other things, this pigment helps color the eye's iris. Different combinations of alleles result in small differences in eye color. So the frequency distribution for the eye-color trait appears to be continuous over the range from black to light blue.

dividing the full range of different phenotypes into measurable categories. Next, count the individual students in each category. This gives you the relative frequencies of phenotypes that are distributed across the range of measurable values.

Figure 11.16*b* is a bar chart that plots the proportion of students in each category against the range of measured phenotypes. Vertical bars that are shortest represent the categories with the least number of students. The bar that is tallest represents the category with the greatest number of students. Draw a graph line around all of the bars and you get a "bell-shaped" curve. Such curves are typical of populations that show continuous variation in a trait.

number of individuals																		
1	4	8	10	16	16	15	15	14	13	13	11	9	8	8	5	1	2

60 (5 feet)	61	62	63	64	65	66	67	68	69	70	71	72	73	74	75	76	77

height (inches)

a Students at Brigham Young University in a splendid example of continuous variation in height.

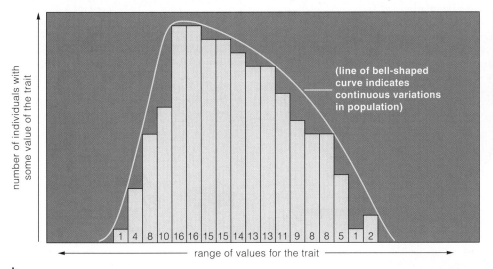

number of individuals with some value of the trait

(line of bell-shaped curve indicates continuous variations in population)

1	4	8	10	16	16	15	15	14	13	13	11	9	8	8	5	1	2

range of values for the trait

b

c Idealized bell-shaped curve for populations showing continuous variation in some trait.

Figure 11.16 Continuous variation in a trait that is characteristic of the human population. (**a**) Suppose you want to know the frequency distribution for height in a group of 169 biology students at Brigham Young University. You decide on how finely the range of possible heights should be divided. Then you measure and assign each student to the proper category. Finally, you divide the number in each category by the total number of all students in all categories.

(**b**) Often a bar graph is used to depict continuous variation. Here, the proportion of students in each category is plotted against the range of measured phenotypes. Notice the curved line above the bars. It is a specific example of the kind of "bell-shaped" curve that emerges for populations showing continuous variation in some trait. The bell-shaped curve in (**c**) is a generalized model for this phenomenon.

11.8 EXAMPLES OF ENVIRONMENTAL EFFECTS ON PHENOTYPE

We have mentioned in passing that environmental conditions can contribute to variable gene expression among individuals of a population. Before leaving this chapter, consider just two examples of environmentally induced variations in phenotype.

The environment influences plant growth. You may have observed water buttercups growing half in and half out of a pond. The submerged leaves have a quite different appearance, compared to the leaves above the water's surface (Figure 11.17). In this type of plant, the genes responsible for leaf shape produce different phenotypes under the two environmental conditions.

Or think of Himalayan rabbits and Siamese cats. Both carry the Himalayan allele (*ch*), which specifies a heat-sensitive version of one of the enzymes used in melanin production. At the surface of warm body regions, the enzyme is less active. There, fur grows in lighter than it does at cooler regions, including the ears and other extremities. The experiment described in Figure 11.18 provides observable evidence of the variation.

And so we conclude this chapter, which has dealt with the heritable and environmental aspects of phenotypic variations. What is the take-home lesson? Simply this:

Owing to gene mutations, cumulative gene interactions, and environmental effects on genes, individuals of a population show degrees of variation in their traits.

Figure 11.17 Effect of different environmental conditions on gene expression in plants. Leaves growing from submerged stems of the water buttercup (*Ranunculus aquatilis*) are more finely divided than leaves growing in air. The variation occurs even in the same leaf if it develops half in and half out of water.

Figure 11.18 Effect of different environmental conditions on gene expression in animals. A Himalayan rabbit normally has black hair only on its long ears, nose, tail, and lower leg limbs. In one experiment, a patch of a rabbit's white fur was plucked clean, then an icepack was secured over the hairless patch. While the cold conditions were maintained, the hairs that grew back were black.

Himalayan rabbits are homozygous for the *ch* allele of a gene that codes for tyrosinase (an enzyme needed to produce melanin). The allele specifies a heat-sensitive version of the enzyme that functions only when temperatures are below about 33°C. When hairs grow under warmer conditions, they appear light (no melanin is produced). Light fur normally covers warm body regions. Ears and other slender extremities are cooler; they tend to lose metabolic heat more rapidly.

SUMMARY

1. A gene is a unit of information about a heritable trait. The alleles of a gene are slightly different versions of that information. Through experimental crosses with pea plants, Mendel gathered evidence that diploid organisms have two genes for each trait and that genes retain their identity when transmitted to offspring.

2. Mendel performed monohybrid crosses (between two true-breeding plants showing different versions of a single trait). The crosses provided evidence that a gene can have different molecular forms (alleles), some of which are dominant over other, recessive forms.

3. Homozygous dominant individuals have two dominant alleles (AA) for the trait being studied. Homozygous recessives have two recessive alleles (aa). Heterozygotes have two nonidentical alleles (Aa).

4. In Mendel's monohybrid crosses ($AA \times aa$), all F_1 offspring were Aa. Crosses between F_1 plants resulted in these combinations of alleles in F_2 offspring:

	A	a	
A	AA	Aa	AA (dominant)
			Aa (dominant)
a	Aa	aa	Aa (dominant)
			aa (recessive)

This produced the expected phenotypic ratio of 3:1.

5. Results from Mendel's monohybrid crosses led to the formulation of a theory of segregation. In modern terms, diploid organisms have pairs of genes, on pairs of homologous chromosomes. The two genes of each pair segregate from each other during meiosis, such that each gamete formed ends up with one or the other.

6. Mendel also performed dihybrid crosses (between two true-breeding plants showing different versions of two traits). Results from several experiments were close to a 9:3:3:1 phenotypic ratio:

9 dominant for both traits
3 dominant for A, recessive for b
3 dominant for B, recessive for a
1 recessive for both traits

7. Mendel's dihybrid crosses led to the formulation of a theory of independent assortment. In modern terms, the gene pairs of two homologous chromosomes tend to be sorted into one gamete or another independently of how the gene pairs of other chromosomes are sorted out.

8. Four factors can influence gene expression. First, degrees of dominance exist between some gene pairs. Second, gene pairs can interact to influence the same trait. Third, a single gene can influence many seemingly unrelated traits. Fourth, environmental conditions can affect gene expression.

Review Questions

1. State the theory of segregation. Does segregation occur during mitosis or meiosis? *173*

2. Define the difference between these terms. *171*
   a. gene and allele
   b. dominant allele and recessive allele
   c. homozygote and heterozygote
   d. genotype and phenotype

3. Define true-breeding. What is a hybrid? *170, 172*

4. Distinguish between monohybrid and dihybrid crosses. What is a testcross, and why is it useful in genetic analysis? *172–174*

5. State the theory of independent assortment. Does independent assortment occur during mitosis or meiosis? *175*

Self-Quiz *(Answers in Appendix IV)*

1. Alleles are _____ .
   a. different molecular forms of a gene
   b. different molecular forms of a chromosome
   c. self-fertilizing, true-breeding homozygotes
   d. self-fertilizing, true-breeding heterozygotes

2. A heterozygote has _____ for the trait being studied.
   a. a pair of identical alleles
   b. a pair of nonidentical alleles
   c. a haploid condition, in genetic terms
   d. a and c

3. The observable traits of an organism are its _____ .
   a. phenotype   c. genotype
   b. sociobiology   d. pedigree

4. Offspring of a monohybrid cross $AA \times aa$ are _____ .
   a. all AA d. 1/2 AA and 1/2 aa
   b. all aa e. none of the above
   c. all Aa

5. Second-generation offspring from a cross are the _____ .
   a. F_1 generation   c. hybrid generation
   b. F_2 generation   d. none of the above

6. Assuming complete dominance, offspring of the cross $Aa \times Aa$ will show a phenotypic ratio of _____ .
   a. 3:1   c. 9:1
   b. 1:2:1   d. 9:3:3:1

7. Crosses between F_1 individuals resulting from the cross $AABB \times aabb$ lead to F_2 phenotypic ratios close to _____ .
   a. 1:2:1   c. 3:1
   b. 1:1:1:1   d. 9:3:3:1

8. Match each genetic term appropriately.
   ___ dihybrid cross   a. bb
   ___ monohybrid cross   b. $AaBb \times AaBb$
   ___ homozygous condition   c. Aa
   ___ heterozygous condition d. $Aa \times Aa$

Genetics Problems *(Answers in Appendix III)*

1. One gene has alleles *A* and *a*. Another has alleles *B* and *b*. For each genotype listed, what type(s) of gametes will be produced? (Assume independent assortment occurs.)

   a. *AA BB*   c. *Aa bb*
   b. *Aa BB*   d. *Aa Bb*

2. Still referring to Problem 1, what will be the genotypes of offspring from the following matings? With what frequency will each genotype show up?

   a. *AA BB* × *aa BB*   c. *Aa Bb* × *aa bb*
   b. *Aa BB* × *AA Bb*   d. *Aa Bb* × *Aa Bb*

3. In one experiment, Mendel crossed a pea plant that bred true for green pods with one that bred true for yellow pods. All the F_1 plants had green pods. Which form of the trait (green or yellow pods) is recessive? Explain how you arrived at your conclusion.

4. At one gene location on a human chromosome, a dominant allele controls whether you can curl the sides of your tongue upward *(see photo)*. People homozygous for the recessive allele cannot roll their tongue. At a different gene location, a dominant allele controls whether earlobes are attached or detached (see Figure 11.1). These two gene pairs assort independently.

Suppose a tongue-rolling woman with detached earlobes marries a man who has attached earlobes and can't roll his tongue. Their first child has attached earlobes and can't roll its tongue.

   a. What are the genotypes of the mother, father, and child?
   b. What is the probability that a second child will have detached earlobes and won't be a tongue roller?

5. Go back to Problem 1, and assume you now study a third gene having alleles *C* and *c*. For each genotype listed, what type(s) of gametes will be produced?

   a. *AA BB CC*   c. *Aa BB Cc*
   b. *Aa BB cc*   d. *Aa Bb Cc*

6. Mendel crossed a true-breeding tall, purple-flowered pea plant with a true-breeding dwarf, white-flowered plant. All F_1 plants were tall and purple-flowered. If an F_1 plant self-fertilizes, what is the probability that a randomly selected F_2 plant will be heterozygous for the genes specifying height and flower color?

7. Assume that a new gene has been identified in mice. One of its alleles specifies yellow fur color. A second allele specifies brown fur color. Suppose you are asked to determine whether the relationship between the two alleles is one of simple dominance, incomplete dominance, or codominance. What types of crosses would give you the answer? On what types of observations would you base your conclusions?

8. The ABO blood-typing system has been used to settle cases of disputed paternity. Suppose, as a geneticist, you must testify during a case in which the mother has type A blood, the child has type O blood, and the alleged father has type B blood. How would you respond to the following statements?

   a. *Attorney of the alleged father:* "The mother has type A blood, so the child's type O blood must have come from the father. Because my client has type B blood, he could not have fathered this child."

   b. *Mother's attorney:* "Further tests prove this man is heterozygous, so he must be the father."

9. As in Labrador retrievers (page 178), fur color in mice is governed by genes concerned with producing and distributing melanin. At one gene location, a dominant allele (B), specifies dark brown and a recessive allele (b) specifies light brown, or tan. At another gene location, a dominant allele (C) permits melanin production and a recessive allele (c) shuts it down and results in albinism.

    a. A homozygous bb cc albino mouse mates with a homozygous BB CC brown mouse. State the probable genotypic and phenotypic ratios for the F_1 and F_2 offspring.

    b. If an F_1 mouse from Problem 9a is backcrossed with its albino parent, what phenotypic and genotypic ratios would you expect?

10. A dominant allele W confers black fur on guinea pigs. If the guinea pig is homozygous recessive (ww), it has white fur. Fred would like to know if his pet black-furred guinea pig is homozygous WW or heterozygous (Ww). How might he determine his pet's genotype?

11. In snapdragons, red-flowering plants are homozygous for a certain allele (R^1R^1). White-flowering plants are homozygous for a different allele (R^2R^2). Heterozygotes (R^1R^2) bear pink flowers. What phenotypes are likely to appear among the F_1 offspring of the following crosses? And what are the expected proportions for each phenotype?

    a. $R^1R^1 \times R^1R^2$
                 c. $R^1R^2 \times R^1R^2$
    b. $R^1R^1 \times R^2R^2$
                 d. $R^1R^2 \times R^2R^2$

(*Note: in cases of incomplete dominance, it is inappropriate to refer to either allele of a pair as dominant or recessive. When the phenotype of a heterozygote is halfway between those of the two homozygotes, then there is no dominance. Such alleles are usually designated with superscript numerals, as shown above, rather than by uppercase letters for dominance and lowercase letters for recessiveness.*)

12. An allele codes for a mutant form of hemoglobin (Hb^S instead of Hb^A). Homozygotes (Hb^SHb^S) are affected by sickle cell anemia. Heterozygotes (Hb^AHb^S) show few outward symptoms. Suppose a female whose father was homozygous for the Hb^S allele marries a heterozygous male, and they consider having children. For *each* pregnancy,

    a. What is the probability of having a child homozygous for the Hb^S allele?

    b. What is the probability of having a child homozygous for the Hb^A allele?

    c. What is the probability of having a child heterozygous Hb^AHb^S?

13. An inability to produce melanin results in albinism. A recessive allele, a, is responsible for this phenotype. State the possible genotypes of both parents and offspring of the following crosses.

    a. Both parents have normal phenotypes but have both albino and normal children.

    b. Both parents are albino and have only albino children.

    c. A normal woman and an albino man who have two albino and two normal children.

14. In chickens two pairs of genes affect the comb type (p. 179). When both genes are in the recessive form, the chicken will have a single (or normal) comb. Pea comb is determined by the dominant allele of one of these genes, P. The dominant allele of the other gene, R, causes rose comb. An epistatic interaction occurs if a chicken has at least one of both dominants, $P\_ R\_$, producing a walnut comb. Predict the offspring ratios of a cross between two walnut-combed chickens that are heterozygous for both genes ($PpRr$).

15. If their selection of mates were truly random, human populations might show continuous variation for skin pigmentation. Hypothetically speaking, assume that this is the case and that two pairs of genes control the trait. Assume further that alleles A and B contribute equally to the production of the pigment melanin, and that alleles a and b are incapable of contributing to its formation. $AABB$ would be associated with the greatest amount of pigmentation, $aabb$ with the least amount, and all other allelic combinations with intermediate amounts.

    a. Predict the amount of pigmentation of the offspring from a mating of two individuals with the genotypes $AABB \times aabb$.

    b. Predict the genotypes of offspring if both parents are $AaBb$.

Selected Key Terms

Readings

Cummings, M. 1994. *Human Heredity*. Third edition, St. Paul: West Publishing Company.

Griffiths, A., et al. 1993. *An Introduction to Genetic Analysis*. Fourth edition. New York: Freeman.

Mendel, G. 1966. "Experiments on Plant Hybrids." Translation in C. Stern and E. Sherwood, *The Origin of Genetics: A Mendel Source Book*. New York: Freeman.

Orel, V. 1984. *Mendel*. New York: Oxford University Press.

12 CHROMOSOMES AND HUMAN GENETICS

Too Young To Be Old

Imagine being ten years old, trapped in a body that with each passing day becomes a bit more shriveled, more frail, *old*. You are just tall enough to peer over the top of the kitchen counter, and you weigh less than thirty-five pounds. Already you are bald, and your nose is crinkled and beaklike. Possibly you have only a few more years to live. Yet, like Mickey Hayes and Fransie Geringer (Figure 12.1), you still play, laugh, and hug your friends.

Of every 8 million newborns, one is destined to grow old far too soon. That rare individual possesses a mutated gene on just one of forty-six chromosomes, inherited either from its mother or father. Through hundreds, thousands, then many millions of DNA replications and mitotic cell divisions, terrible information encoded in that gene was methodically distributed to every cell in the embryo, then in the newborn's body. The outcome will be accelerated aging and a greatly

reduced life expectancy. This is the defining feature of *Hutchinson-Gilford progeria syndrome*. There is no cure.

The mutation leads to disruptions in the gene interactions underlying cell division, growth, and development. Outward symptoms start to emerge when the child is less than two years old. The skin becomes thinner. Muscles become flabby. Limb bones that should lengthen and become stronger start to soften instead. Hair loss is pronounced and typically ends in baldness.

Most progeriacs die in their early teens from strokes or heart attacks. These are the final insults, brought on by a hardening of the arteries—a condition that is typical of advanced age.

There are no documented cases of progeria running in families, which suggests that the gene must mutate spontaneously, at random. The disorder can develop with equal frequency in either boys or girls, so the gene cannot be on a sex chromosome. Because the characteristic phenotype always develops, the mutated gene must be dominant over its partner on the homologous chromosome.

We began this unit of the book by looking at cell division, the starting point of inheritance. Then we started thinking about how chromosomes—and the genes they carry—are shuffled at meiosis and fertilization. In this chapter we delve more deeply into the patterns of chromosomal inheritance, with emphasis on humans. At times the methods of analysis might seem abstract. But keep in mind that we are talking about messages of inheritance in yourself and in other human individuals. When Mickey Hayes turned eighteen, he was the oldest living progeriac. Fransie was seventeen when he died.

Figure 12.1 Two boys, both less than ten years old, who met during a gathering of progeriacs at Disneyland, California. Progeria is a heritable (genetic) disorder. It is characterized by accelerated aging and extremely reduced life expectancy.

KEY CONCEPTS

1. One gene follows another along the length of a chromosome. Each has its own position in that sequence.

2. A chromosome's gene sequence does not necessarily remain intact through meiosis and gamete formation. Through crossing over, some genes leave the sequence, and genes from the homologous chromosome replace them.

3. A chromosome's structure may change, as when a segment is deleted, duplicated, inverted, or moved to a new location. Also, improper separation of duplicated chromosomes during meiosis or mitosis may lead to changes in chromosome number.

4. Allele shufflings, changes in chromosome structure, and changes in chromosome number contribute to variation in traits. Often they lead to genetic abnormalities or disorders.

12.1 RETURN OF THE PEA PLANT

The year was 1884. Mendel's paper on pea plants had been gathering dust in a hundred libraries for nearly two decades, and Mendel himself had just passed away. Ironically, the experiments described in that forgotten paper were about to be devised once again.

Improvements in microscopy had rekindled efforts to locate the hereditary material within cells. By 1882, Walther Flemming had observed threadlike bodies—chromosomes—in the nucleus of dividing cells. By 1884, a question was taking shape: Could chromosomes be the hereditary material?

Then researchers realized each gamete has half the number of chromosomes of a fertilized egg. In 1887, August Weismann proposed that a special division process must reduce the chromosome number by half before gametes form. Sure enough, in that same year meiosis was discovered. Weismann began to promote his theory of heredity: The chromosome number is halved during meiosis, then restored at fertilization. Thus a cell's hereditary material is half paternal in origin, and half maternal. His theory was hotly debated, and it prompted a flurry of experimental crosses—just like the ones Mendel had carried out.

Finally, in 1900, researchers came across Mendel's paper while checking literature related to their own genetic crosses. To their surprise, their results merely confirmed what Mendel had already proposed. Diploid cells have two units (genes) for each heritable trait, and the units segregate before gametes form.

Researchers learned a great deal about chromosomes during the decades that followed. Let's start with a few high points of their work, which will serve as background for understanding human inheritance.

12.2 THE CHROMOSOMAL BASIS OF INHERITANCE—AN OVERVIEW

From the preceding chapters in this unit, you already are familiar with some concepts concerning the structure of chromosomes and their behavior during meiosis. Let's now integrate them with a few concepts that will help explain patterns of human inheritance:

---

1. **Genes** are units of information about heritable traits. Each kind of gene has a particular location in a particular type of chromosome.

2. For each eukaryotic species, the different genes are distributed among different types of chromosomes.

3. Many cells that give rise to gametes are diploid (2*n*), with pairs of **homologous chromosomes**. In humans, one chromosome of each type is inherited from the mother, and its homologue from the father.

4. Each type of chromosome has the same length, shape, and gene sequence as its homologous partner, and interacts with it during meiosis. Different sex chromosomes (*X* and *Y*) are homologous in one small region only.

5. A given pair of homologous chromosomes may be carrying identical or nonidentical alleles at corresponding locations in their gene sequence. **Alleles** are slightly different molecular forms of the same gene, and they arise through mutation.

6. There is no pattern to the way that maternal and paternal chromosomes become attached to the microtubular spindle just before metaphase I of meiosis, so different combinations are possible at anaphase I. As a result, gametes and then new individuals receive a mix of alleles of maternal and paternal chromosomes.

7. Genes close together on the same chromosome *tend* to stay together through meiosis, but the chromosome breaks and exchanges corresponding segments with its homologous partner. This event, called **crossing over**, puts new combinations of alleles into chromosomes.

8. A chromosome's structure may change, as when a segment is deleted, duplicated, inverted, or moved to a new location. Also, improper separation of chromosomes that have been duplicated in advance of meiosis or mitosis may lead to changes in chromosome number in gametes and the new individual.

9. For any given trait, the particular combination of alleles inherited by the new individual may have neutral, beneficial, or harmful effects.

Focus on Science

Preparing a Karyotype Diagram

In laboratories throughout the world, karyotype diagrams help answer questions about an individual's chromosomes. You can use the diagram in Figure *a* as a basis for understanding this chapter's examples of karyotypes that point to human genetic disorders.

Chromosomes are most condensed and easiest to identify in dividing cells, particularly those at metaphase of mitosis. Technicians don't count on finding a cell that happens to be dividing in the body when they go looking for it. Instead, they culture cells *in vitro* (literally, "in glass"). They put a sample of blood, skin, bone marrow, amniotic fluid (which bathes embryos), or some other tissue in a glass container. In that container is a solution that stimulates cell growth and mitotic division for many generations.

The dividing cells can be arrested at metaphase by adding colchicine to the culture medium. Colchicine is an extract of the autumn crocus (*Colchicum autumnale*). Technicians and researchers use it to block formation of microtubular spindles—and to prevent duplicated chromosomes from separating during nuclear division. With the proper colchicine concentration and exposure time, many metaphase cells can accumulate, and this increases the chances of finding good candidates for karyotype diagrams.

Next, the culture medium is transferred to tubes of a centrifuge, a motor-driven spinning device. The cells have greater mass and density than the surrounding solution, so the spinning force moves them farthest away from the

Autosomes and Sex Chromosomes

Generally speaking, a pair of homologous chromosomes is exactly alike in length, shape, and gene sequence. In the late 1800s, however, microscopists discovered an exception to this. In humans and many other species, a distinctive chromosome is present in females *or* males, but not in both. The *Focus* essay shows a lineup of the twenty-three pairs of human chromosomes. In males only, the last two in the series are physically different. A long, X-shaped chromosome is paired with a shorter one, shaped like an upside-down Y. These two are called the **X chromosome** and **Y chromosome**. Despite the physical difference, the two are able to synapse in a small region and function as homologues during meiosis. Unlike the males (XY), females are XX; they have two X chromosomes.

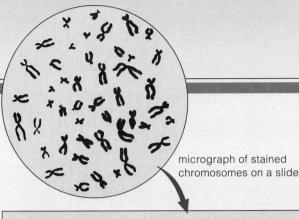

micrograph of stained chromosomes on a slide

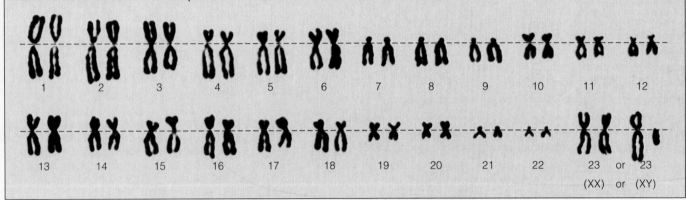

a Karyotype of human somatic cells. Such cells have 22 pairs of autosomes and 1 pair of sex chromosomes (XX or XY), for a diploid chromosome number of 46. These chromosomes are all duplicated.

1 2 3 4 5 6 7 8 9 10 11 12

13 14 15 16 17 18 19 20 21 22 23 or 23
(XX) or (XY)

motor, to the bottom of the attached tubes. (Separation in response to the spinning force is called *centrifugation*.) Afterward, the cells are transferred to a saline solution. They swell (by osmosis), then move apart—and so do their metaphase chromosomes. The cells are ready to be dropped onto a microscope slide, fixed (as by air-drying), and stained.

Chromosomes take up some stains (orcein and Giemsa dyes) uniformly. This is enough to allow identification of their size and shape, as in Figure *a*. With new staining procedures, horizontal bands appear in human chromosomes that fluoresce when viewed in ultraviolet light. Examples of these appear in several chapters of the book.

Finally, a microscope image of the chromosomes is photographed and enlarged, so that the chromosomes can be cut apart individually. Cutouts can be arranged by size, shape, and length of arms, then the homologous pairs can be aligned horizontally by their centromeres (Figure *a*).

Human X and Y chromosomes are examples of **sex chromosomes**. Inheritance of one type or the other governs gender—that is, whether a new individual will be male or female. All other chromosomes, which are the same in both sexes, are designated **autosomes**.

Karyotype Analysis

Today, microscopists routinely analyze a cell's sex chromosomes and autosomes. Chromosomes, recall, are most highly condensed at metaphase of mitosis. At that time, each has a characteristic size, length, and centromere location. Also, the chromosomes of many species show distinct banding patterns when they are stained in certain ways. Regions that are the most con-

densed often take up more stain and form darker bands. Figure 10.4 shows an example of this.

The number of metaphase chromosomes and their defining characteristics are called the **karyotype** of an individual (or species). The *Focus* essay explains how to construct a karyotype diagram. Such a diagram is a cut-up, rearranged photograph in which all the autosomes are lined up, largest to smallest, and the sex chromosomes are positioned last.

A sex chromosome is one whose presence determines a new individual's gender. An autosome is any chromosome other than a sex chromosome.

12.3 SEX DETERMINATION IN HUMANS

Karyotype analysis of human gametes provides evidence that each egg produced by a female carries one X chromosome. Half of the sperm cells produced by a male carry an X chromosome and half carry a Y. If an X-bearing sperm fertilizes an X-bearing egg, the new individual will develop into a female. Conversely, if the fertilizing sperm carries a Y chromosome, the individual will develop into a male (Figure 12.2).

Among the very few genes on the Y chromosome is a "male-determining gene." As Figure 12.3 indicates, expression of this gene leads to the formation of testes, which are the primary male reproductive organs. In the absence of this gene, ovaries form automatically. Ovaries are the primary female reproductive organs. Testes and ovaries both produce hormones that govern the development of particular sexual traits.

A human X chromosome probably carries more than 300 genes. Like other chromosomes, it carries some genes associated with sexual traits, such as the distribution of body fat. But most of its genes deal with *nonsexual* traits, such as blood-clotting functions.

A certain gene on the human Y chromosome dictates that a new individual will develop into a male. In the absence of the Y chromosome (and the gene), a female develops.

Figure 12.3   Boys, girls, and the Y chromosome.

For about the first four weeks of its existence, a human embryo is neither male nor female, even though it normally carries XY or XX chromosomes. However, ducts and other structures start forming that can go either way. In XY embryos, the primary male reproductive organs (testes) start to form during the next four to six weeks (**a–c**). A gene region on the Y chromosome seems to govern a fork in the developmental road that can lead either to maleness or to femaleness. In XX embryos, the primary female reproductive organs (ovaries) start to form. They form automatically in the absence of a Y chromosome.

The testes start producing testosterone and other sex hormones that influence the development of the male reproductive system. The ovaries also start producing sex hormones that influence the development of the female reproductive system.

The master gene for sex determination is named SRY (short for *Sex-determining Region of the Y* chromosome). So far, the same gene has been identified in DNA from male humans, chimpanzees, rabbits, pigs, horses, cattle, and tigers. None of the females tested had the gene. Tests with mice indicate that the gene region becomes active about the time testes start developing.

The SRY gene resembles regions of DNA that are known to specify regulatory proteins. Such proteins bind to certain parts of DNA and so turn genes on and off. Apparently, the SRY gene product regulates a cascade of reactions that are necessary for sex determination.

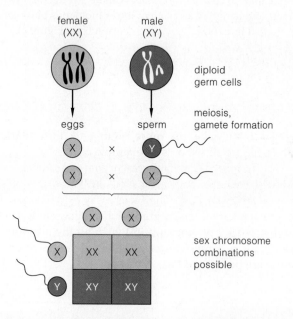

Figure 12.2   Pattern of sex determination in humans. Males transmit their Y chromosome to sons but not to daughters. Males receive their X chromosome only from their mother.

umbilical cord (lifeline between embryo and maternal tissues)

embryo floating in protective, fluid-filled sac (the amnion)

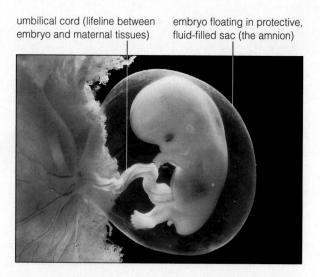

a   A human embryo, eight weeks old. Male reproductive organs have already started to develop.

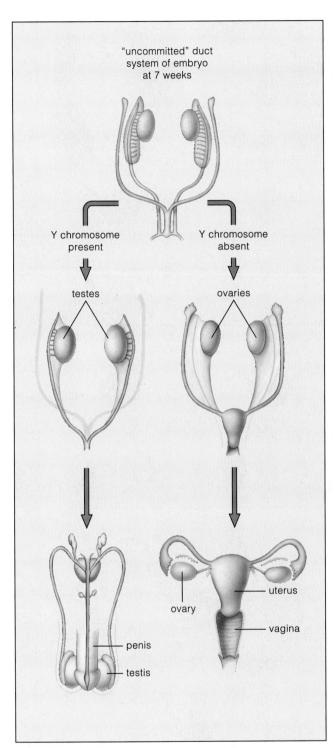

b Duct system in the early embryo that develops into a male or female reproductive system. Compare pages 774 and 778.

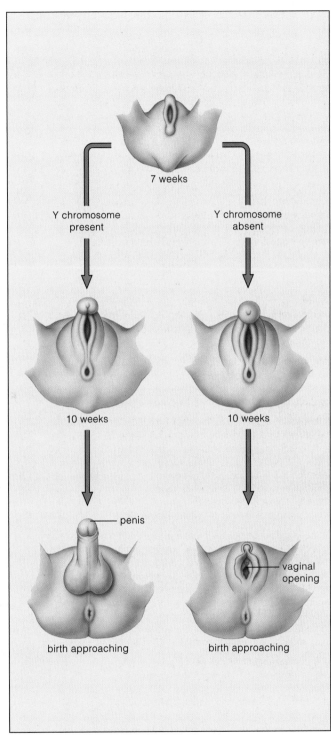

c External appearance of the developing reproductive organs in embryos.

Figure 12.4 X-linked genes: clues to inheritance patterns.

In the early 1900s, the embryologist Thomas Morgan was studying inheritance patterns. During those studies, he and his coworkers discovered an apparent genetic basis for the connection between gender and certain nonsexual traits. For example, human males and females both have blood-clotting mechanisms. Yet hemophilia (a blood-clotting disorder) shows up most often in the males, not females, of a family lineage. This gender-specific outcome was not like anything Mendel saw in his hybrid crosses between pea plants. (Either parent plant could carry a recessive allele. It made no difference; the resulting phenotype was the same.)

Morgan studied eye color and other nonsexual traits of *Drosophila melanogaster.* These fruit flies can be grown in bottles on bits of cornmeal, molasses, and agar. A female lays hundreds of eggs in a few days, and new flies that develop from the eggs can themselves reproduce in less than two weeks. Morgan could track hereditary traits through nearly thirty generations of thousands of flies in a year's time.

At first, all the flies were wild-type for eye color; they had brick-red eyes, as in (**a**). (*Wild-type* simply means the normal or most common form of a trait in a population.) Then, through an apparent mutation in a gene controlling eye color, a white-eyed male appeared (**b**).

Morgan established true-breeding strains of white-eyed males and females. Then he did a series of *reciprocal* crosses. These are pairs of crosses. In the first, one parent displays the trait in question. In the second, the other parent displays it.

White-eyed males were mated with homozygous red-eyed females. All the F_1 offspring of the cross had red eyes. But of the F_2 offspring, only some of the males had white eyes. Then white-eyed females were mated with true-breeding red-eyed males. Of the F_1 offspring of that second cross, half were red-eyed females and half were white-eyed males. Of the F_2 offspring, 1/4 were red-eyed females, 1/4 white-eyed females, 1/4 red-eyed males, and 1/4 white-eyed males!

The seemingly odd results implied a relationship between the eye-color gene and gender. Probably the gene was located on a sex chromosome. But which one? Because females (XX) could be white-eyed, the recessive allele would have to be on one of their X chromosomes. Suppose white-eyed males (XY) also carry the recessive allele on their X chromosome—and suppose there is no corresponding eye-color allele on the Y chromosome. The males would have white eyes, for they have no dominant allele to mask the effect of the recessive one.

(**c**) This diagram shows the expected results when the idea of an X-linked gene is combined with Mendel's concept of segregation. By proposing that a specific gene is located on the X chromosome but not on the Y, Morgan was able to explain the outcome of his reciprocal crosses. The results of the experiments matched the predicted outcomes.

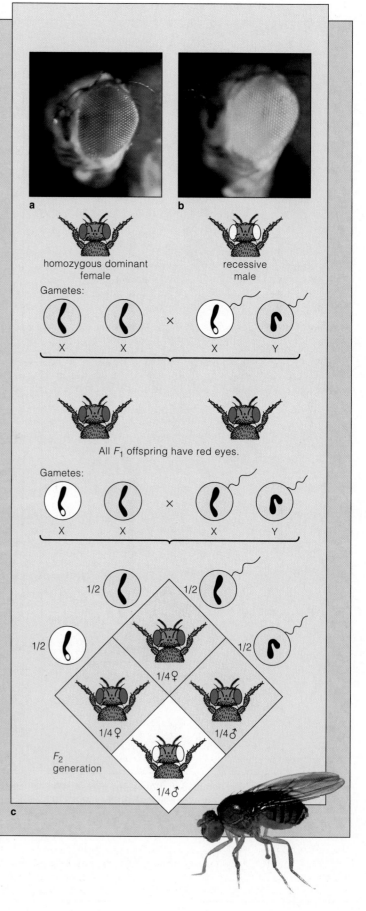

12.4 EARLY QUESTIONS ABOUT GENE LOCATIONS

X-Linked Genes: Clues to Patterns of Inheritance

Some time ago, researchers only suspected that each gene has a specific location on a chromosome. Through a series of hybridization experiments with fruit flies (*Drosophila melanogaster*), Thomas Hunt Morgan and his coworkers helped confirm this. For example, as described in Figure 12.4, they found strong evidence of a gene for eye color on the *Drosophila* X chromosome. For a time, genes located on sex chromosomes were called "sex-linked genes." Today researchers use the more precise terms, **X-linked** and **Y-linked genes**.

Linkage Groups and Crossing Over

It seemed, during the early *Drosophila* experiments, that the genes "linked" to one chromosome or another travel together and end up in the same gamete. In time, researchers identified a large number of genes, and the genes located on each type of chromosome came to be called **linkage groups**. For example, *D. melanogaster* has four linkage groups, which correspond to its four pairs of homologous chromosomes. Indian corn (*Zea mays*) has ten linkage groups, corresponding to ten pairs of homologous chromosomes.

As we now know, linkage groups are vulnerable to crossing over (Figure 12.5). We also know that crossing over is not a rare event. In most eukaryotic organisms, humans included, meiosis cannot be completed properly unless *every* pair of homologous chromosomes takes part in at least one crossover.

Imagine any two genes at two different locations on the same chromosome. The probability of a crossover disrupting their linkage is proportional to the distance separating them. Suppose genes *A* and *B* are twice as far apart as two other genes, *C* and *D*:

We would expect crossing over to disrupt the linkage between *A* and *B* much more often.

Two genes are very closely linked when the distance between them is small; they nearly always end up in the same gamete. Linkage is more vulnerable to crossing over when the distance between two genes is greater. When two genes are very far apart, crossing over disrupts their linkage so often that the genes assort independently of each other into gametes.

The farther apart two genes are on a chromosome, the greater will be the frequency of crossing over and recombination between them.

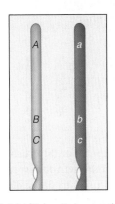 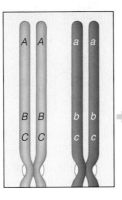 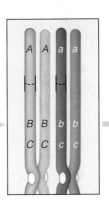 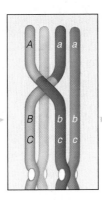 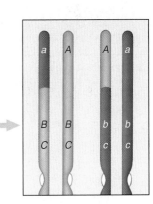

a Diploid (2*n*) cells have pairs of genes, at corresponding locations, on pairs of homologous chromosomes. The two genes of each pair may or may not be identical. In this example, nonidentical alleles are present at all three locations (*A* with *a*, *B* with *b*, and *C* with *c*).

b The same gene regions at interphase, after DNA replication at interphase. Both of these homologous chromosomes are now in the duplicated state.

c At prophase I of meiosis, two of the nonsister chromatids break while aligned very tightly together. (compare Figure 10.6).

d The nonsister chromatids swap segments, then enzymes seal the broken ends. The breakage and exchange represent one crossover event.

e The outcome of the crossover is genetic recombination between two of four chromatids. (They are shown here following meiosis, as separate, unduplicated chromosomes.)

Figure 12.5 Crossing over, an event that influences inheritance. A preceding chapter started us thinking about crossing over by showing the time of its occurrence in meiosis (Figure 10.6). Here we see how it disrupts part of a linkage group—that is, the genes that otherwise would stay together on the same chromosome during meiosis and gamete formation.

12.5 RECOMBINATION PATTERNS AND CHROMOSOME MAPPING

Even without knowing about chromosomes, Mendel almost certainly would have recognized Figure 12.6a as an illustration of the theory that the genes on one chromosome are assorted into gametes independently of the genes on a different chromosome. Had he known about crossing over, he would have been able to make sense of Figure 12.6b, which is an example of the resulting genetic recombination.

For a specific example, consider an experimental cross between two watermelon plants. One plant is true-breeding for green-rind, round melons, the dominant phenotypes. The other is true-breeding for stripe-rind, oblong melons, the recessive phenotypes. The gene for rind color is closely linked to the gene for melon shape. As you might expect, all of the F_1 offspring of the cross produced green, round melons.

In a testcross, F_1 plants were hybridized with the homozygous recessive parent. Knowing that the genes for the two traits are closely linked, you predict that the phenotypic ratio of the F_2 offspring will be 1:1:

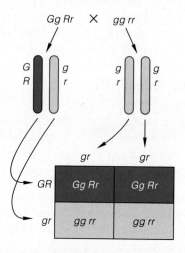

Yet the actual F_2 phenotypes were not as expected:

46 green, round	4 green, oblong
47 striped, oblong	3 striped, round

How can you explain these results? You can assume that 7 percent of the offspring (the 4 green, oblong and the 3 striped, round) inherited a chromosome that had undergone a crossover between the gene locus for rind color and the gene locus for rind shape:

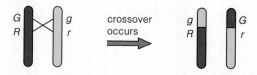

The results of many such experimental crosses tell us that genes undergo recombination in fairly regular patterns. The patterns have been used to determine the positions of genes relative to one another along the length of a chromosome, an activity called linkage mapping. For example, of the several thousand known genes in the four types of *Drosophila* chromosomes, the positions of about a thousand have been mapped.

What about the estimated 50,000 to 100,000 genes on the twenty-three types of human chromosomes? These must be mapped by different methods, of the sort described in Chapter 16. Humans, unlike watermelon plants, do not lend themselves to experimental crosses. And with this jarring thought fresh in your mind, you are ready to begin the next section.

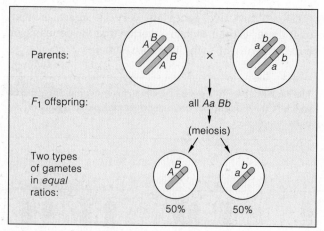

a Complete linkage (no crossing over)

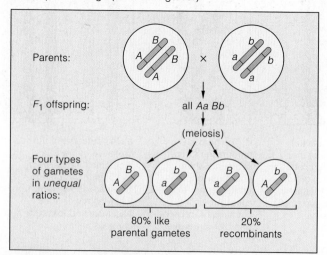

b Incomplete linkage owing to crossing over

Figure 12.6 How crossing over can affect gene linkage, using two genes on the same chromosome as the example.

12.6 HUMAN GENETIC ANALYSIS

Some organisms, including pea plants and fruit flies, lend themselves to genetic analysis. They grow and reproduce rapidly in small spaces, under controlled conditions. It doesn't take long to track a trait through many generations.

Humans are another story. We live under variable conditions in diverse environments. We select our own mates and reproduce if and when we want to. Human subjects live as long as the geneticists who study them, so tracking traits through generations can be rather tedious. Most human families are so small, there aren't enough offspring for easy inferences about inheritance.

To get around some of the problems associated with analyzing human inheritance, geneticists put together **pedigrees**. A pedigree is a chart that is constructed, by standardized methods, to show the genetic connections among individuals. Figure 12.7 includes an example, along with definitions of a few of the standardized symbols used to represent individuals.

When analyzing pedigrees, geneticists rely on their knowledge of probability and Mendelian inheritance patterns, which may yield clues to a trait's genetic basis. For instance, they might determine that the responsible allele is dominant or recessive, or that it is located on an autosome or a sex chromosome. Gathering many family pedigrees increases the numerical base for their analysis. When a trait clearly shows a simple Mendelian inheritance pattern, a geneticist may be justified in predicting the probability of its occurrence among children of prospective parents. We will return to this topic later in the chapter.

For many genes, pedigree analysis may reveal simple Mendelian inheritance patterns that permit inferences about the probability of their transmission to children.

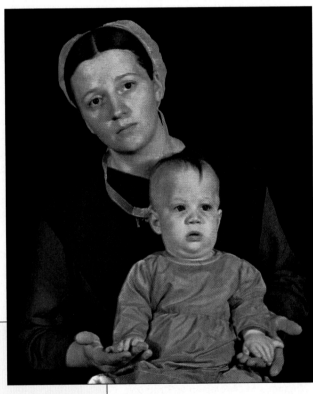

Figure 12.7 (**a**) Some symbols used in constructing pedigree diagrams. (**b**) This is an example of a pedigree for *polydactyly*. An individual with this condition has extra fingers, extra toes, or both. Expression of the gene governing polydactyly can vary from one individual to the next. Here, black numerals designate the number of fingers on each hand. Blue ones designate the number of toes on each foot.

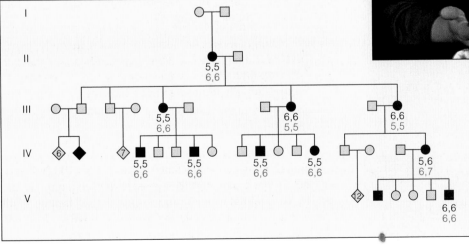

12.7 REGARDING HUMAN GENETIC DISORDERS

Table 12.1 lists some heritable traits that have been studied in detail. A few of these are abnormalities, or deviations from the average condition. Said another way, a **genetic abnormality** is simply a rare or less common version of a trait, as when a person is born with six toes on each foot instead of five. Whether we view an abnormal trait as disfiguring or merely interesting is subjective. There is nothing inherently life-threatening or even ugly about it.

By contrast, a **genetic disorder** is an inherited condition that results in mild to severe medical problems. Alleles underlying severe genetic disorders do not abound in populations, for they put individuals at great risk. They do not disappear entirely for two reasons. First, rare mutations put new copies of the alleles in the population. Second, in heterozygotes, such an allele is paired with a normal one that may cover its functions, so it can be passed to offspring.

You may hear someone call a genetic disorder a disease, but the terms are not always interchangeable. "Diseases" result from infection by bacteria, viruses, or some other environmental agent that invades and multiplies inside the body. Illness follows only if the agent's activities damage tissues and interfere with normal body functions. When a person's genes increase susceptibility or weaken the response to an infection, the resulting illness might be called a genetic disease.

With these qualifications in mind, let's turn to some patterns of inheritance in the human population.

A genetic abnormality simply is a rare or less common version of an inherited trait. A genetic disorder is an inherited condition that results in mild to severe medical problems.

Table 12.1	Examples of Human Genetic Disorders
Disorder or Abnormality*	Main Consequences
Autosomal Recessive Inheritance:	
Albinism 178	Absence of pigmentation
Sickle-cell anemia 176	Severe tissue, organ damage
Galactosemia 196	Brain, liver, eye damage
Phenylketonuria 204	Mental retardation
Autosomal Dominant Inheritance:	
Achondroplasia 197	One form of dwarfism
Amyotrophic lateral sclerosis 199	Motor neurons deteriorate, muscles waste away
Camptodactyly 180	Rigid, bent little fingers
Familial hypercholesterolemia 33, 197	High cholesterol levels in blood, clogged arteries
Huntington's disorder 197	Nervous system degenerates progressively, irreversibly
Polydactyly 195	Extra fingers, toes, or both
Progeria 187, 197	Drastic premature aging
X-Linked Dominant Inheritance:	
Faulty enamel trait 198	Problems with teeth
X-Linked Recessive Inheritance:	
Hemophilia A 198	Deficient blood-clotting
Duchenne muscular dystrophy 198	Muscles waste away
Testicular feminization 617	XY individual but female traits, sterility
Red-green color blindness 606	Reds, greens appear the same
Changes in Chromosome Structure:	
Cri-du-chat 200	Retardation, skewed larynx
Fragile X syndrome 200	Mental retardation
Changes in Chromosome Number:	
Down syndrome 202	Mental retardation, heart defects
Turner syndrome 203	Sterility, abnormal ovaries and sexual traits
Klinefelter syndrome 203	Sterility, retardation
XYY condition 203	Mild retardation in some; no symptoms in others

*Number indicates the page(s) on which the disorder is described.

12.8 PATTERNS OF AUTOSOMAL INHERITANCE

Autosomal Recessive Inheritance

Two clues tell us that a recessive allele on an autosome is responsible for a trait:

1. If both parents are heterozygous, there is a 50 percent chance each child will be heterozygous and a 25 percent chance it will be homozygous recessive. Figure 12.8 is a diagram of this outcome.

2. If both parents are homozygous recessive, any child of theirs will be, also.

Galactosemia is a genetic disorder arising from autosomal recessive inheritance. About 1 in 100,000 newborns are homozygous recessive for an enzyme that helps prevent a breakdown product of lactose (milk sugar) from accumulating to toxic levels. Normally, lactose is converted to glucose and galactose, then to glucose-1-phosphate, which can be broken down by

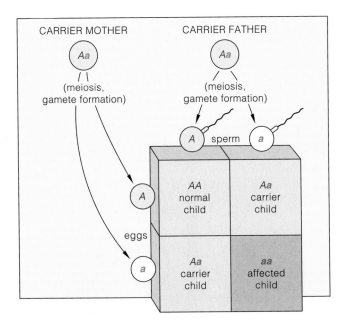

Figure 12.8 One pattern for autosomal recessive inheritance. This example shows the phenotypic outcomes possible when both parents are heterozygous carriers of the recessive allele (shown in *red*).

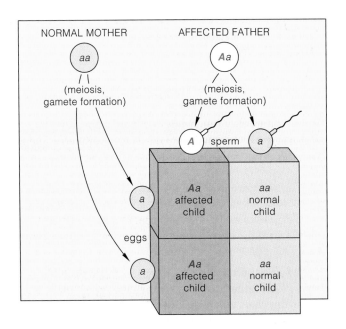

Figure 12.9 One pattern for autosomal dominant inheritance. In this example, assume the dominant allele (shown in *red*) is fully expressed in the carriers.

glycolysis or converted to glycogen (page 134). The defective enzyme blocks the full conversion:

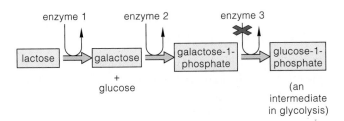

A high blood level of galactose can damage the eyes, liver, and brain. Malnutrition, diarrhea, and vomiting are early symptoms. Untreated galactosemics often die in childhood. However, the telling symptom—a high galactose level—can be detected in urine samples. When affected individuals are placed on a diet that includes milk substitutes and excludes dairy products, they can grow up symptom-free.

Autosomal Dominant Inheritance

Figure 12.9 shows an example of what can happen when a dominant allele on an autosome is responsible for a trait. Two clues point to this condition:

1. Usually the trait appears in each generation, for the allele usually is expressed even in heterozygotes.

2. If one parent is heterozygous and the other homozygous recessive, there is a 50 percent chance that any child of theirs will be heterozygous.

In Chapter 3, you read about a genetic disorder that can lead to cholesterol-clogged arteries. The disorder, called *familial hypercholesterolemia*, arises from an autosomal dominant allele. Affected individuals can be placed on a diet low in cholesterol and saturated fats. Also, certain drugs can lower cholesterol levels.

Even though a few dominant alleles cause severe genetic disorders, they persist in populations. Some are perpetuated by spontaneous mutations. This is the case for progeria, the rare aging disorder described in the introduction to this chapter.

In other cases, a dominant allele may not prevent reproduction. Consider *achondroplasia*, which affects about 1 in 10,000 humans. Usually, the homozygous dominant condition results in a stillbirth. However, heterozygotes are able to reproduce. They cannot form cartilage properly when limb bones are growing, and this leads to abnormally short arms and legs. Adults are less than 4 feet, 4 inches tall. The dominant allele often has no other phenotypic effects than this. In others, severe symptoms may not appear until after a person reproduces and so transmits the allele.

In still other cases, affected persons have children before their symptoms become severe. In *Huntington's disorder*, a progressive deterioration of the nervous system, symptoms may not even begin until about age forty. Most people have already reproduced by then.

12.9 PATTERNS OF X-LINKED INHERITANCE

X-Linked Recessive Inheritance

Certain clues may tell you that a trait is specified by an allele on the X chromosome. When the X-linked allele is recessive, you will detect this pattern:

1. The recessive phenotype shows up far more often in males than females. (A recessive allele can be masked in females, who may inherit a dominant allele on their other X chromosome. It cannot be masked in males, who have only one X chromosome.)

2. A son cannot inherit the recessive allele from his father. A daughter can. If she is heterozygous, there is a 50 percent chance that each son of hers will inherit the allele (Figure 12.10).

Hemophilia A is an example of X-linked recessive inheritance. In most people, a blood-clotting mechanism quickly stops bleeding from minor injuries. The reactions that lead to clot formation require the products of several genes. If any of the genes is mutated, its defective product can cause affected persons to bleed for an abnormal time.

A mutated gene for "clotting factor VIII" causes hemophilia A. It affects about 1 in 7,000 males, who run the risk of dying from untreated bruises, cuts, or internal bleeding. Blood-clotting time is more or less normal in heterozygous females.

The frequency of hemophilia A was unusually high among the males in royal families of nineteenth-century Europe. As Figure 12.11 indicates, Queen Victoria of England was a carrier. At one time, the recessive allele was present in eighteen of her sixty-nine descendants. One hemophilic great-grandchild, Crown Prince Alexis, was a focus of political intrigue that helped usher in the Russian revolution.

Duchenne muscular dystrophy is another example of X-linked recessive inheritance. Although rare, it is the most common of several heritable disorders in which certain groups of muscles slowly degenerate over time. The muscles themselves enlarge as fat and connective tissue is deposited in them, but the muscle cells atrophy (waste away).

At first, children with this form of dystrophy may appear to be normal. But somewhere between their second and tenth birthday, their muscles start to weaken. Such children become progressively clumsier and fall frequently. Usually, after their twelfth birthday, they no longer can walk. In time, muscles of the chest and head degenerate. Affected individuals typically die of respiratory failure in their early twenties. There is no cure.

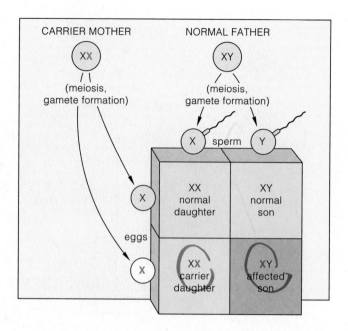

Figure 12.10 One pattern for X-linked inheritance. This example shows the phenotypic outcomes possible when the mother carries a recessive allele on one of her X chromosomes (shown in *red*).

X-Linked Dominant Inheritance

The *faulty enamel trait* is one of very few known examples of X-linked dominant inheritance. With this disorder, the hard, thick enamel coating that normally protects teeth fails to develop properly.

The inheritance pattern is similar to that for X-linked recessive alleles, except that the trait is expressed in heterozygous females. The expression tends to be less pronounced in females than in males. A son cannot inherit the dominant allele responsible for the trait from an affected father, but all of his daughters will. If a woman is heterozygous, she will transmit the allele to half of her offspring, regardless of their sex.

A Few Qualifications

Don't take the preceding examples of autosomal and X-linked traits too seriously. We include them not to turn you into an armchair geneticist, but rather to give you a general idea of the kinds of clues that hold meaning for trained geneticists.

Before diagnosing a case, geneticists often find it necessary to pool together many pedigrees. Typically they make detailed analyses of clinical data and keep

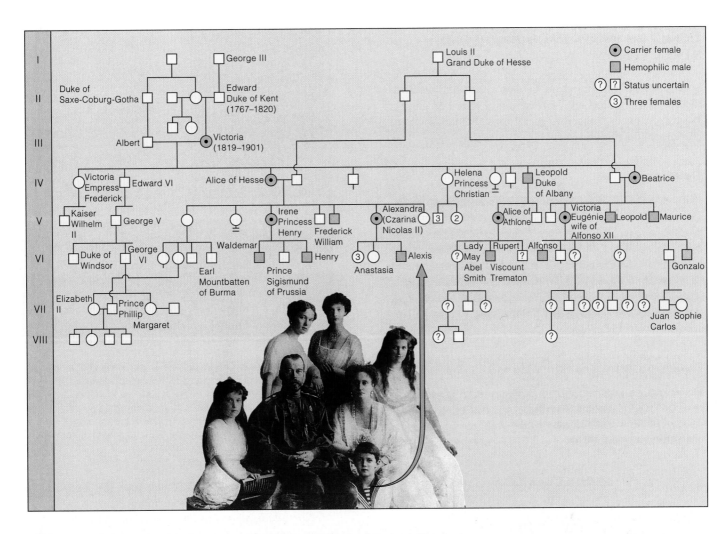

Figure 12.11 Partial pedigree of the descendants of Queen Victoria of England. The chart shows carriers and affected males that possessed the X-linked allele conferring the disorder hemophilia A. Many individuals of later generations are not included. The photograph shows the Russian royal family members. The mother was a carrier of the mutated allele. Crown Prince Alexis was hemophilic.

abreast of current research. Why? Consider that more than one type of gene may be responsible for a given phenotype. Geneticists already know of dozens of conditions that can arise from a mutated gene on an autosome *or* a mutated gene on the X chromosome. They know of genes on autosomes that show dominance in males and recessiveness in females—so they appear to be due to X-linked recessive inheritance, even though they are not.

Besides this, don't assume that errant genes automatically relegate a person to the sidelines of life. More than twenty years ago, Stephen Hawking noticed his muscles were starting to weaken. In time it became difficult for him to speak, to swallow, to use his hands. Motor neurons in his brain and spinal cord were deteriorating, and scar tissue was forming along his spinal

cord. Without normal control signals from his nervous system, his skeletal muscles began to waste away. Such are the symptoms of a rare, ultimately fatal disorder, *amyotrophic lateral sclerosis.* Possibly a virus issues the death warrant, but an autosomal dominant allele serves as the executioner.

And yet, through his acclaimed research in astrophysics and his accessible, eloquent writings, Hawking has changed the way we view time and the universe. His affliction has not immobilized his mind, which dances freely around the most challenging of questions—*When did time begin? Does the universe have boundaries? If we know the past, why can't we know the future?* He lives to move beyond the horizons of our knowledge, believing that when we find the ultimate answers, we will know the mind of God.

12.10 CHANGES IN CHROMOSOME STRUCTURE

So far, we have focused on genetic abnormalities and disorders associated with autosomal or X-linked genes. We turn now to some rare conditions that arise when chromosomes undergo structural alterations.

Deletions

A **deletion** is the loss of a chromosome region by irradiation, viral attack, chemical action, or other environmental factors. One or more genes may be lost, and this nearly always causes problems. A certain deletion from human chromosome 5 causes mental retardation and an abnormally shaped larynx. When an affected infant cries, he or she produces meowing sounds. Hence the name of the disorder, *cri-du-chat* (cat-cry). Figure 12.12*a* shows an affected child.

Inversions and Translocations

An **inversion** is a segment that separated from a chromosome and then was inserted at the same place—but in reverse. The reversal alters the position and order of the chromosome's genes:

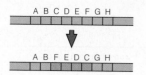

Most often, a **translocation** is part of one chromosome that has exchanged places with a corresponding part of another, *non*homologous chromosome. As an example, chromosome 14 may end up with a segment of chromosome 8 (and chromosome 8 with a segment of 14). Controls over the segment's genes are lost at the new location, and a form of cancer results.

Duplications

Even a normal chromosome contains **duplications**, which are gene sequences that are repeated several to many times. Often the same gene sequence is repeated thousands of times. Some duplications are built into the DNA of the species; others are not.

Years ago, geneticists noticed there were twice as many males as females in mental institutions. They thought X-linked recessive inheritance might be one reason why. (Can you say why?) Later, they found an abnormal constricted region on the X chromosomes of mentally impaired, related males (Figure 12.12*b*). Abnormally expanded repeats cause the constriction.

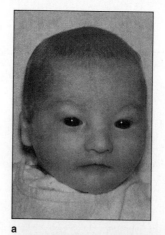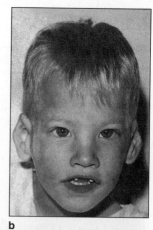

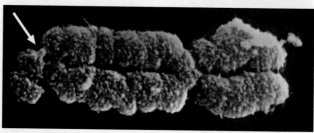

Figure 12.12 (**a,b**) Cri-du-chat syndrome, a result of a deletion on the short arm of chromosome 5. The left photograph shows the patient just after birth. The right photograph, taken four years later, shows how facial features change. By this age, patients no longer make the mewing sounds typical of the syndrome. (**c**) On the long arm of the X chromosome, the constricted region (arrow) responsible for the fragile X chromosome.

They give rise to the *fragile X syndrome*, the second most common form of mental impairment.

Not all duplications are harmful. Many have had roles in evolution. Cells require specific gene products, so mutations that alter most genes are probably selected against. But *duplicates* of some gene sequences that do no harm could be retained and would be free to mutate, for the normal gene would still provide the required product. In time, duplicated and then slightly modified sequences could yield products with related functions or even new ones. This apparently happened in gene regions coding for the polypeptide chains of the hemoglobin molecule. In humans and other primates, those regions have multiple copies of strikingly similar gene sequences, and they produce whole families of slightly different chains.

Together with duplications, some inversions and translocations probably helped put the primate ancestors of humans on a unique evolutionary road. Of the twenty-three pairs of human chromosomes, eighteen are nearly identical to their counterparts in chimpanzees and gorillas. The other five pairs differ at inverted and translocated regions.

On rare occasions, a segment of a chromosome may be lost, inverted, moved to a new location, or duplicated.

12.11 CHANGES IN CHROMOSOME NUMBER

Various abnormal cellular events can put too many or too few chromosomes into gametes. New individuals end up with the wrong chromosome number. The effects range from minor physical changes to lethal disruption of organ systems. More often, affected individuals are miscarried, or spontaneously aborted before birth.

Categories of Change

New individuals may end up with one extra or one less chromosome. This condition, called **aneuploidy**, is a major cause of human reproductive failure, affecting possibly half of all fertilized eggs. Most miscarried and autopsied human embryos were aneuploids.

New individuals also may end up with three or more of each type of chromosome. This condition is called **polyploidy**. About half of all flowering plant species are polyploid (page 294). So are some insects, fishes, and other animals. Polyploidy is lethal for humans. It may disrupt interactions between the genes of autosomes and sex chromosomes at key steps in the complex pathways of development and reproduction. All but 1 percent of human polyploids die before birth, and the rare newborns die within a month.

Mechanisms of Change

A chromosome number can change during mitotic or meiotic cell divisions or during the fertilization process. Suppose a cell cycle proceeds through DNA duplication and mitosis, then is arrested before the cytoplasm divides. This polyploid cell is "tetraploid," with four of each type of chromosome. Or suppose one or more pairs of chromosomes fail to separate properly during mitosis or meiosis. Such events are called **nondisjunction**. As Figure 12.13 shows, some or all of the resulting cells end up with too many or too few chromosomes. Researchers routinely induce nondisjunction by exposing cells to colchicine (compare the *Focus* essay on page 188).

What if a gamete with an extra chromosome ($n + 1$) unites with a normal gamete at fertilization? The new individual will be "trisomic," with three of one type of chromosome ($2n + 1$). If the gamete is missing a chromosome, then the individual will be "monosomic" ($2n - 1$). The following pages provide specific examples of such changes.

Compared to the parental chromosome number, aneuploids have one extra or one less chromosome. Polyploids have three or more of each type of chromosome.

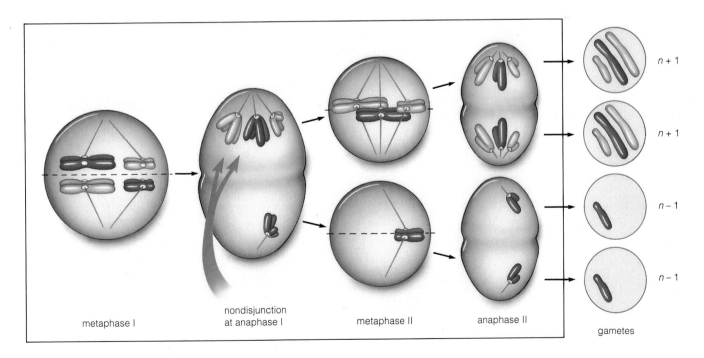

metaphase I   nondisjunction at anaphase I   metaphase II   anaphase II $n + 1$ $n + 1$ $n - 1$ $n - 1$ gametes

Figure 12.13 Nondisjunction. In this example, chromosomes fail to separate during anaphase I of meiosis and so change the chromosome number in the resulting gametes. As another example, make a similar sketch of nondisjunction at anaphase II of meiosis. What will the chromosome numbers be in the resulting gametes?

12.12 WHEN THE NUMBER OF AUTOSOMES CHANGES

Most changes in the number of autosomes arise through nondisjunction during the formation of gametes. Here we consider the most common of the resulting disorders.

Chromosome 21 is one of the smallest chromosomes in human cells. Someone who inherits three of them is categorized as a trisomic 21 and will show the effects of *Down syndrome*. ("Syndrome" simply means a set of symptoms that characterize a disorder.) Figure 12.14*a* shows a karyotype of an affected girl.

Symptoms vary considerably. However, most affected individuals show moderate to severe mental impairment. About 40 percent develop heart defects. As a result of abnormal skeletal development, older children have shortened body parts, loose joints, and poorly aligned bones of the hips, fingers, and toes. Muscles and muscle reflexes are weaker than normal, and development of speech and other motor functions is quite slow.

Nevertheless, with special training, trisomic 21 individuals often take part in normal activities. As a group, they seem truly to enjoy life (Figure 12.15). These are cheerful, impish, affectionate people who characteristically derive great pleasure from music and dancing.

Down syndrome is one of many genetic disorders that can be detected by prenatal diagnosis, as described in the *Focus* essay that concludes this chapter. Before detection procedures were widespread, about 1 in 700 newborns of all ethnic groups were trisomic 21. Today the number is closer to 1 of every 1,100. The risk is much greater when women are over thirty-five years old (Figure 12.14*b*).

The genes responsible for the disorder seem to reside in a certain band region on the long arm of chromosome 21. One of the genes codes for the precursor of *beta*-amyloid protein. Excessive amounts of this protein are present in plaques and tangles in the brain of trisomic 21 patients. The same is true of *Alzheimer disease*, which is characterized by confusion, memory loss, increasing inability to perform simple tasks and, in time, the loss of all motor functions. Intriguingly, past thirty years of age, trisomic 21 patients show many of the same symptoms. They also show similar forms of deterioration in the same regions of the brain.

Figure 12.14 Down syndrome. (**a**) Karyotype of an affected girl. The *red* arrows identify the trisomy of chromosome 21. (**b**) Relationship between the frequency of Down syndrome and the mother's age at the time her child was born. Results are from a study of 1,119 children with the disorder who were born in Victoria, Australia, between 1942 and 1957.

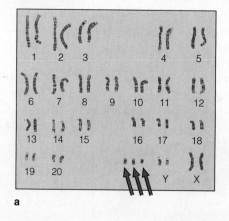

a

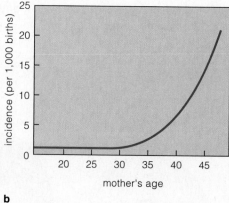

b

Figure 12.15 A few children with Down syndrome. The photographs at center and to the right show two of the enthusiastic participants in the Special Olympics.

12.13 WHEN THE NUMBER OF SEX CHROMOSOMES CHANGES

Most sex chromosome abnormalities arise through nondisjunction during gamete formation, as shown in Figure 12.16. Let's look at a few phenotypic outcomes.

Turner Syndrome

Inheritance of one X chromosome without a partner X or Y chromosome gives rise to *Turner syndrome*, which affects about 1 in 2,500 to 10,000 newborn girls. About 75 percent of these cases are results of nondisjunction in the father.

Turner syndrome is not as common as other sex chromosome abnormalities. Probably this is because at least 98 percent of all X0 zygotes are spontaneously aborted early in pregnancy. And X0 embryos represent 20 percent of all spontaneous abortions of embryos with chromosome abnormalities.

Despite the near-lethality, X0 survivors are not as disadvantaged as other aneuploids. They grow up well proportioned, albeit short (4 feet, 8 inches tall, on the average). Generally, their behavior is normal during childhood. But most Turner females are infertile. They do not have functional ovaries and so cannot produce eggs or sex hormones. Without sex hormones, breast enlargement and the development of other secondary sexual traits cannot occur. Possibly as a result of their arrested sexual development and small size, females often become passive and are easily intimidated by peers during their teens. Some patients have benefited from hormone therapy and corrective surgery.

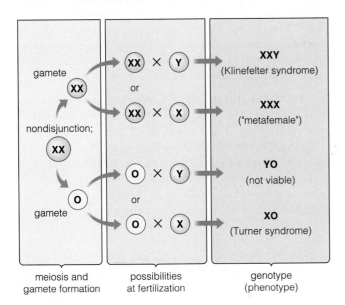

Figure 12.16 Examples of genetic disorders resulting from nondisjunction of X chromosomes followed by fertilization involving normal sperm.

Klinefelter Syndrome

About 1 in 500 to 2,000 liveborn males inherits two X chromosomes as well as one Y chromosome. About 67 percent of the time, the XXY condition results from nondisjunction in the mother, and the other 33 percent in the father.

Symptoms of the resulting *Klinefelter syndrome* do not develop until after the onset of puberty. XXY males are taller than average, and they are sterile or show low fertility. Their testes usually are much smaller than average, although the scrotum and penis are normal in size. Facial hair is often sparse, and there may be some breast enlargement. Injections of the hormone testosterone can reverse the feminized traits but not the low fertility. Some XXY males show mild mental impairment, but many fall within the normal range of intelligence. Except for their low fertility, many show no outward symptoms at all.

XYY Condition

About 1 in every 1,000 males has one X and two Y chromosomes. These XYY males tend to be taller than average. Some may be mildly retarded, but most are phenotypically normal.

At one time, XYY males were thought to be genetically predisposed to become criminals. This erroneous conclusion was based on small numbers of cases in highly selected groups, such as inmates in prisons. The investigators often knew in advance who the XYY males were, and this may have biased their evaluations. There were no double-blind studies, in which data on karyotypes and on personal histories were gathered independently by entirely different investigators, then matched up only after both sets of data were completed. Fanning the stereotype was a sensationalized report in 1968 that Richard Speck, a mass-murderer of young nurses, was an XYY male. He wasn't.

In 1976, a Danish geneticist reported on a large-scale study based on the records of 4,139 tall males, twenty-six years old, who had reported to their draft board. Besides the results of a physical examination and intelligence test, the records also provided clues to socioeconomic status, educational history, and any criminal convictions. Twelve of the males were XYY, which left more than 4,000 for the control group. The only significant finding was that tall, mentally impaired males who engage in criminal activity are more likely to get caught—irrespective of karyotype.

Nondisjunction during gamete formation accounts for most changes in chromosome number.

Prospects and Problems in Human Genetics

With the arrival of a newborn, parents typically want to know whether it is a girl or boy. Then most apprehensively ask, "Is the baby normal?" Quite naturally, they want their baby to be free of genetic disorders, and most of the time it is. Chapter 45 describes the story of human reproduction and development when all goes well. But what are the options when it does not?

We do not approach diseases and heritable disorders the same way. "Diseases" are the outcome of infection by bacteria, viruses, and other agents from the outside. We eliminate or control infectious agents with antibiotics and other weapons. By contrast, how do we attack an "enemy" within our genes?

Do we institute regional, national, or global programs to identify people carrying harmful alleles? Do we tell them they are "defective" and run a risk of bestowing their disorder on their children? Who decides which alleles are "harmful"? Should society bear the cost of treating Down syndrome and other genetic disorders? If so, should society also have a say in whether affected embryos will be born at all, or aborted? These questions are only the tip of an ethical iceberg. And answers have not been worked out in universally acceptable ways.

Phenotypic Treatments Genetic disorders cannot be cured. But often their symptoms can be suppressed or minimized. For example, this can be done by controlling diets, making environmental adjustments, or intervening surgically or with hormone therapy.

Diet control works for several genetic disorders, including galactosemia (page 196). *Phenylketonuria* (or PKU for short) is another case in point. A certain gene codes for an enzyme that converts one amino acid to another (phenylalanine to tyrosine). In people who are homozygous for a recessive mutated form of the gene, phenylalanine accumulates. If the excess is diverted to other pathways, phenylpyruvate and other compounds may be produced. A high level of phenylpyruvate can lead to mental impairment.

But suppose affected persons restrict the amount of phenylalanine in their diet. Their body does not have to dispose of excess amounts, so they can lead normal lives. Phenylketonurics are usually aware that diet soft drinks and many other products often are artificially sweetened with aspartame, which contains phenylalanine. Such products carry warning labels alerting them to this.

Environmental adjustments help counter symptoms of some disorders. Sickle-cell anemics (page 176) can avoid strenuous activity when oxygen levels are low, as at high altitudes. Albinos (page 178) can avoid direct sunlight.

Surgical reconstructions can correct or minimize many phenotypic problems. In one form of *cleft lip,* a vertical fissure cuts through the lip midsection and often extends into the roof of the mouth. Surgery usually corrects the lip's appearance and function.

Genetic Screening Some genetic disorders can be detected early enough to start preventive measures before symptoms develop. "Genetic screening" refers to large-scale programs to detect affected persons or carriers in a population. For example, most hospitals in the United States routinely screen all newborns for PKU, so it is now less common to see people with symptoms of this disorder.

Genetic Counseling Sometimes, prospective parents suspect that they are likely to produce a severely afflicted child. (Their first child or a close relative may suffer from a genetic disorder.) Parents at risk may request information from clinical psychologists, geneticists, and social workers.

Counseling starts with accurate diagnosis of parental genotypes. This may reveal the risk of a specific disorder. Biochemical tests can be used to detect many metabolic disorders. Detailed family pedigrees may be constructed to aid the diagnosis.

For disorders showing simple Mendelian inheritance, geneticists can predict the chances of having an affected child. But as this chapter made clear, not all disorders follow Mendelian patterns. Even ones that do can be influenced by other factors. Even when the risk has been defined with some confidence, prospective parents must know that the risk is the same for each pregnancy. For example, if a pregnancy has one chance in four of producing a child with a genetic disorder, the same odds apply to every subsequent pregnancy, also.

Prenatal Diagnosis Suppose a woman who is forty-five years old and pregnant wants to know if her fetus is trisomic 21 and so will develop Down syndrome. Prenatal ("before birth") diagnostic methods can detect this genetic disorder and more than a hundred others.

In **amniocentesis**, a fluid sample is drawn from fluid inside the amnion, a sac that surrounds the developing fetus (page 190). Fetal cells in the sample are cultured, then they are analyzed by karyotyping and other tests (Figure *a*). In a different procedure, **chorionic villi sampling** (CVS), cells are drawn from fluid in the chorion, a sac around the amnion.

Like amniocentesis, CVS is risky. Either procedure may accidentally cause infection or puncture the fetus. Besides this, amniocentesis is performed during the fifth month of pregnancy, when a mother already feels kicks and movements inside her. CVS can be performed earlier, between the ninth and twelfth weeks. However, even though the

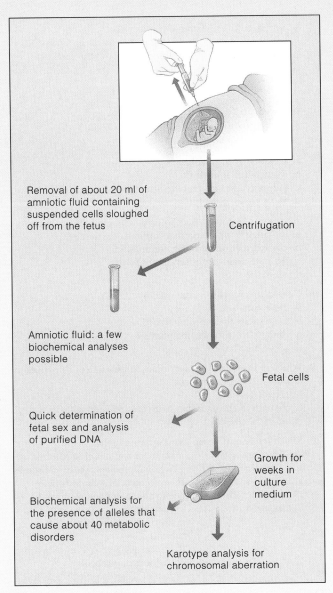

Removal of about 20 ml of amniotic fluid containing suspended cells sloughed off from the fetus

Centrifugation

Amniotic fluid: a few biochemical analyses possible

Fetal cells

Quick determination of fetal sex and analysis of purified DNA

Growth for weeks in culture medium

Biochemical analysis for the presence of alleles that cause about 40 metabolic disorders

Karotype analysis for chromosomal aberration

a Steps in amniocentesis.

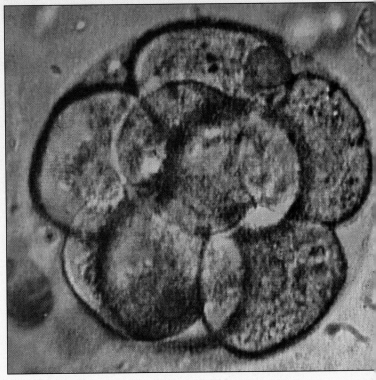

b Eight-cell stage of human development.

fetus is only about half as long as the little finger, its major organs have already started to form.

What choice do prospective parents make if either procedure reveals a devastating genetic disorder? Do they opt for induced **abortion**—an induced expulsion of the embryo from the uterus? This can be an agonizing decision. They must weigh awareness of the crushing severity of the disorder against ethical and religious beliefs. Worse, they must play out their personal tragedy on a larger stage, dominated now by a nationwide battle between fiercely vocal "pro-life" and "pro-choice" factions.

In 1992, clinical trials of preimplantation diagnosis proved successful. This new procedure relies on **in vitro fertilization**. Sperm and eggs donated by prospective parents are put in an enriched medium in a petri dish. There, one or more eggs may become fertilized. Two days later, cell divisions convert each fertilized egg into a ball of eight cells (Figure *b*). The tiny ball might be considered a prepregnancy stage. Like unfertilized eggs flushed monthly from a woman's body, it is free-floating; it is not connected to the uterus.

All cells in that ball have the same genes, and they are not differentiated. They are not yet committed to giving rise to specialized cells of the heart, toes, and other tissues. Researchers pluck one of the cells from each ball and analyze its genes for suspected disorders. Only a ball that is free of genetic defects is inserted back into the uterus.

At this writing, several couples who are at risk of passing on muscular dystrophy, cystic fibrosis, and other severe disorders have opted for the procedure. The procedure is still highly experimental, and it is costly. Yet a number of "test-tube" babies have been born. All are in good health—and free of the harmful genes.

SUMMARY

1. Genes, the units of instruction for heritable traits, are arranged one after the other along chromosomes.

2. Human cells are diploid (2n), with twenty-three pairs of homologous chromosomes that interact during meiosis. Except for X and Y chromosomes, each pair has the same length, shape, and gene sequence.

3. Human females have a pair of X chromosomes. Males have X paired with Y. All other pairs of chromosomes are autosomes (the same in both females and males). A gene on the Y chromosome determines gender.

4. Clues to inheritance often show up in pedigrees (charts of genetic connections through lines of descent). Certain patterns are characteristic of dominant or recessive alleles on autosomes or on the X chromosome.

5. Genes on the same chromosome represent a linkage group. Crossing over (the breakage and exchange of segments between homologues) disrupts linkages. The farther apart two genes are, the greater will be the frequency of crossovers between them.

6. A chromosome's structure may change on rare occasions. A segment may be deleted, inverted, moved to a new location, or duplicated.

7. The chromosome number may change on rare occasions. New individuals may end up with one more or one less chromosome than the parents (aneuploidy). Or they may end up with three or more of each type of chromosome (polyploidy). Nondisjunction during meiosis and gamete formation accounts for most of these chromosome abnormalities.

8. Crossing over and changes in chromosome number or in a chromosome's structure may influence the course of evolution. The changes in genotype (genetic makeup) lead to variations in phenotype (observable traits) among members of a population, so that evolution is possible.

Self-Quiz (Answers in Appendix IV)

1. _____ segregate during _____ .
   a. Homologues; mitosis
   b. Genes on one chromosome; meiosis
   c. Homologues; meiosis
   d. Genes on one chromosome; mitosis

2. The genes of one chromosome and the genes of the homologous chromosome end up in separate _____ .
   a. body cells
   b. gametes
   c. nonhomologous chromosomes
   d. offspring
   e. both b and d are possible

3. Genes on the same chromosome tend to remain together during _____ and end up in the same _____ .
   a. mitosis; body cell
   b. mitosis; gamete
   c. meiosis; body cell
   d. meiosis; gamete
   e. both a and d

4. The probability of a crossover occurring between two genes on the same chromosome is _____ .
   a. unrelated to the distance between them
   b. increased if they are closer together on the chromosome
   c. increased if they are farther apart on the chromosome
   d. impossible

5. Chromosome structure can be altered by _____ .
   a. deletions
   b. duplications
   c. inversions
   d. translocations
   e. all of the above

6. Nondisjunction can be caused by _____ .
   a. crossing over in meiosis
   b. segregation in meiosis
   c. failure of chromosomes to separate during meiosis
   d. multiple independent assortments

7. A gamete affected by nondisjunction would have _____ .
   a. a change from the normal chromosome number
   b. one extra or one missing chromosome
   c. the potential for a genetic disorder
   d. all of the above

8. Genetic disorders can be caused by _____ .
   a. gene mutations
   b. changes in chromosome structure
   c. changes in chromosome number
   d. all of the above

9. Which of the following contributes to variation in a population?
   a. independent assortment
   b. crossing over
   c. changes in chromosome structure and number
   d. all of the above

10. Match the chromosome terms appropriately.
    ____ crossing over
    ____ deletion
    ____ nondisjunction
    ____ translocation
    ____ karyotype
    a. number and defining features of individual's metaphase chromosomes
    b. movement of a chromosome segment to a nonhomologous chromosome
    c. disrupts gene linkages during meiosis
    d. causes gametes to have abnormal chromosome numbers
    e. loss of a chromosome segment

Genetics Problems (Answers in Appendix III)

1. Human females are XX and males are XY.
   a. Does a male inherit his X chromosome from his mother or father?
   b. With respect to an X-linked gene, how many different types of gametes can a male produce?
   c. If a female is homozygous for an X-linked gene, how many different types of gametes can she produce with respect to that gene?
   d. If a female is heterozygous for an X-linked gene, how many different types of gametes can she produce with respect to that gene?

2. One allele of a presumed Y-linked gene results in nonhairy ears in males (*see photograph*). Another allele of the same gene results in *hairy pinnae* (relatively long hairs at the outer ear).

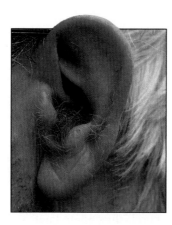

    a. Why would you *not* expect females to have hairy pinnae?

    b. A hairy-eared male's son will have hairy ears, but no daughter will. Explain why.

3. Suppose that you have two linked genes with alleles *A,a* and *B,b* respectively. An individual is heterozygous for both genes, as in the following:

$$
\begin{array}{cc}
A & B \\
A & B \\
\\
a & b \\
a & b
\end{array}
$$

If the crossover frequency between these two genes is 0 percent, what genotypes would be expected among gametes from this individual, and with what frequencies?

4. *Linkage maps* are plots of the relative distance between genes that undergo crossing over and other chromosomal rearrangements. They are based on the patterns in which such genes are distributed into gametes. They don't show *exact* physical distances between genes, because the probability of crossing over isn't equal along a chromosome's length. The distances are measured in map units that are based on the frequency of recombination between genes. One genetic map unit equals 1 percent recombination. Thus, for example, if the crossover frequency for two genes is 10 percent, then ten map units separate them.

    Consider two linked genes on a *D. melanogaster* chromosome. One gene, which influences eye color, has a red (dominant) allele and a purple (recessive) allele. Another gene dictates wing length. A dominant allele at this locus codes for long wings; a recessive allele codes for vestigial (short) wings. Suppose a fully homozygous dominant female having red eyes and long wings mates with

a male having purple eyes and vestigial wings. Suppose next that the F_1 females mate with males having purple eyes and vestigial wings. The second cross results in offspring with the following traits:

252 red eyes, long wings
276 purple eyes, vestigial wings
 42 red eyes, vestigial wings
 30 purple eyes, long wings

600 offspring total

Based on these data, how many map units separate the two linked genes?

5. Suppose you cross a homozygous dominant long-winged fruit fly with a homozygous recessive vestigial-winged fly. Shortly after mating, the fertilized eggs are exposed to a level of x-rays known to cause mutation and chromosomal deletions. When these fertilized eggs subsequently develop into adults, most of the flies are long-winged and heterozygous. However, a few are vestigial-winged. Provide possible explanations for the unexpected appearance of these vestigial-winged adults.

6. Individuals affected by Down syndrome typically have an extra chromosome 21, so their cells have a total of 47 chromosomes. However, in a few cases of Down syndrome, 46 chromosomes are present. Included in this total are two normal-appearing chromosomes 21 and a longer-than-normal chromosome 14. Interpret this observation and indicate how these few individuals can have a normal chromosome number.

7. The mugwump, a type of tree-dwelling mammal, has a reversed sex-chromosome condition. The male is XX and the female is XY. However, perfectly good sex-linked genes are found to have the same effect as in humans. For example, a recessive, X-linked allele *c* produces red-green color blindness. If a normal female mugwump mates with a phenotypically normal male mugwump whose mother was color blind, what is the probability that a son from that mating will be color blind? A daughter?

8. One type of childhood muscular dystrophy is a recessive, X-linked trait in humans. A slowly progressing loss of muscle function leads to death, usually by age twenty or so. Unlike color blindness, this disorder is restricted to males, not ever having been found in a female. Suggest why.

Readings

Cummings, M. 1994. *Human Heredity: Principles and Issues*. Third edition. St. Paul, Minnesota: West.

Edlin, G. 1988. *Genetic Principles: Human and Social Consequences*. Second edition. Portola Valley, California: Jones & Bartlett.

Fuhrmann, W., and F. Vogel. 1986. *Genetic Counseling*. Third edition. New York: Springer-Verlag.

Holden, C. 1987. "The Genetics of Personality." *Science* 237:598–601. For students interested in human behavioral genetics.

13 DNA STRUCTURE AND FUNCTION

Cardboard Atoms and Bent-Wire Bonds

Linus Pauling in 1951 did something no one had done before. Through his training in biochemistry, a talent for model building, and a few educated guesses, he deduced the three-dimensional structure of a protein (collagen). His discovery electrified the scientific community. If the secrets of proteins could be pried open, why not other biological molecules? Further, wouldn't structural details provide clues to molecular functions? And who would go down in history as having discovered the molecule that contains instructions for reproducing parental traits in offspring—*the very secrets of inheritance*? Scientists around the world started scrambling after that ultimate prize.

Maybe hereditary instructions were encoded in the structure of some unknown class of proteins. After all, heritable traits are spectacularly diverse. Surely the molecules encoding information about those traits were structurally diverse also. Proteins are put together from potentially limitless combinations of amino acid subunits, so they almost certainly could function as the sentences (genes) in each cell's book of inheritance.

Yet there was something about another substance—DNA—that excited many researchers. Among them were James Watson, a young postdoctoral student from Indiana University, and Francis Crick, an energetic researcher at Cambridge University. How could DNA, a molecule consisting of only four kinds of subunits, hold genetic information? Watson and Crick spent long hours arguing over everything they had read about the size, shape, and bonding requirements of the subunits of DNA. They fiddled with cardboard cutouts of the subunits. They badgered chemists to identify potential bonds they might have overlooked. They assembled models from bits of metal held together with wire "bonds" bent at appropriate angles.

In 1953, they put together a model that fit all the pertinent biochemical rules and all the facts about DNA they had gleaned from other sources (Figure 13.1). They had discovered the structure of DNA. The breathtaking simplicity of that structure also enabled them to solve a long-standing riddle—*how life can show unity at the molecular level and yet give rise to so much diversity at the level of whole organisms.*

With this chapter, we turn to investigations that led to our current understanding of DNA structure and function. They are revealing of how ideas are generated in science. On the one hand, having a shot at fame and fortune quickens the pulse of men and women in any profession, and scientists are no exception. On the other hand, science proceeds as a community effort, with individuals sharing not only what they can explain but also what they do not understand. Even if an experiment "fails," it may turn up information that others can use or lead to questions that others can answer. Unexpected results, too, might be clues to something important about the natural world.

Figure 13.1 James Watson and Francis Crick posing in 1953 by their newly unveiled model of the structure of DNA. The photograph on the facing page is a more recent model. This computer-generated model is more sophisticated in appearance, yet it is basically the same as the prototype that Watson and Crick put together nearly four decades before.

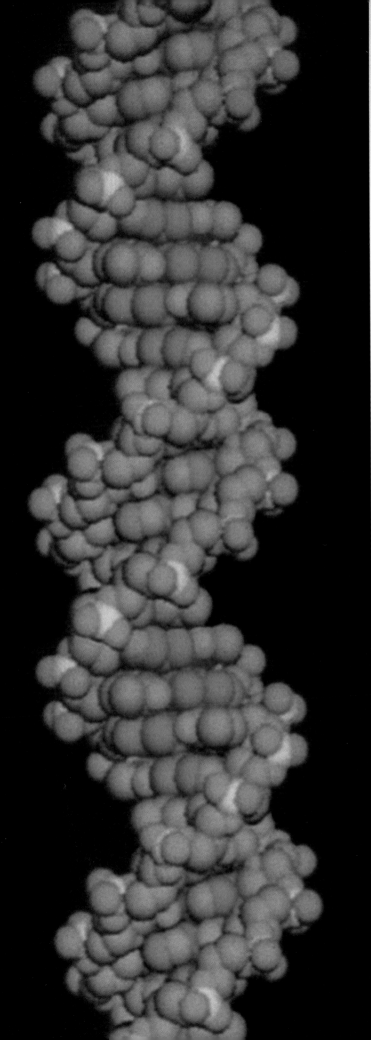

KEY CONCEPTS

1. In living cells, DNA is the storehouse of information about heritable traits. The hereditary information is encoded in the nucleotide subunits that make up the DNA molecule.

2. DNA consists of four kinds of nucleotides that differ in only one component, a nitrogen-containing base. The four bases are adenine, guanine, thymine, and cytosine. Great numbers of these occur one after another in a DNA molecule, and the order in which one kind follows another is different for each species of organism.

3. In a DNA molecule, two strands of nucleotides twist together like a spiral stairway; they form a double helix. Hydrogen bonds connect the bases of one strand to bases of the other. As a rule, adenine pairs (hydrogen-bonds) only with thymine, and guanine only with cytosine.

4. Before a cell divides, its DNA is replicated with the help of enzymes and other proteins. Each double-stranded DNA molecule starts unwinding. A new, complementary strand is assembled on the exposed bases of each parent strand according to the base-pairing rule.

5. There is only one DNA molecule in an unduplicated chromosome. Except in bacteria, great numbers of proteins are attached to the DNA and function in its structural organization.

One might have wondered, in the spring of 1868, why Johann Friedrich Miescher was collecting cells from the pus of open wounds and, later, from the sperm of a fish. Miescher, a physician, wanted to identify the chemical composition of the nucleus. He was interested in those cells because they are composed mostly of nuclear material, with very little cytoplasm. Miescher succeeded in isolating an acidic substance, one with a notable amount of phosphorus. He called it "nuclein." He had discovered what came to be known as deoxyribonucleic acid, or **DNA**.

The discovery caused scarcely a ripple through the scientific community. At the time, no one knew much about the physical basis of inheritance—that is, *which* chemical substance in cells actually encodes the instructions for reproducing parental traits in offspring. Only a few researchers suspected that the nucleus might hold the answer. In fact, seventy-five years passed before DNA was recognized as having profound biological importance.

13.1 DISCOVERY OF DNA FUNCTION

Early Clues

In 1928, an Army medical officer, Fred Griffith, was attempting to create a vaccine against *Streptococcus pneumoniae*, which causes a type of pneumonia. Many vaccines are preparations of killed or weakened bacterial cells that, when introduced into the body, can mobilize defenses against a real attack. Griffith never did create a vaccine, but his work unexpectedly opened a door to the molecular world of heredity.

Griffith isolated and cultured two strains of the bacterium. He noticed that colonies of one strain had a rough surface appearance and those of the other strain appeared smooth. He designated the strains *R* and *S* and used them in a series of four experiments:

1. Laboratory mice were injected with live R cells. As Figure 13.2 indicates, they did not develop pneumonia. *The R strain was harmless.*

2. Mice were injected with live S cells. The mice died. Blood samples from them teemed with live S cells. *The S strain was pathogenic* (disease-causing).

3. S cells were killed by exposure to high temperature. Mice injected with these cells did not die.

4. Live R cells were mixed with heat-killed S cells and injected into mice. The mice died—and blood samples from them teemed *with* live S cells!

What was going on in the fourth experiment? Maybe heat-killed S cells in the mixture weren't really dead. But if that were true, then mice injected with heat-killed S cells alone (experiment 3) would have died. Maybe harmless R cells in the mixture had mutated into a killer form. But if that were true, then mice injected with the R cells alone (experiment 1) would have died.

The simplest explanation was this: *Heat did kill the S cells but did not destroy their hereditary material, including the part that specified "how to cause infection."* Somehow, that material had been transferred from dead S cells to living R cells—where it was put to use.

Further experiments showed that harmless cells had indeed picked up information about infection and had been permanently transformed into pathogens. Hundreds of generations of bacteria descended from the transformed cells also caused infections!

Griffith's unexpected results intrigued a microbiologist, Oswald Avery, and his colleagues. They were able to transform harmless cells with extracts of killed pathogens. In 1944, they reported that the hereditary substance in the extracts was probably DNA, not pro-

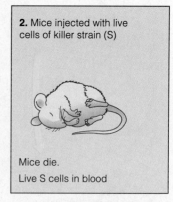
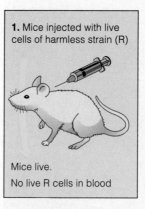
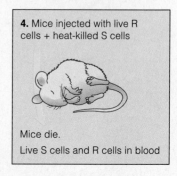
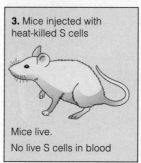

1. Mice injected with live cells of harmless strain (R)

Mice live.
No live R cells in blood

2. Mice injected with live cells of killer strain (S)

Mice die.
Live S cells in blood

3. Mice injected with heat-killed S cells

Mice live.
No live S cells in blood

4. Mice injected with live R cells + heat-killed S cells

Mice die.
Live S cells and R cells in blood

Figure 13.2 Summary of results from Griffith's experiments with a harmless strain and a disease-causing strain of *Streptococcus pneumoniae*, as described in the text.

teins, as was widely believed. They had added protein-digesting enzymes to some extracts, but cells were transformed anyway. Then they added an enzyme that breaks apart DNA but not proteins—and that enzyme blocked hereditary transformation.

Yet how were Avery's impressive findings received? Many (if not most) biochemists refused to give up on the proteins. His experimental results, they said, probably applied only to bacteria.

Confirmation of DNA Function

By the early 1950s, Max Delbrück, Alfred Hershey, Salvador Luria, and other molecular detectives were using certain viruses as their experimental subjects. The viruses, called **bacteriophages**, infect bacterial cells such as *Escherichia coli*. Viruses are about as biochemically simple as you can get. They are not alive, but they do contain hereditary information for building new virus particles. At some point after they infect a host cell, viral enzymes take over a portion of the cell's metabolic machinery—which starts churning out the substances required to construct new virus particles.

By 1952, researchers knew that some bacteriophages consist only of DNA and protein. Also, electron micrographs revealed that the main part of these viruses

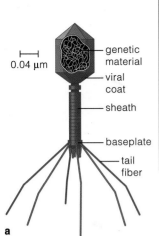

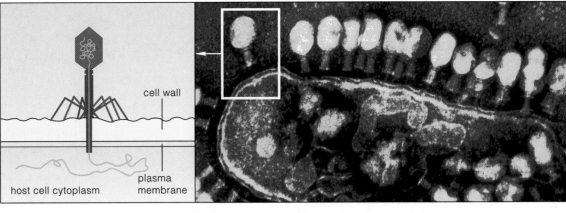

b

Figure 13.3 (**a**) Structural organization of a T4 bacteriophage. (**b**) Electron micrograph of T4 virus particles infecting a host bacterial cell (*Escherichia coli*). The diagram shows DNA, the genetic material of this bacteriophage, being injected into the cell cytoplasm.

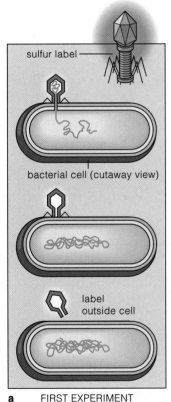

a FIRST EXPERIMENT

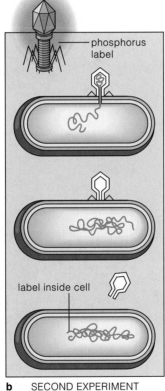

b SECOND EXPERIMENT

Figure 13.4 Two examples of the landmark experiments pointing to DNA as the substance of heredity. In the 1940s, Alfred Hershey and his colleague, Martha Chase, were studying the biochemical basis of inheritance. They were aware that certain bacteriophages were composed of proteins and DNA. Did the proteins, DNA, or both contain the viral genetic information?

To find out, Hershey and Chase designed two experiments, based on the following biochemical facts: First, proteins incorporate sulfur (S) but not phosphorus (P). Second, DNA incorporates phosphorus but not sulfur.

(**a**) In one experiment, bacterial cells were grown on a culture medium that included the radioisotope ^{35}S and no other form of sulfur. To synthesize proteins, the bacterial cells would have to use the radioisotope—which would serve as a tracer. After the cells became labeled with the tracer, bacteriophages were allowed to infect them. Subsequently, viral proteins were synthesized inside the host cells. These proteins also became labeled with ^{35}S. So did the new generation of virus particles.

Next, the labeled bacteriophages were allowed to infect a new batch of unlabeled bacteria that were suspended in a fluid culture medium. Afterward, Hershey and Chase whirred the fluid in a kitchen blender. Whirring dislodged the infectious particles from the cells. The particles became suspended in the fluid medium. Analysis revealed the presence of labeled protein in the fluid—*not* inside the bacterial cells.

(**b**) In the second experiment, new bacterial cells were cultured. The only phosphorus available for synthesizing DNA was the radioisotope ^{32}P. Bacteriophages were allowed to infect the cells. As predicted, the viral DNA assembled inside the infected cells became labeled, as did the new generation of virus particles. Next, the labeled particles were allowed to infect bacteria in a fluid medium, and then were dislodged from the host cells. Analysis revealed that the labeled DNA was not in the fluid. It remained *inside* the host cells, where its instructions had to be put to use. Here was evidence that DNA is the genetic material of these bacteriophages.

remains *outside* of the cells they are infecting (Figure 13.3). Such viruses probably were injecting genetic material alone *into* host cells. If that were true, was the material DNA, protein, or both?

Through many experiments, researchers came up with convincing evidence that DNA, not proteins, functions as the molecule of heredity. Figure 13.4 describes two of these landmark experiments.

Information for producing the heritable traits of single-celled and multicelled organisms is encoded in DNA.

DNA STRUCTURE

Components of DNA

Long before the bacteriophage studies were under way, biochemists knew that DNA contains only four types of nucleotides, the building blocks of nucleic acids. A **nucleotide** consists of a five-carbon sugar (which is deoxyribose in DNA), a phosphate group, and one of the following nitrogen-containing bases:

adenine	guanine	thymine	cytosine
(A)	**(G)**	**(T)**	**(C)**

As you can see from Figure 13.5, all four types of nucleotides in DNA have these components joined together in much the same way. But T and C are pyrimidines, which are single-ring structures. A and G are purines, which are double-ring structures. These are larger, bulkier molecules.

By 1949, the biochemist Erwin Chargaff had shared two crucial insights into the composition of DNA with the scientific community. First, the amount of adenine relative to guanine differs from one species to the next. Second, the amount of adenine always equals the amount of thymine, and the amount of guanine always equals the amount of cytosine. We may show this as:

$$A = T \quad \text{and} \quad G = C$$

The relative proportions of the four different nucleotides were tantalizing clues to their arrangement in DNA. In some way, the four had to follow one after another along the length of the molecule.

The first convincing evidence of that arrangement came from Maurice Wilkins's laboratory in England. One of Wilkins's coworkers, Rosalind Franklin, had obtained good **x-ray diffraction images** of DNA fibers.

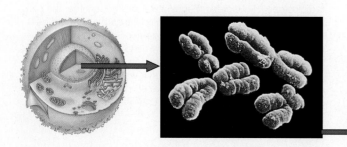

Figure 13.5 The four kinds of nucleotide subunits of DNA. Small numerals on the structural formulas identify the carbon atoms to which other parts of the molecule are attached.

ADENINE (A) — double-ring structure

GUANINE (G) — double-ring structure

THYMINE (T) — single-ring structure

CYTOSINE (C) — single-ring structure

All chromosomes contain DNA. What does DNA contain? Four kinds of nucleotides. Each nucleotide has a five-carbon sugar (shaded red). That sugar has a phosphate group attached to the fifth carbon atom of its ring structure. It also has one of four kinds of nitrogen-containing bases (shaded blue) attached to its first carbon atom. The nucleotides differ only in which base is attached to that atom.

DNA does not readily lend itself to x-ray diffraction. However, the researchers rapidly spun a suspension of DNA molecules, spooled them onto a rod, and gently pulled them into gossamer fibers, like cotton candy. If the atoms in DNA were arranged in a regular order, an x-ray beam directed at a fiber would scatter in a regular pattern that could be captured on film. The pattern would consist only of dots and streaks. It would not, in itself, reveal molecular structure. But it could be used to calculate the positions of atoms in DNA.

According to calculations in Wilkins's laboratory, DNA had to be long and thin, with a 2-nanometer diameter. Some molecular configuration was being repeated every 0.34 nanometer along its length, and another one, every 3.4 nanometers.

Could the sequence of nucleotide bases be twisting, like a circular stairway? Certainly Pauling thought about this possibility. After all, he was the one who discovered that a collagen chain has a helical shape, held in place by hydrogen bonds. He wrote to Wilkins, requesting copies of the x-ray diffraction images. The response was lukewarm, perhaps a delaying tactic. Why should Wilkins help a formidable competitor win the race to glory?

As it turned out, neither won the race. Watson and Crick were about to close in on the answer.

Patterns of Base Pairing

As Watson and Crick perceived, DNA consists of *two* strands of nucleotides, held together at their bases by hydrogen bonds. These bonds form when the two strands run in opposing directions and twist together into a double helix (Figure 13.6). Only two kinds of base pairings form along the entire length of the molecule: A–T and G–C.

This bonding pattern permits variation in the order of bases in any given strand. For example, in even a tiny stretch of DNA from a rose, gorilla, human, or any other organism, the sequence might be:

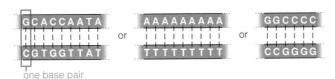

one base pair

And so DNA molecules show constancy and variation from one species to the next. This is the molecular foundation for the unity and diversity of life.

Base pairing between the two nucleotide strands in DNA is *constant* for all species (A with T, and G with C).

The base sequence (that is, which base follows another in a nucleotide strand) is *different* from species to species.

Figure 13.6 Representations of a DNA double helix. Notice how the two sugar-phosphate backbones run in *opposing* directions. (Think of the ribose units of one strand as being upside down.) By comparing the numerals used to identify each carbon atom of the ribose molecule (1', 2', 3', and so on), you see that one strand runs in the 5'→ 3' direction and the other runs in the 3'→ 5' direction.

2-nanometer diameter, overall

distance between each pair of bases = 0.34 nanometer

In all these respects, the Watson-Crick model of DNA structure is consistent with the known biochemical and x-ray diffraction data.

each full twist of the DNA double helix = 3.4 nanometers

The pattern of base pairing (A only with T, and G only with C) is consistent with the known composition of DNA (A = T, and G = C)

13.3 DNA REPLICATION AND DNA REPAIR

The discovery of DNA structure was a turning point in studies of inheritance. Until then, no one could explain **DNA replication**—that is, how hereditary material is duplicated prior to cell division. The Watson-Crick model suggested at once how this might be done.

Enzymes can readily break hydrogen bonds between the two nucleotide strands of a DNA molecule. When they act at a given site, one strand unwinds from the other, thereby exposing some nucleotide bases. Cells have stockpiles of free nucleotides, and these pair with exposed bases. Each parent strand remains intact, and a companion strand is assembled on each one, according to the base-pairing rule (A to T, and G to C). As soon as a stretch of the parent strand has its new, partner strand, the two twist into a double helix.

Because the parent strand is conserved, each "new" DNA molecule is really half old, half new (Figure 13.7). That is why biologists sometimes refer to this process as *semiconservative* replication.

unwinding enzyme ——

binding proteins (stabilize the DNA in single-strand form) ——

primer-synthesizing enzyme ——

replication fork

a Replication of eukaryotic DNA begins at many short, specific base sequences called *origins*. Organized complexes of enzymes and other proteins unwind the two strands of the parent DNA molecule, prevent rewinding, and assemble new strands on each one. They do this only in *replication forks*. These are limited, V-shaped regions that advance in both directions away from an origin. Enzymes rewind the half-old, half-new molecules while they are being completed, behind each advancing replication fork.

origin

DNA replication requires a large team of molecular workers. For example, as Figure 13.8 indicates, one kind of enzyme unwinds the two nucleotide strands, and many proteins bind to them and hold them apart. **DNA polymerases** are key players; they attach free nucleotides to a growing strand. **DNA ligases** seal new short stretches of nucleotides into one continuous strand. Chapter 16 will describe how some of these enzymes have uses in recombinant DNA technology.

DNA polymerases, DNA ligases, and other enzymes also engage in a process called **DNA repair**. If the sequence of bases in one strand of a double helix becomes altered, DNA polymerases "read" the complementary sequence on the other strand. With the aid of other repair enzymes, they restore the original sequence. Figure 13.9 gives an example of what can happen when the excision-repair function is impaired.

Prior to cell division, the double-stranded DNA molecule unwinds and is replicated. Each parent strand remains intact—it is conserved—and enzymes assemble a new, complementary strand on each one.

Certain enzymes also correct base-pairing errors in the nucleotide sequence of DNA.

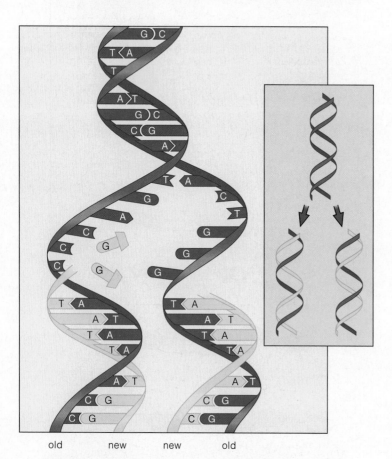

old　　　new　　　new　　　old

Figure 13.7 Semiconservative nature of DNA replication. The original two-stranded DNA molecule is shown in *blue*. Each parent strand remains intact, and a new strand (*yellow*) is assembled on each one.

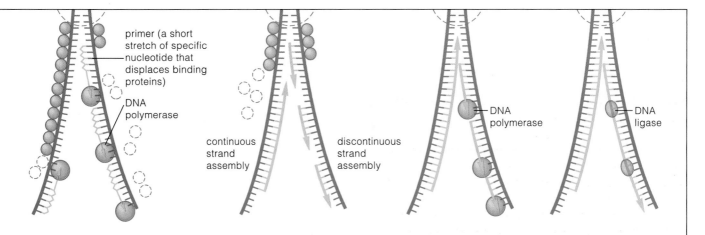

b At each replication fork, some binding proteins are displaced when enzymes synthesize primers at intervals along both of the parent templates. Other enzymes (DNA polymerases) recognize these primers as "start" tags. They join nucleotide units together behind primers. Exposed bases of the parent templates dictate which kind of nucleotide is added in sequence, according to the base-pairing rule.

c As discovered by Reiji Okazaki, strand assembly is *continuous* on one parent template. It is *discontinuous* on the other, where nucleotides must be assembled in short stretches only. Why? They can be joined together in the 5′ → 3′ direction only, because this leaves one of the −OH groups of the sugar-phosphate backbone exposed (**f**). This exposed group is the only site where nucleotide units can be joined together.

d Primers are removed. DNA polymerases fill in the gaps but cannot make the final connections between stretches of nucleotides.

e DNA ligases seal the tiny nicks that remain, the result being continuous strands on the parent templates.

Figure 13.8 A closer look at the functions of enzymes and other proteins in DNA replication. The free nucleotides brought up for strand assembly also provide energy to drive the process. Each has three phosphate groups attached. DNA polymerase splits away two of these and uses some of the released energy to attach the nucleotide to a growing strand.

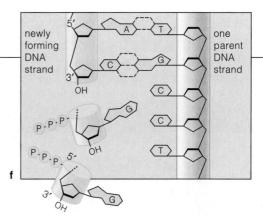

Figure 13.9 When DNA can't be fixed. A newborn destined to develop *xeroderma pigmentosum* faces a dim future, literally. Individuals affected by this genetic disorder cannot be exposed to sunlight, even briefly, without risking disfiguring skin tumors and possible early death from cancer. The disorder is a result of faulty DNA repair in the body's cells. One or more of the genes thought to govern repair processes have become damaged.

The ultraviolet wavelengths in light from the sun, tanning lamps, and other sources can cause molecular changes in DNA (page 635). Among other things, the wavelengths can promote covalent bonding between two adjacent thymine bases in a nucleotide strand. Thus the two nucleotides to which the bases belong are combined into an abnormal, bulky molecule called a "thymine dimer."

Normally, a DNA repair mechanism gets rid of such bulky lesions. The mechanism requires at least seven gene products. Mutation in one or more of the required genes can skew the repair machinery. Thymine dimers can accumulate in skin cells of individuals with such mutations. The accumulation triggers development of lesions, including skin cancer. The photograph shows a common skin cancer, called *basal cell carcinoma*.

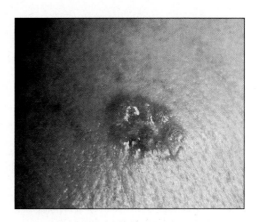

basal cell carcinoma

13.4 ORGANIZATION OF DNA IN CHROMOSOMES

Each chromosome has one DNA molecule. If the DNA of all forty-six chromosomes in one of your somatic cells were stretched out end to end, the string might extend from your shoulder to your fingertips. What keeps all that DNA from becoming a tangled mess? Proteins.

Eukaryotic DNA has many **histones** and other protein molecules bound tightly to it (Figure 13.10). Some histones are like spools for winding up small stretches of DNA. Each histone-DNA spool is a **nucleosome**. Another histone stabilizes the spools.

Histone-DNA interactions can make a chromosome coil back on itself again and again. The coiling greatly increases its diameter. Further folding results in a series of loops. Proteins other than histones serve as a structural "scaffold" for the loops.

Apparently, scaffold regions intervene between the genes, not in regions that contain information for building proteins. Are the scaffold proteins organizing the chromosome into functional "domains"? Would that organization make it easier for DNA replication and protein synthesis to proceed? These are just two of the possibilities being probed by the new generation of molecular detectives.

Figure 13.10 Levels of organization of DNA in a eukaryotic chromosome.

SUMMARY

1. Hereditary information of cells and multicelled organisms is encoded in DNA (deoxyribonucleic acid).

2. DNA is composed of nucleotide subunits. Each of these has a five-carbon sugar (deoxyribose), a phosphate group, and one of four nitrogen-containing bases (adenine, thymine, guanine, or cytosine).

3. A DNA molecule consists of two nucleotide strands twisted together into a double helix. The bases of one strand pair (hydrogen-bond) with bases of the other.

4. The bases of the two strands in a DNA double helix pair in constant fashion. Adenine always pairs with thymine (A to T), and guanine with cytosine (G to C). *Which* base pair follows the next (A–T, T–A, G–C, or C–G) varies along the length of the strands.

5. Overall, the DNA of one species includes a number of unique stretches of base pairs that set it apart from the DNA of all other species.

6. During DNA replication, enzymes unwind the two strands of a double helix and assemble a new strand of complementary sequence on each one. Two double-stranded molecules result. One strand of each molecule is "old" (it is conserved) and the other is "new."

7. Many histones and other proteins are tightly bound to eukaryotic DNA. The structural organization of chromosomes arises from DNA-protein interactions.

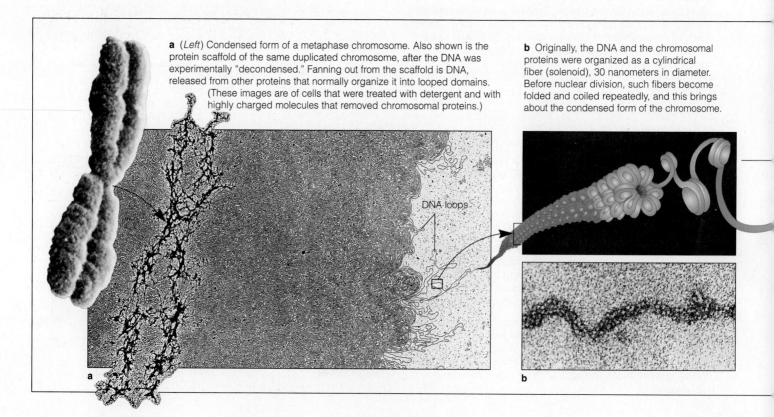

a (*Left*) Condensed form of a metaphase chromosome. Also shown is the protein scaffold of the same duplicated chromosome, after the DNA was experimentally "decondensed." Fanning out from the scaffold is DNA, released from other proteins that normally organize it into looped domains. (These images are of cells that were treated with detergent and with highly charged molecules that removed chromosomal proteins.)

b Originally, the DNA and the chromosomal proteins were organized as a cylindrical fiber (solenoid), 30 nanometers in diameter. Before nuclear division, such fibers become folded and coiled repeatedly, and this brings about the condensed form of the chromosome.

DNA loops

Review Questions

1. Name the three molecular parts of a nucleotide in DNA. Name the four different kinds of nitrogen-containing bases that occur in the nucleotides of DNA. *212*

2. What kind of bond holds two DNA chains together in a double helix? Which nucleotide base-pairs with adenine? Which pairs with guanine? *212, 213*

3. The four bases in DNA may differ greatly in relative amounts from one species to the next—yet the relative amounts are always the same among members of a single species. How does base pairing explain these twin properties—the unity and diversity—of DNA molecules? *213*

4. When regions of a double helix are unwound during DNA replication, do the two unwound strands join back together again after a new DNA molecule has formed? *214, 215*

Self-Quiz *(Answers in Appendix IV)*

1. Which is *not* a nucleotide base in DNA?
   a. adenine   c. uracil   e. guanine
   b. thymine   d. cytosine

2. What are the base-pairing rules for DNA?
   a. A–G, T–C   c. A–U, C–G
   b. A–C, T–G   d. A–T, C–G

3. A DNA strand with the sequence C–G–A–T–T–G would be complementary to the sequence _____ .
   a. C–G–A–T–T–G   c. T–A–G–C–C–T
   b. G–C–T–A–A–G   d. G–C–T–A–A–C

4. One species' DNA differs from others in its _____ .
   a. sugars   c. base sequence
   b. phosphate groups d. all of the above

5. When DNA replication begins, _____ .
   a. the two DNA strands start to unwind from each other
   b. the two DNA strands condense for base transfers
   c. two DNA molecules bond
   d. old strands move to find new strands

6. DNA replication results in _____ .
   a. two half-old, half-new molecules
   b. two molecules, one with old strands and one with newly assembled strands
   c. three double-stranded molecules, one with new strands and two that are discarded
   d. none of the above

7. DNA replication requires _____ .
   a. a supply of free   c. many enzymes and other
      nucleotides   proteins
   b. new hydrogen bonds   d. all of the above

8. Match the DNA concepts appropriately.
   ____ base sequence variation   a. two nucleotide strands
   ____ metaphase chromosome   twisted together
   ____ constancy in   b. A = T, G = C
        base pairing   c. duplication of hereditary
   ____ replication   material
   ____ DNA double   d. accounts for diversity
        helix   among species
                                   e. condensed DNA with
                                      protein scaffold

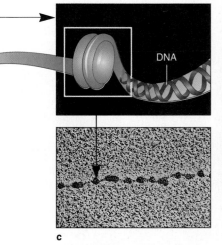

c A chromosome immersed in a salt solution loosens up to a beads-on-a-string organization. The "string" is DNA. Each "bead" is a nucleosome (**d**). Short stretches of the chromosome may look like this when DNA is being replicated or when a gene's instructions are being read. Each nucleosome consists of a double loop of DNA around a core of proteins (eight histone molecules). Another histone (H1) stabilizes the arrangement.

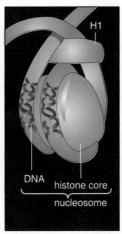

c   d

Selected Key Terms

bacteriophage *210*   DNA replication *214*
DNA *209*   histone *216*
DNA ligase *214*   nucleosome *216*
DNA polymerase *214*   nucleotide *212*
DNA repair *214*   x-ray diffraction image *212*

Readings

Cairns, J., G. Stent, and J. Watson, eds. 1966. *Phage and the Origins of Molecular Biology.* Cold Spring Harbor, New York: Cold Spring Harbor Laboratories. Gives a sense of the insights, wit, humility, and personalities of the founders of and converts to molecular genetics.

Radman, M., and R. Wagner. August 1988. "The High Fidelity of DNA Duplication." *Scientific American* 259(2): 40–46.

Watson, J. 1978. *The Double Helix.* New York: Atheneum. Highly personal view of scientists and their methods, interwoven into an account of how DNA structure was discovered.

Wolfe, S. 1993. *Molecular and Cellular Biology.* Belmont, California: Wadsworth. Comprehensive, current, and accessible.

14 FROM DNA TO PROTEINS

Beyond Byssus

Picture a mussel, of the sort shown in Figure 14.1. Hard-shelled but soft of body, it is using its muscular foot to probe a wave-scoured rock. At any moment, pounding waves can whack the mussel into the water, hurl it repeatedly against the rock with shell-shattering force, and so offer up a gooey lunch for gulls.

By chance, the mussel's foot comes across a crevice in the rock. The foot moves, broomlike, and sweeps the crevice clean. It presses down, forcing air out from under it, then arches up. The result is a vacuum-sealed chamber, rather like the one that forms when a plumber's rubber plunger is being squished down and up to unclog a drain. Into this vacuum chamber the mussel spews a fluid composed of keratin and other proteins, which bubbles into a sticky foam. And now, by curling its foot into a small tubular shape and pump-ing the foam through it, the mussel forms sticky threads about as wide as a human whisker. As a final touch, it varnishes the threads with another protein and thereby ends up with an adhesive called byssus—which anchors the mussel to the rock.

Byssus is the world's premier underwater adhesive. Nothing that humans have manufactured comes close. (Sooner or later, water chemically degrades or deforms synthetic adhesives.) Byssus fascinates biochemists, dentists, and surgeons looking for better ways to do tissue grafts and to rejoin severed nerves. Genetic engi-neers are inserting mussel DNA into yeast cells, which reproduce in large numbers and serve as "factories" for translating mussel genes into useful amounts of pro-teins. This exciting work, like the mussel's own byssus building, starts with one of life's universal concepts.

Every protein is synthesized according to instructions contained in DNA.

You are about to trace the steps leading from DNA to protein. Many enzymes are players in this pathway, as are molecules of RNA. The same steps produce *all* of the world's proteins, from mussel-inspired adhesives to the keratin in your fingernails to the insect-digesting enzymes of a Venus flytrap.

Figure 14.1 Mussels, busily demonstrating the importance of proteins for survival. When mussels come across a suitable anchoring site, they use their foot like a plumber's plunger to create a vacuum chamber. In this chamber they manufacture the world's best underwater adhesive from a mix of proteins. The adhesive anchors them to substrates in their wave-swept habitat.

KEY CONCEPTS

1. Life cannot exist without enzymes and other proteins. Proteins consist of polypeptide chains, which consist of amino acids. The sequence of amino acids corresponds to a gene—which is a sequence of nucleotide bases in DNA. And the path leading from genes to proteins has two steps, called transcription and translation.

2. In transcription, the double-stranded DNA molecule is unwound at a gene region, then an RNA molecule is assembled on the exposed bases of one of the strands.

3. In translation, RNA directs the linkage of one amino acid after another, in the sequence required to produce a specific kind of polypeptide chain.

4. With few exceptions, the genetic "code words" by which DNA instructions are translated into proteins are the same in all species of organisms.

5. A gene's base sequence is vulnerable to permanent changes, called mutations. Such changes are the original source of genetic variation. Mutations lead to alterations in protein structure, protein function, or both, and these alterations may lead to differences in traits among the individuals of a given population.

Discovering the Connection Between Genes and Proteins

Garrod's Hypothesis In the early 1900s a physician, Archibald Garrod, was puzzling over certain illnesses. They appeared to be heritable, for they kept recurring in the same families. They also appeared to be metabolic disorders, for blood or urine samples from the affected patients contained abnormally high levels of a substance that was known to be produced at a certain step in a metabolic pathway. Most likely, the enzyme at the *next* step in the pathway was defective in a way that prevented it from using the substance as a substrate. If this were true, then the pathway would be blocked from that step onward, as this diagram indicates:

$$A \longrightarrow B \longrightarrow \overset{\displaystyle C\ C\ C}{\underset{\displaystyle C}{C}} \times D$$

*pathway
is blocked*

This hypothesis could explain why molecules of a particular substance were accumulating in excess amounts in affected individuals. Garrod suspected that his patients differed from normal individuals in only one respect: they had inherited one metabolic defect. Thus, Garrod concluded, *specific "units" of inheritance (genes) function through the synthesis of specific enzymes.*

Beadle, Tatum, and a Bread Mold Thirty-three years later, George Beadle and Edward Tatum were experimenting with a bread mold (*Neurospora crassa*). This fungus is a common spoiler of baked goods (Figure *a*), but it also lends itself to laboratory experiments. *N. crassa* can be grown easily on an inexpensive culture medium that contains only sucrose, mineral salts, and biotin (one of the B vitamins). It synthesizes all other nutrients it requires, including other vitamins.

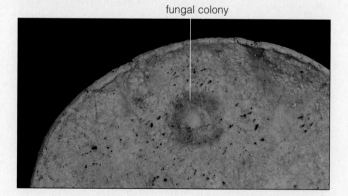

fungal colony

a Colony of the red bread mold (*Neurospora crassa*) growing on a stale tortilla.

Suppose an enzyme of a synthesis pathway is defective as a result of a gene mutation. The researchers suspected this had happened in some strains of *N. crassa*. One strain grew only when supplied with vitamin B_6, another with vitamin B_1, and so on. Chemical analysis of cell extracts revealed a different defective enzyme in each mutant strain. In other words, *each inherited mutation corresponded to a defective enzyme.* Here was evidence favoring the "one-gene, one-enzyme" hypothesis.

Gel Electrophoresis Studies The one-gene, one-enzyme hypothesis was refined through studies of *sickle-cell anemia*. This heritable disorder arises from the presence of a protein, hemoglobin, in red blood cells. The abnormal hemoglobin is designated HbS instead of HbA.

In 1949 Linus Pauling and Harvey Itano subjected molecules of HbS and HbA to **gel electrophoresis**. This laboratory procedure uses an electric field to move molecules through a viscous gel and separate them according to their size, shape, and net surface charge. Often the gel is sandwiched between glass or plastic plates to form a viscous slab. The two ends of the slab are suspended in two salt solutions, which are connected by electrodes to a power source (Figure *b*). When voltage is applied to the appara-

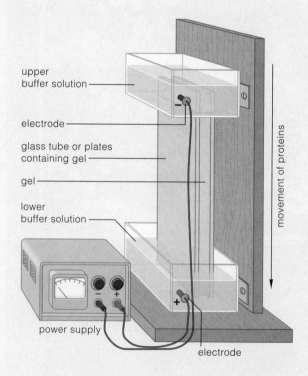

upper buffer solution

electrode

glass tube or plates containing gel

gel

lower buffer solution

power supply

electrode

movement of proteins

b One type of apparatus used in gel electrophoresis.

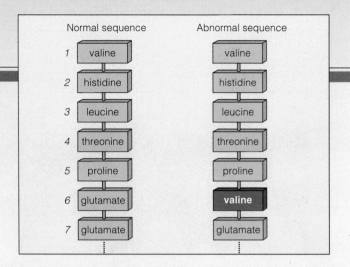

Normal sequence	Abnormal sequence
1 valine	valine
2 histidine	histidine
3 leucine	leucine
4 threonine	threonine
5 proline	proline
6 glutamate	**valine**
7 glutamate	glutamate

c The single amino acid substitution that gives rise to abnormal HbS hemoglobin molecules.

tus, molecules present in the gel will migrate through the electric field according to their individual charge. The molecules move away from one another in the gel and can be identified by staining the gel after a predetermined period of electrophoresis.

Pauling and Itano carefully layered a mixture of HbS and HbA molecules onto the gel at the top of the slab. The molecules gradually migrated down through the gel and separated into distinct bands. The band moving fastest carried the greatest surface charge, and this turned out to be composed of HbA molecules. HbS molecules moved more slowly.

One Gene, One Polypeptide   Later, Vernon Ingram pinpointed the difference between HbS and HbA. Hemoglobin, recall, consists of four polypeptide chains (page 44). Two are designated alpha and the other two, beta. An abnormal HbS chain arises from a single gene mutation that affects protein synthesis. The mutation causes the addition of valine instead of glutamate as the sixth amino acid of the beta chain (Figure *c*). Whereas glutamate carries an overall negative charge, valine has no net charge—and so HbS behaved differently in the electrophoresis studies.

As a result of this mutation, HbS hemoglobin has a "sticky" (hydrophobic) patch. In blood capillaries, where oxygen concentrations are at their lowest, hemoglobin molecules interact at the sticky patches. They aggregate into rods and distort red blood cells, and this affects organs throughout the body (page 176).

The discovery of the genetic difference between the alpha and beta chains of hemoglobin suggested that *two* genes code for hemoglobin—one for each kind of polypeptide chain. More importantly, it suggested further that genes must code for proteins in general, not just for enzymes.

And so a more precise hypothesis emerged: *The amino acid sequences of polypeptide chains—the structural units of proteins—are encoded in genes.*

TRANSCRIPTION AND TRANSLATION: AN OVERVIEW

DNA is like a book of instructions in each cell. The alphabet used to create the book is simple enough: A, T, G, and C. But how do we get from an alphabet to a protein? The answer starts with the structure of DNA.

DNA, recall, is a double-stranded molecule (Figure 13.6). The strands consist of four types of nucleotides, which differ in one component only. That component, a nitrogen-containing base, may be adenine, thymine, guanine, or cytosine. Which base follows the next along the length of a strand—that is, the **base sequence**—differs from one kind of organism to the next.

Before a cell divides, its DNA is replicated and the two strands unwind entirely from each other. At other times in a cell's life, the two strands unwind only in certain regions to expose certain genes. *And those genes contain instructions for building proteins.*

Researchers discovered this connection between genes and proteins through many studies, of the sort described in the *Focus* essay.

It takes two steps, **transcription** and **translation**, to carry out a gene's protein-building instructions. In eukaryotic cells, transcription proceeds in the nucleus. In this step, the nucleotide bases of DNA serve as a structural pattern (template) for assembling a strand of **ribonucleic acid** (**RNA**) from the cell's pool of free nucleotides. Afterward, the RNA moves to the cytoplasm, where translation proceeds. In this second step, RNA directs the assembly of amino acids into polypeptide chains. Later, the chains become folded into the three-dimensional shapes of proteins.

In short, DNA guides the synthesis of RNA, then RNA guides the synthesis of proteins:

DNA   *transcription*   RNA   *translation*   protein

The new proteins will have structural and functional roles in cells. Some even will have roles in building more DNA, RNA, and proteins.

From this overview, you may have the impression that protein synthesis requires only one class of RNA molecules. Actually, it requires three. Transcription of most genes produces **messenger RNA** (**mRNA**), the only class of RNA that carries protein-building instructions. Transcription of some other genes produces **ribosomal RNA** (**rRNA**), which are the main components of ribosomes. Ribosomes, recall, are the sites where polypeptide chains are assembled. Transcription of still other genes produces **transfer RNA** (**tRNA**).

As you will see, tRNA delivers amino acids one by one to a ribosome, in the order specified by mRNA.

14.2 TRANSCRIPTION OF DNA INTO RNA

How RNA Is Assembled

An RNA molecule is almost, but not quite, like a single strand of DNA. It is composed of only four types of nucleotides, each consisting of a five-carbon sugar (ribose), a phosphate group, and a base. The bases are adenine, cytosine, guanine, and uracil (Figure 14.2). **Uracil** is like the thymine in DNA; it pairs with adenine. Thus a new RNA strand can be put together on a DNA region according to the base-pairing rule:

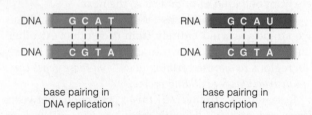

base pairing in DNA replication

base pairing in transcription

Transcription resembles DNA replication in another respect. Enzymes add the nucleotide units to a growing RNA strand one at a time, in the $5' \rightarrow 3'$ direction. (Here you may wish to refer to Figure 13.8.)

Transcription *differs* from DNA replication in three key respects. First, only a certain stretch of one DNA strand—not the whole molecule—serves as the template. Second, different enzymes, called **RNA polymerases**, catalyze the addition of nucleotides to the 3' end of a growing strand. Third, transcription results in a *single* strand of nucleotides in RNA.

Transcription starts at a **promoter**, a base sequence that signals the start of a gene. Proteins help position an RNA polymerase on the DNA so that it binds with the promoter. The enzyme moves along the DNA, joining

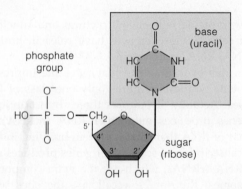

Figure 14.2 Structure of one of the four nucleotides of RNA. The other three have a different base (adenine, guanine, or cytosine instead of the uracil shown here). Compare Figure 13.5, which shows the nucleotides of DNA.

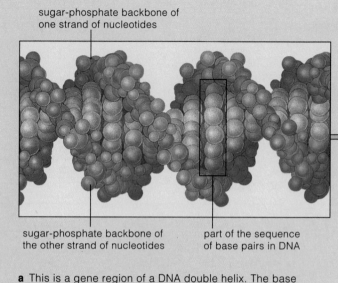

a This is a gene region of a DNA double helix. The base sequence of one of the nucleotide strands in this region is about to be transcribed into an RNA molecule.

Figure 14.3 Gene transcription: the synthesis of an RNA molecule on a DNA template.

nucleotides together (Figure 14.3). When it reaches a base sequence that serves as a stop signal, the RNA is released as a free transcript.

Finishing Touches on mRNA Transcripts

Newly formed mRNA is an unfinished molecule, one that must be modified before its protein-building instructions can be put to use. Just as a dressmaker might snip off some threads or add bows on a dress before it leaves the shop, so does a eukaryotic cell tailor this pre-mRNA.

Very quickly, enzymes attach a cap to the 5' end of the pre-mRNA molecule. The cap is a nucleotide that has a methyl group and phosphate groups bonded to it. Enzymes also attach about 100 to 200 adenine-containing nucleotides to the 3' end of most pre-mRNA molecules (hence the name, "poly-A tail"). The tail becomes wound up with proteins. Later on, in the cytoplasm, the cap will promote the binding of mRNA to a ribosome. There also, enzymes will gradually destroy the wound-up tail from the tip on up, then the mRNA. Clearly, such tails "pace" enzyme access to mRNA. Perhaps they help keep protein-building messages intact for as long as the cell requires them.

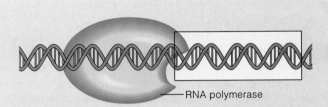

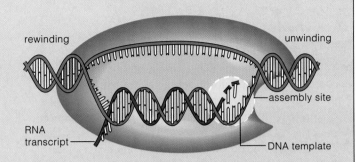

b An RNA polymerase molecule binds to a promoter in the DNA. It will use the base sequence positioned downstream from that site as a template for linking nucleotides into a strand of RNA.

d Throughout transcription, the DNA double helix unwinds just in front of the RNA polymerase. Short lengths of the newly forming RNA strand temporarily wind up with the DNA template strand. Then they unwind from it—and the two strands of DNA wind up together again.

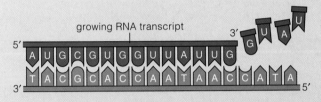

c During transcription, RNA nucleotides are based-paired, one after another, with the exposed bases on the DNA template.

new RNA transcript

e At the end of the gene region, the last stretch of the RNA is unwound from the DNA template and so is released.

Besides this, the mRNA message itself is modified. Most eukaryotic genes have one or more **introns**, which are base sequences that do not get translated into an amino acid sequence. Introns intervene between **exons**, the only parts of mRNA that become translated into protein. As Figure 14.4 shows, introns are transcribed right along with exons, but enzymes snip them out before the mRNA leaves the nucleus in mature form.

Some introns may be evolutionary junk, the left-overs of past mutations that led nowhere. However, other introns are sites where instructions for building a particular protein can be snipped apart and spliced back together in different ways. Alternative splicing allows different cells in your body to produce modified versions of an mRNA transcript—and modified versions of the resulting protein—from the same gene. We will return to this topic in the next chapter.

In gene transcription, a sequence of bases in one of the two strands of a DNA molecule serves as the template for assembling an RNA strand. The assembly follows base-pairing rules (adenine with uracil, cytosine with guanine).

Before leaving the nucleus, new RNA transcripts undergo modification into final form.

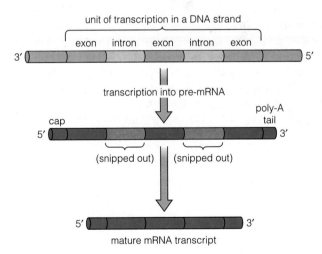

Figure 14.4 Transcription and modification of newly formed mRNA in the nucleus of eukaryotic cells. The cap simply is a nucleotide with functional groups attached. The poly-A tail is a string of adenine nucleotides.

14.3 FROM mRNA TO PROTEINS

The Genetic Code

Like a DNA strand, mRNA is a linear sequence of nucleotides. What are the protein-building "words" encoded in its sequence? Gobind Khorana, Marshall Nirenberg, and others came up with the answer. They deduced that enzymes recognize nucleotide bases *three at a time*, as triplets. In mRNA, the base triplets are called **codons**. Figure 14.5 shows how the order of different codons in an mRNA molecule dictates the order in which particular amino acids are added to a growing polypeptide chain. Count the codons listed in Figure 14.6, and you see there are sixty-four kinds.

Most of the twenty kinds of amino acids correspond to more than one codon. (Glutamate corresponds to the code words GAA *or* GAG, for example.) The codon AUG also establishes the reading frame for translation. That is, enzymes start their "three-bases-at-a-time"

selections at an AUG that serves as the "start" signal in an mRNA strand. Three codons (UAA, UAG, UGA) are stop signals. They stop enzymes from adding any more amino acids to a new polypeptide chain.

The set of sixty-four different codons is the **genetic code**. It the basis of protein synthesis in all organisms.

Protein-building instructions are encoded in the nucleotide sequence of DNA and mRNA. The genetic code is a set of sixty-four base triplets (nucleotide bases, read in blocks of three). A codon is a base triplet in mRNA.

Different combinations of codons specify the amino acid sequence of different polypeptide chains, start to finish.

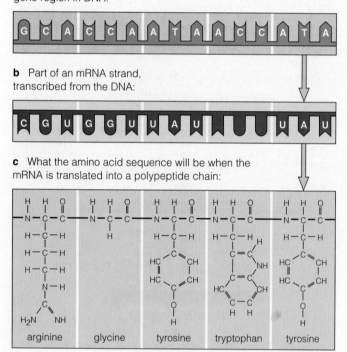

a Base sequence of a gene region in DNA:

G C A C C A A T A A C C A T A

b Part of an mRNA strand, transcribed from the DNA:

C G U G G U U A U U A U

c What the amino acid sequence will be when the mRNA is translated into a polypeptide chain:

arginine — glycine — tyrosine — tryptophan — tyrosine

Figure 14.5 The steps from genes to proteins. (**a**) This region of a DNA double helix was unwound during transcription. (**b**) Exposed bases on one strand served as a template for assembling an mRNA strand. In the mRNA, every three nucleotide bases equaled one codon. Each codon called for one amino acid in this polypeptide chain. (**c**) Referring to Figure 14.6, can you fill in the blank codon for tryptophan in the chain?

First Letter	Second Letter				Third Letter
	U	C	A	G	
U	phenylalanine	serine	tyrosine	cysteine	U
	phenylalanine	serine	tyrosine	cysteine	C
	leucine	serine	stop	stop	A
	leucine	serine	stop	tryptophan	G
C	leucine	proline	histidine	arginine	U
	leucine	proline	histidine	arginine	C
	leucine	proline	glutamine	arginine	A
	leucine	proline	glutamine	arginine	G
A	isoleucine	threonine	asparagine	serine	U
	isoleucine	threonine	asparagine	serine	C
	isoleucine	threonine	lysine	arginine	A
	(start) methionine	threonine	lysine	arginine	G
G	valine	alanine	aspartate	glycine	U
	valine	alanine	aspartate	glycine	C
	valine	alanine	glutamate	glycine	A
	valine	alanine	glutamate	glycine	G

Figure 14.6 The genetic code. The codons in mRNA are nucleotide bases, read in blocks of three. Sixty-one of these base triplets correspond to specific amino acids. Three others serve as signals that stop translation. The left column of the diagram shows the first of the three nucleotides in each codon in mRNA. The middle columns show the second nucleotide. The right column shows the third. Reading from left to right, for instance, the triplet UGG corresponds to tryptophan. Both UUU and UUC correspond to phenylalanine.

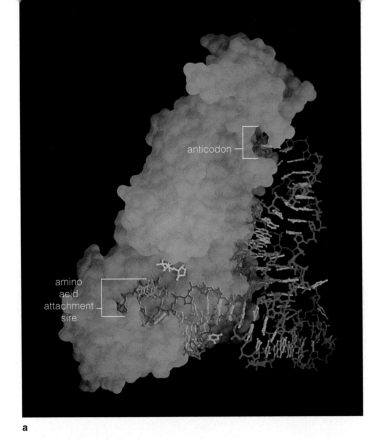

a

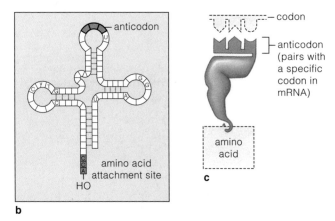

b

c

Figure 14.7 (**a**) Computer-generated, three-dimensional model of one type of tRNA molecule. The tRNA (*reddish brown*) is shown attached to a bacterial enzyme (*green*), along with an ATP molecule (*gold*). This particular enzyme attaches amino acids to tRNAs. (**b**) Structural features common to all tRNAs. (**c**) Simplified model of tRNA that is used in subsequent illustrations. The "hook" at one end is the site to which a specific amino acid can become attached.

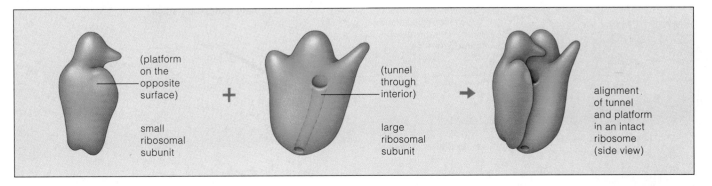

Figure 14.8 Model of eukaryotic ribosomes. Polypeptide chains are assembled on the small subunit's platform. Newly forming chains may move through the large subunit's tunnel.

Roles of tRNA and rRNA

Cells have pools of free amino acids and free tRNA molecules in the cytoplasm. tRNAs have a molecular "hook," an attachment site for amino acids. They also have an **anticodon**, a nucleotide triplet that can base-pair with codons (Figure 14.7). When tRNAs bind to codons, they automatically position their attached amino acids in the order specified by mRNA.

A cell has more than sixty kinds of codons but fewer kinds of tRNAs. How do they match up? By the base-pairing rules, adenine must pair with uracil, and cytosine with guanine. However, for codon-anticodon interactions, the rules loosen up for the third base. For example, CCU, CCC, CCA, and CCG all specify proline. Such freedom in codon-anticodon pairing at the third base is called the "wobble effect."

tRNAs interact with mRNA at binding sites on the surface of **ribosomes**. Each ribosome has two subunits

(Figure 14.8). These are assembled in the nucleus from rRNA and proteins, then are shipped separately to the cytoplasm. There they will combine as functional units only during translation, as described next.

---

mRNAs are the only molecules that carry protein-building instructions from DNA to the cytoplasm.

tRNAs bind to codons and so position their attached amino acids in the order specified by mRNA. Thus they translate the mRNA into a corresponding sequence of amino acids.

rRNAs are components of ribosomes, the sites where amino acids are assembled into polypeptide chains.

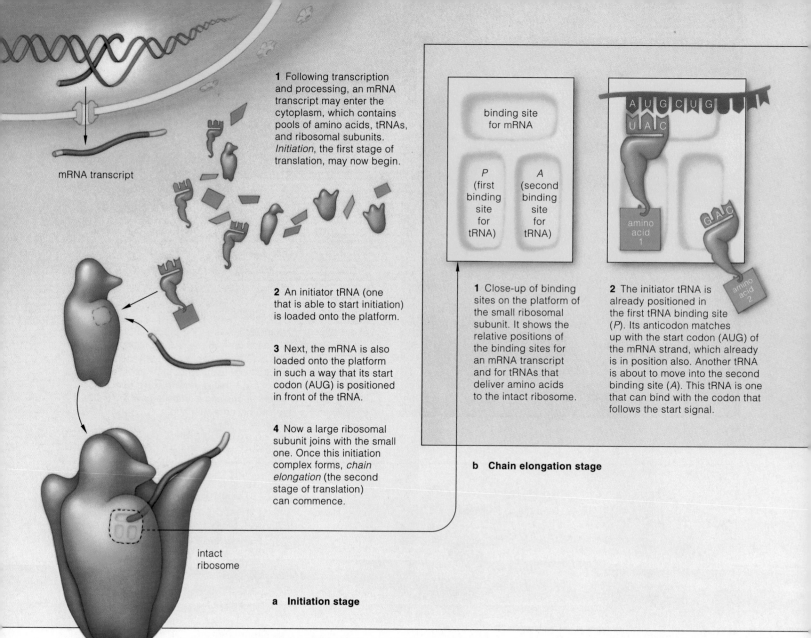

1 Following transcription and processing, an mRNA transcript may enter the cytoplasm, which contains pools of amino acids, tRNAs, and ribosomal subunits. *Initiation*, the first stage of translation, may now begin.

mRNA transcript

2 An initiator tRNA (one that is able to start initiation) is loaded onto the platform.

3 Next, the mRNA is also loaded onto the platform in such a way that its start codon (AUG) is positioned in front of the tRNA.

4 Now a large ribosomal subunit joins with the small one. Once this initiation complex forms, *chain elongation* (the second stage of translation) can commence.

binding site for mRNA

P (first binding site for tRNA)

A (second binding site for tRNA)

1 Close-up of binding sites on the platform of the small ribosomal subunit. It shows the relative positions of the binding sites for an mRNA transcript and for tRNAs that deliver amino acids to the intact ribosome.

AUGCUG
UAC

amino acid 1

GAC

amino acid 2

2 The initiator tRNA is already positioned in the first tRNA binding site (*P*). Its anticodon matches up with the start codon (AUG) of the mRNA strand, which already is in position also. Another tRNA is about to move into the second binding site (*A*). This tRNA is one that can bind with the codon that follows the start signal.

b   Chain elongation stage

intact ribosome

a   Initiation stage

Figure 14.9   Translation, the second step of protein synthesis.

14.4   STAGES OF TRANSLATION

Translation occurs in the cytoplasm. It requires three stages, called initiation, elongation, and termination.

In *initiation*, a tRNA that can start transcription and an mRNA transcript become loaded onto an intact ribosome. First, the initiator tRNA binds with the small ribosomal subunit. So does the start codon (AUG) of the mRNA transcript. After this, a large ribosomal subunit binds with the small one (Figure 14.9a). The next stage can begin.

In *elongation*, a polypeptide chain forms as the mRNA strand passes between the ribosomal subunits, like a thread being moved through the eye of a needle. Some proteins in the ribosome that function as enzymes join amino acids together in the sequence dictated by mRNA's codons. As Figure 14.9b indicates, they catalyze the formation of a peptide bond between every

two amino acids delivered to the ribosome. Here you may also wish to refer to an earlier diagram of peptide bond formation (Figure 3.16).

In *termination*, a stop codon is reached and there is no corresponding anticodon. Now the enzymes bind to certain proteins (release factors). The binding causes the ribosome as well as the polypeptide chain to detach from the mRNA (Figure 14.9c). The detached chain may join the pool of free proteins in the cytoplasm. Or it may enter the cytomembrane system, starting with the compartments of rough ER. As indicated on page 64, many newly formed chains take on their final form in that system before being shipped to their destinations.

Often, enzymes and tRNAs repeatedly translate the same mRNA transcript in a given period. The transcript threads through many ribosomes, which are arranged

 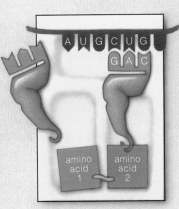 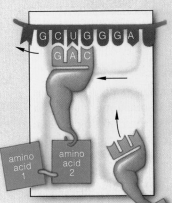 

3 Through enzyme action, the bond between the initiator tRNA and the amino acid hooked to it is broken. At the same time, an enzyme catalyzes the formation of a peptide bond between the two amino acids. After these bonding events are over, the initiator tRNA will be released from the ribosome.

4 Now the first amino acid is attached only to the second one—which is still hooked to the second tRNA. This tRNA will move into the *P* site on the ribosomal platform, sliding the mRNA with it by one codon. When it does so, the third codon will become aligned above the *A* site.

5 A third tRNA is about to move into the *A* site. Its anticodon is capable of base-pairing with the third codon of the mRNA transcript. Next, through enzyme action, a peptide bond will form between amino acids 2 and 3.

6 Steps 3 through 5 are repeated again and again. The polypeptide chain continues to grow this way until enzymes reach a stop codon in the mRNA transcript. Now *termination*, the last stage of transcription, can begin, as shown in (**c**).

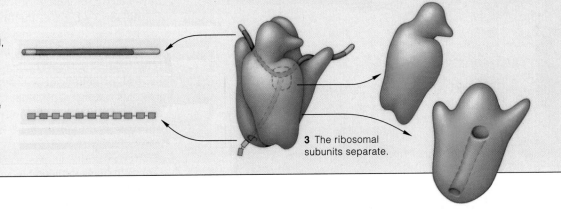

1 Once a stop codon is reached, the mRNA transcript is released from the ribosome.

2 The newly formed polypeptide chain also is released.

c Chain termination stage

3 The ribosomal subunits separate.

one after another, closely together in assembly-line fashion. Figure 14.10 shows one of these arrangements, which are called **polysomes**. Their presence indicates that a cell is producing many copies of a polypeptide chain from the same transcript.

Translation is initiated by the convergence of a small ribosomal unit, an initiator tRNA, an mRNA transcript, then a large ribosomal subunit.

Next, anticodons of tRNAs base-pair with mRNA codons. A polypeptide chain grows through peptide bond formation between every two amino acids delivered to the ribosome.

Translation ends when a stop codon triggers events that cause the chain and the mRNA to detach from the ribosome.

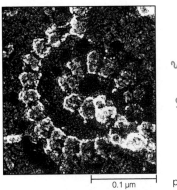 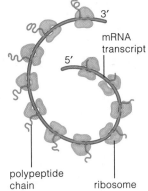

0.1 µm

Figure 14.10 From a eukaryotic cell, a polysome (many ribosomes simultaneously translating the same mRNA molecule).

14.5 MUTATION AND PROTEIN SYNTHESIS

Whenever a cell puts its genetic code into action, it is synthesizing precisely those proteins that it requires for its structure and functioning. If something changes a gene's code words, we can expect the resulting protein to change. If that protein is central to cell architecture or metabolism, we can expect the outcome to be a dead or damaged cell.

Every so often, genes do change. Maybe one base is substituted for another in the DNA sequence. Or maybe an extra base is inserted or a base is lost. These small-scale changes in the nucleotide sequence of a DNA molecule are **gene mutations**.

Some mutations result from exposure to **mutagens**. These are agents that increase the risk of heritable alterations in the structure of DNA. Ultraviolet light, ionizing radiation, and certain substances in tobacco smoke are common mutagens. So are free radicals, those rogue molecular fragments described on page 92.

Even in the absence of mutagens, however, mutations arise in cells. For example, spontaneous mutations may follow replication errors, as when adenine wrongly pairs with a cytosine unit on a DNA template strand. DNA repair enzymes may simply detect an error, then "fix" the wrong base (Figure 14.11).

Regardless of whether it is spontaneous or induced by a mutagen, "base-pair substitutions" of this sort may lead to the substitution of one amino acid for another during protein synthesis. Earlier in the chapter, you saw how one of these substitutions results in sickle-cell anemia—a genetic disorder that has wide-ranging structural and physiological consequences.

Or consider the "frameshift mutation," in which one to several base pairs are inserted into a DNA molecule or deleted from it. Remember, polymerases read a nucleotide sequence in blocks of three. As Figure 14.12 shows, an insertion or deletion in a gene region can shift this reading frame. Because the gene is not read correctly, an abnormal protein is synthesized.

As Barbara McClintock discovered, spontaneous mutations also result when "jumping genes" are on the move. These DNA regions also are called transposable elements. They can move from one location to another in the same DNA molecule or in a different one. Often they inactivate genes into which they are inserted. As Figure 14.13 suggests, the unpredictability of the jumps can give rise to interesting changes in phenotype.

In terms of individual lifetimes, gene mutations are rather rare events. Whether they prove to be harmful, neutral, or beneficial will depend on how the resulting proteins interact with other genes and with the environment. And the outcome may have evolutionary consequences, as the *Commentary* makes clear.

Gene mutations are heritable, small-scale alterations in the nucleotide sequence of DNA. They may be harmful, neutral, or beneficial, depending on how the proteins they specify interact with other genes and with the environment.

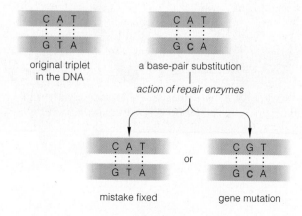

Figure 14.11 Example of a base-pair substitution.

Figure 14.12 Outcomes of three frameshift mutations. In this hypothetical example, three different base insertions occur in the same nucleotide sequence of DNA. The word *cat* represents a code word (base triplet) in mRNA. In the first mutation, an extra nucleotide (arrow) is inserted into a gene. When mRNA is transcribed off the gene, the insertion puts the reading frame out of phase and changes the code words. Thus the wrong amino acids are called up during translation. The second base insertion does not improve matters. The third one restores most of the code words, so only part of the resulting protein will be defective.

Insertions of extra bases into a DNA molecule often give rise to mutant phenotypes. Studies of such mutations in bacteriophages led to the discovery of the genetic code.

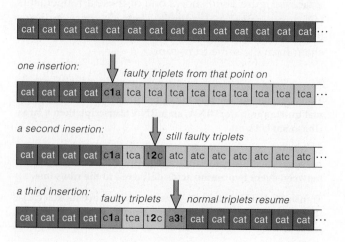

Mutation, Gene Products, and Evolution

In the natural world, gene mutations are rare, chance events. We cannot predict exactly when or in which individual they will appear. Even so, each gene does have a characteristic *mutation rate*, which simply is the probability of its mutating between or during DNA replications. The mutation rate for a typical gene is one in a million (10^6) replications.

Genes also mutate independently of one another. To predict the probability of any two mutations occurring in a given cell, we would have to multiply the individual mutation rates for two of its genes. For example, if the rate for the first gene is one in a million per cell generation and the rate for the second is one in a billion (10^9), then there is only one chance in a million billion (10^{15}) that both genes will mutate in the same cell. Many trillions of replications occur during human growth and development, so chances are that at least one cell will contain a new mutation. If the cell happens to be a reproductive cell, the mutation may be passed on to offspring.

What will be the phenotypic outcome of a given gene mutation? That depends on how the resulting protein functions in the individual. Many mutations turn out to be harmful. No matter what the species, each individual generally inherits a combination of many finely tuned genes that work well together under a certain range of operating conditions. A mutation that has drastic effects on essential traits increases the odds that an individual will not survive or reproduce. Such *lethal* genes obviously do not accumulate in the DNA. Yet many other mutations have little, if any, effect on traits. As you will read in the next unit, these

neutral mutations have accumulated in the DNA of all species.

Natural selection has played a major role in perpetuating *beneficial* genes as well as the mechanisms that protect them. For example, we can readily understand the adaptive advantage of genes that specify effective repair enzymes, ones that have little tolerance of mismatched base pairs. Such enzymes have helped protect the overall stability of all the DNA molecules that have been replicated countless times for nearly 4 billion years.

Reflect on some of the concepts you have read about so far in this book. You have an idea that all living things on earth share the same chemical heritage. Your DNA has the same kinds of substances, and follows the same base-pairing rules, as the DNA of earthworms in Missouri and grasses on the Mongolian steppes. Your DNA is replicated in much the same way as theirs, and the same genetic code is followed in translating its messages into proteins. In the evolutionary view, the reason you don't look like an earthworm or a grass plant is largely a result of natural selection of different mutations that originated in different lines of descent. Thus the sequence of base pairs in DNA has come to be different in you, the plant, and the worm.

And so we have three concepts of profound importance. *First, DNA is the source of the unity of life. Second, gene mutations are the original source of life's diversity. Finally, the changing environment is the testing ground for the success or failure of the proteins specified by each novel gene product that appears on the evolutionary scene.*

Figure 14.13 Mutation in different kernels of Indian corn (*Zea mays*). All of the kernel cells have the same pigment-coding genes. Yet some kernels are colorless or spottily colored.

In an ancestor of this plant, a transposable element (jumping gene) in one of its cells left its position in a DNA molecule, invaded another, and shut down a pigment gene. This plant inherited the mutations. As cell divisions proceeded in the growing plant, none of the mutated cell's descendants could produce pigment molecules. They gave rise to colorless kernel tissue.

Later, in some cells, the transposable element slipped out of the pigment gene. All of the descendants of *those* cells were able to produce pigment. They gave rise to colored kernel tissue.

SUMMARY

1. A gene is a sequence of nucleotide bases in DNA. Most genes contain protein-building instructions; their nucleotide sequence corresponds to the sequence of amino acids in a polypeptide chain. Proteins are composed of one or more polypeptide chains.

2. Some genes contain instructions for building RNA molecules, which are necessary for protein synthesis. There are three classes of RNA molecules:

   a. Messenger RNA (mRNA) carries protein-building instructions from DNA to the cytoplasm. It is the only class of RNA molecules to do this.

   b. Transfer RNA (tRNA) is the translator that converts the sequential message of mRNA into a corresponding sequence of amino acids.

   c. Ribosomal RNA (rRNA) is a component of ribosomes, the actual sites where amino acids are assembled into polypeptide chains.

3. As summarized in Figure 14.14, the path leading from genes to proteins has two steps. In the first step, called transcription, a DNA template is used to synthesize RNA. In the second step, called translation, three classes of RNA interact in ways that lead to the synthesis of polypeptide chains, which later will become folded into the three-dimensional shape of proteins:

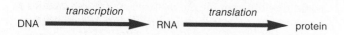

4. Here are the key points concerning transcription:

   a. A double-stranded DNA molecule is unwound at a particular gene region, then an RNA molecule is assembled on the exposed bases of one of the strands, which serves as a template.

   b. Transcription of DNA into RNA follows the same base-pairing rule that applies to DNA replication. However, uracil takes the place of thymine in an RNA strand. It pairs with adenine:

   c. In eukaryotic cells, new RNA transcripts undergo modification into final form before being shipped from the nucleus. Typically, enzymes attach a cap at the 5' end and a poly-A tail at the 3' end. For mRNAs, the noncoding portions (introns) are excised and the coding portions (exons) are spliced together. Only mature mRNA transcripts are translated in the cytoplasm.

5. Here are the key points concerning translation:

   a. mRNA interacts with tRNAs and ribosomes in such a way that amino acids become linked one after another, in the sequence required to produce a specific kind of polypeptide chain.

   b. In all species, translation is based on the genetic code, a set of sixty-four base triplets (nucleotide bases, which enzymes read in blocks of three).

   c. In mRNA, a base triplet is called a codon. An anticodon is a complementary triplet in tRNA. Some combination of codons specifies the amino acid sequence of a polypeptide chain, start to finish.

6. Translation proceeds through three stages:

   a. Initiation. A small ribosomal subunit binds with an initiator tRNA, then with an mRNA transcript. The small subunit then binds with a large ribosomal subunit to form the initiation complex.

   b. Chain elongation. tRNAs deliver amino acids to the ribosome. Each has an anticodon, a base triplet that is complementary to a codon in mRNA and that can base-pair with it. Each amino acid brought to the ribosome by tRNAs is linked (by a peptide bond) to the growing polypeptide chain.

   c. Chain termination. A stop codon triggers events that cause the polypeptide chain and mRNA to detach from the ribosome.

7. Gene mutations are heritable, small-scale alterations in the nucleotide sequence of DNA. They may be harmful, neutral, or beneficial, depending on how the proteins they specify interact with other genes and with the environment.

   a. Mutations can be induced by exposure to mutagens, which are agents that cause heritable alterations in the structure of DNA. They also may arise spontaneously, as through replication errors and movements of transposable elements.

8. DNA is the source of the unity of life. Gene mutations are the original source of life's diversity. The changing environment is the testing ground for the success or failure of the proteins specified by each novel gene product that appears on the evolutionary scene.

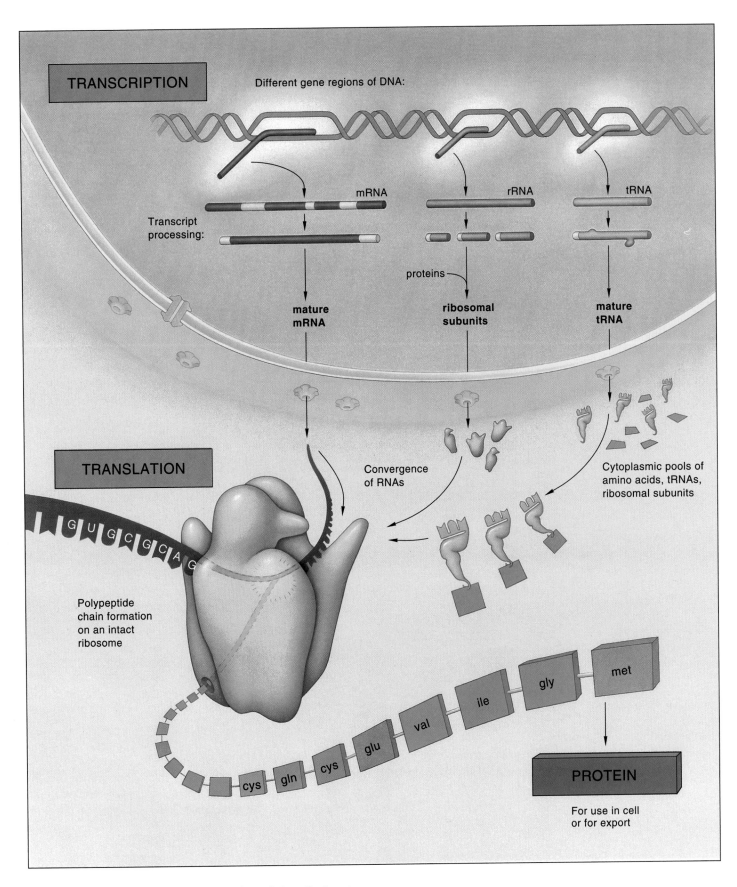

Figure 14.14 Summary of transcription and translation—the two steps leading to protein synthesis—as they proceed in eukaryotic cells.

Review Questions

1. Are the proteins specified by eukaryotic DNA assembled *on* the DNA molecule? If so, state how. If not, tell where they are assembled, and on which molecules. *221*

2. Protein synthesis requires the interaction of three classes of RNA molecules. Name the three classes and define the functions of each. *221, 224, 225*

3. In what key respect does an RNA molecule differ from a DNA molecule? *222*

4. Describe the steps of gene transcription. In what respects does this process resemble DNA replication? In what respects does it differ? *222*

5. Before an mRNA transcript leaves the nucleus, are its introns or exons snipped out? *223*

6. What is the genetic code? Is the code the basis of protein synthesis in all organisms? *224*

7. Distinguish between a codon and an anticodon. *224, 225–227*

8. If sixty-one codons in mRNA actually specify amino acids, and if there are only twenty common amino acids, then more than one codon combination must specify some of the amino acids. How do triplets that code for the same thing usually differ? *225*

9. Describe the events that unfold during the three steps of translation. *226–227*

10. Describe one kind of gene mutation that leads to an altered protein. What determines whether the altered protein will have beneficial, neutral, or harmful effects on the individual? *228, 229*

11. Fill in the blanks on the diagram below. *231*

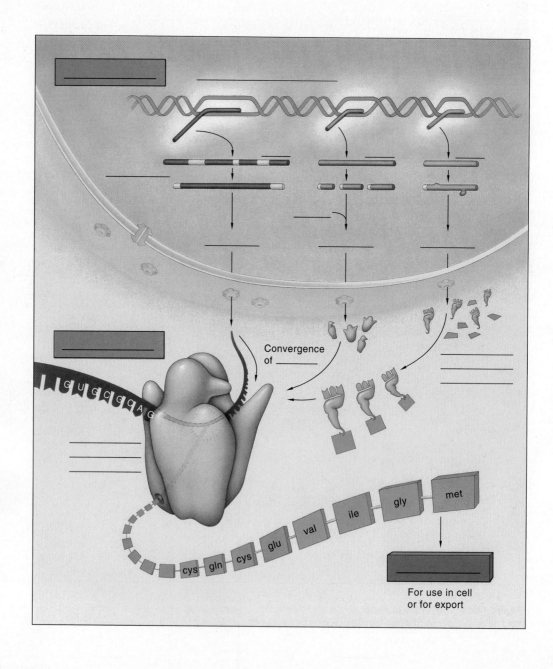

1. Nucleotide bases, read _____ at a time, serve as the "code words" of genes.

2. DNA contains different genes that are transcribed into _____ .
   a. proteins
   b. mRNAs
   c. rRNAs
   d. tRNAs
   e. b, c, and d

3. The RNA molecule is _____ .
   a. a double helix
   b. usually single-stranded
   c. always double-stranded
   d. usually double-stranded

4. mRNA is produced by _____ .
   a. replication
   b. duplication
   c. transcription
   d. translation

5. _____ carries coded instructions for an amino acid sequence to the ribosome.
   a. DNA
   b. rRNA
   c. mRNA
   d. tRNA

6. tRNA _____ .
   a. delivers amino acids to ribosomes
   b. picks up genetic messages from rRNA
   c. synthesizes mRNA
   d. all of the above

7. Each codon calls for a specific _____ .
   a. protein
   b. polypeptide
   c. amino acid
   d. carbohydrate

8. How many amino acids are coded for in this mRNA sequence: CGUUUACACCGUCAC?
   a. three
   b. five
   c. six
   d. seven
   e. more than seven

9. An anticodon pairs with the nitrogen-containing bases of _____ .
   a. mRNA codon
   b. DNA codons
   c. tRNA anticodon
   d. amino acids

10. The loading of mRNA onto an intact ribosome occurs during _____ .
    a. transcription
    b. transcript processing
    c. translation
    d. chain elongation

11. The presence of _____ in a micrograph indicates that a cell is producing many copies of a polypeptide chain from the same transcript.
    a. RNA polymerase
    b. DNA polymerase
    c. polysomes
    d. both a and b

12. Use the genetic code (Figure 14.6) to translate the mRNA sequence UAUCGCACCUCAGGAUGAGAU. Which amino acid sequence is being specified?
    a. tyr-arg-thr-ser-gly-stop-asp . . .
    b. tyr-arg-thr-ser-gly . . .
    c. tyr-arg-tyr-ser-gly-stop-asp . . .
    d. none of the above

13. Match the terms related to protein building.
    _____ disrupts genetic instructions
    _____ genetic code word
    _____ transcription
    _____ translation
    _____ translation stages

    a. initiation, elongation, termination
    b. RNAs convert genetic messages into polypeptide chains
    c. base triplet for an amino acid
    d. one DNA strand is template for the process
    e. gene mutation

Selected Key Terms

anticodon *225*
base sequence *221*
codon *224*
exon *223*
gel electrophoresis *220*
gene mutation *228*
genetic code *224*
intron *223*
messenger RNA (mRNA) *221*
mutagen *228*
polysome *227*
promoter *222*
ribonucleic acid (RNA) *221*
ribosomal RNA (rRNA) *221*
ribosome *225*
RNA polymerase *222*
transcription *221*
transfer RNA (tRNA) *221*
translation *221*
uracil *222*

Readings

Amato, I. January 1991. "Stuck on Mussels." *Science News* 139:8–15.

Grunstein, M. October 1992. "Histones as Regulators of Genes." *American Scientist* 267(4):68–74B.

Liotta, L. February 1992. "Cancer Cell Invasion and Metastasis." *Scientific American* 266(2): 54–63.

Murray, A., and M. Kirschner. March 1991. "What Controls the Cell Cycle?" *Scientific American* 264(3):56–63.

Wolfe, S. 1993. *Molecular and Cellular Biology*. Belmont, California: Wadsworth.

15 CONTROL OF GENE EXPRESSION

Genes, Proteins, and Cancer

Every second, millions of cells in your skin, gut lining, liver, and other body regions divide and replace their worn-out, dead, and dying predecessors. They do not divide willy-nilly. Certain genes specify the enzymes and other proteins that are required for cell growth, DNA replication, spindle formation, chromosome movements, and cytoplasmic division. Certain regulatory proteins control the synthesis and use of these gene products, and they control when the division machinery is put to rest.

On rare occasions, cell division goes out of control. Once controls are lost, divisions cannot stop as long as conditions for growth remain favorable. When cells are not responding to normal controls over growth and division, they form a tissue mass called a **tumor**.

Cells of common skin warts and other *benign* tumors grow rather slowly, in an unprogrammed way, and they still have surface recognition proteins that hold them together in their home tissue. Surgically remove a benign tumor, and you remove its threat to the surrounding tissues.

By contrast, abnormal cells grow and divide more rapidly in a *malignant* tumor, and their physical and metabolic effects on surrounding tissues are destructive. These are grossly disfigured cells (Figure 15.1).

They cannot construct a proper cytoskeleton or plasma membrane, and they cannot synthesize normal versions of recognition proteins. When such cells break loose from their home tissue and enter lymph or blood vessels, they can travel through the body—and become lodged where they don't belong. You may have heard about this process of migration and invasion. It is called **metastasis**.

Basal cell carcinomas, of the sort shown in Figure 13.9, are merely one of more than 200 types of malignant tumors that have been identified to date. All of these malignant tumors are grouped into the general

cytoplasmic
extension of
cancer cell
body

white
blood cell

Figure 15.1 Visible evidence of an absence of gene control. This scanning electron micrograph shows a cancer cell surrounded by a number of the body's defenders—white blood cells—that may or may not be able to destroy it. Cancer arises through the loss of controls that govern certain genes with roles in cell growth and division.

category of **cancer**. Each year in the United States and other developed countries, cancer accounts for 15 to 20 percent of all deaths. It is not just a human affliction. It has been observed in most of the animals that have been studied. Similar abnormalities have even been observed in many kinds of plants.

Researchers have traced most cancers to a certain category of genes, called oncogenes. You will read about oncogenes in this chapter. In some cases of cancer, base substitutions or deletions have altered one of these genes or one of the controls over its transcription. In other cases, a gene has been abnormally duplicated over and over again. In still other cases, a whole gene or part of it moves to a new location in the same chromosome or to an entirely different chromosome. With such alterations, transformations may begin that lead, ultimately, to cancer.

Consider the *myc* gene. It is located in human chromosome 8, and it specifies a regulatory protein that helps control the cell cycle. Sometimes a break occurs in this gene region *and* in a region of chromosome 14. The two chromosomes swap segments, so that part of the *myc* gene ends up in a new location. Possibly the translocated portion escapes the normal controls over its transcription and comes under the influence of new ones. Possibly it leaves behind part of the message concerning how the mRNA transcripts are supposed to be processed. Whatever the case, transformations begin that can lead, ultimately, to a form of *Burkitt's lymphoma*.

Burkitt's lymphoma is a malignant tumor of B lymphocytes. B lymphocytes are the only type of cell that produces the antibody molecules, which are protein weapons against specific agents of infectious disease (page 680). In activated B cells, a gene region that specifies antibody is transcribed at an exceptionally high rate. The translocation between chromosomes 8 and 14 puts the *myc* gene next to this gene region and makes it abnormally active.

Perhaps more than any other example, cancerous transformations bring home the extent to which you and all other organisms depend on gene controls. A dizzying variety of controls govern transcription and translation. They influence when and where gene products will be used. And they are the topic of this chapter.

KEY CONCEPTS

1. In cells, controls govern when, how, and to what extent genes are expressed. The control elements operate in response to changing chemical conditions and to signals from the outside environment.

2. Control is exerted through regulatory proteins and other molecules that interact with DNA, with RNA that has been transcribed off the DNA, or with the gene products. Among most multicelled organisms, the regulatory molecules called hormones have major roles in controlling gene expression.

3. Control elements come into play during transcription and translation. They also act on the resulting polypeptide chains and proteins. In eukaryotic cells, additional controls govern the processing of new RNA transcripts and their shipment out of the nucleus for translation in the cytoplasm.

4. All cells in a multicelled organism have the same genes, but they activate or suppress many of those genes in different ways. The controlled, selective use of genes leads to synthesis of the proteins that give each type of cell its distinctive structure, function, and products.

15.1 THE NATURE OF GENE CONTROL

At this very moment, bacteria are feeding on nutrients in your gut. Your own red blood cells are binding, transporting, or giving up oxygen, and many epithelial cells in your skin are busily synthesizing the protein keratin. Like cells everywhere, they are functioning by virtue of the protein products of genes.

Cells don't express all of their genes all of the time. Some genes and their products are used only once. Others are used some or all of the time or not at all. Which genes are expressed depends on the type of cell, on its responses to chemical conditions and signals from the environment, and on its built-in control systems. These systems consist of molecules, including **regulatory proteins**, that interact with DNA, RNA, or the actual gene products.

For example, nutrient availability shifts rapidly and often for enteric bacteria, which reside in animal intestines. These prokaryotic cells rapidly transcribe genes and synthesize many nutrient-digesting enzyme molecules when nutrients are moving past—then restrict synthesis when they are not. By contrast, the composition and solute concentrations of the fluid bathing your cells do not shift drastically. And few of your cells exhibit such rapid shifts in transcription.

Different kinds of control systems operate during transcription, translation, and after translation, on the gene product. For example, cells use two common control systems to block or enhance transcription. With **negative control systems**, a regulatory protein can bind to DNA at a certain site and block transcription. It can be removed from the DNA by the action of an inducer, a type of molecule that promotes transcription. With **positive control systems**, a regulatory protein binds to the DNA and promotes the initiation of transcription.

The binding of a regulatory protein to DNA can be reversed, as when the conditions that called for the synthesis of a particular protein change.

Summing up, these are the points to keep in mind as you read through the remainder of this chapter:

---

1. Cells exert control over when, how, and to what extent each of their genes is expressed.

2. The expression of a given gene depends on the type of cell and its functions, on chemical conditions, and on signals from the outside environment.

3. Regulatory proteins and other molecules govern gene expression through interactions with DNA, RNA, and gene products (polypeptide chains, and then proteins).

15.2 EXAMPLES OF GENE CONTROL IN PROKARYOTIC CELLS

When nutrients are in good supply and the environment favors growth, bacterial cells tend to grow and divide indefinitely. Controls promote rapid synthesis of the enzymes required for nutrient digestion and other growth-related activities. Transcription is rapid, and translation begins even before mRNA transcripts are completed. (Bacteria have no nucleus; nothing separates the DNA from ribosomes in the cytoplasm.)

If a nutrient-degrading pathway requires several enzymes, all genes for those enzymes are transcribed, often into one continuous mRNA molecule. They are not transcribed when conditions turn unfavorable. Let's look at two examples of this all-or-nothing control of transcription, which is called **cooperative regulation**.

Negative Control of Transcription

Escherichia coli is one of the enteric bacteria residing in the mammalian gut. It lives on glucose, lactose, and other ingested nutrients. Lactose is a sugar in milk. Like other adult mammals, you probably don't drink milk around the clock. When you do drink milk—and only then—*E. coli* cells rapidly transcribe three genes for enzymes with roles in certain breakdown reactions that begin with lactose (Figure 15.2).

A promoter precedes the three genes, which are next to one another. A **promoter**, recall, is a base sequence that signals the start of a gene. Another sequence, an **operator**, intervenes between the promoter and the genes. It is a binding site for a **repressor** protein, which can block transcription. Such an arrangement, in which a promoter and operator service more than one bacterial gene, is an **operon**. Elsewhere in *E. coli* DNA, another gene codes for the repressor, which is able to bind with the operator *or* with a lactose molecule.

When the concentration of lactose is low, a repressor binds with the operator. Being a large molecule, it overlaps the promoter and *prevents* transcription (Figure 15.2b). *Thus lactose-degrading enzymes are not produced when they are not required.*

When the lactose concentration is high, the odds are greater that a lactose molecule will bind with the repressor. Binding alters the repressor's shape, so it cannot bind with the operator. RNA polymerase can now transcribe the genes (Figure 15.2c). *Thus lactose-degrading enzymes are produced only when required.*

Positive Control of Transcription

E. coli cells pay far more metabolic attention to glucose than to lactose. (Can you give a few reasons why?) Even when lactose is available, the lactose operon isn't used

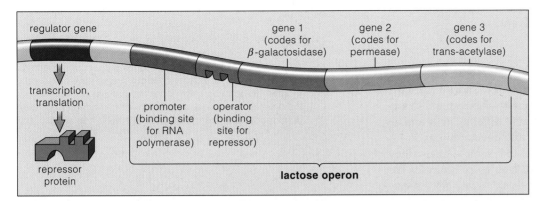

a A repressor protein exerts negative control over three genes of the lactose operon by binding to the operator and inhibiting transcription.

regulator gene

gene 1 (codes for β-galactosidase)

gene 2 (codes for permease)

gene 3 (codes for trans-acetylase)

transcription, translation

promoter (binding site for RNA polymerase)

operator (binding site for repressor)

repressor protein

lactose operon

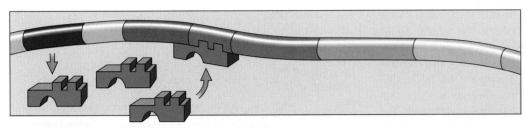

b When the lactose concentration is low, the repressor is free to block transcription. Being a bulky molecule, it overlaps the promoter and prevents binding by RNA polymerase. The enzymes (which are not needed) are not produced.

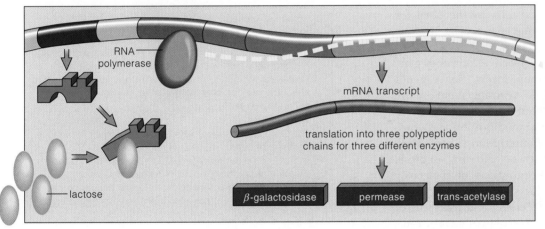

RNA polymerase

mRNA transcript

translation into three polypeptide chains for three different enzymes

lactose

β-galactosidase

permease

trans-acetylase

c At high concentration, lactose is an inducer of transcription. It binds to and distorts the shape of the repressor—which now cannot bind to the operator. The promoter is exposed and the genes can be transcribed.

Figure 15.2 Negative control of the lactose operon. The first gene of the operon codes for an enzyme that splits the disaccharide lactose into its two subunits (glucose and galactose). The second gene codes for an enzyme that transports lactose molecules into the cell. The third enzyme functions in metabolizing certain sugars.

much—*unless glucose is absent*. By contrast, the genes for glucose breakdown are transcribed continuously, at faster rates.

The lactose operon also is subject to *positive* control by an **activator protein**, called CAP. It happens that the operon's promoter is not very efficient at binding RNA polymerase. It does a much better job when CAP adheres to it first. However, CAP will not do this without being activated by a small molecule, called cAMP. Among other things, cAMP is produced from ATP—which *E. coli* produces by breaking down glucose. Not much cAMP is available when glucose is plentiful and

glycolysis is proceeding full bore. At such times, the activator protein is not primed to adhere to the promoter, and transcription of lactose operon genes slows almost to a standstill. When glucose is scarce, cAMP accumulates, the CAP-cAMP complex forms, and the lactose operon genes receive appropriate attention.

Prokaryotic cells, which must respond rapidly to changing conditions, commonly rely on a small number of regulatory proteins that exert rapid, on-off control of transcription.

15.3 GENE CONTROL IN EUKARYOTIC CELLS

In eukaryotic cells, gene control is complicated business. Like prokaryotes, these cells respond to short-term shifts in diet and level of activity. But they also respond to signals that influence gene activity in different cells and so bring about the body's growth and development.

Your own cells inherited the same genes (they descended from the same fertilized egg). Many of the genes specify proteins that are basic to any cell's structure and functions. That is why the protein subunits of ribosomes are the same from one cell to the next, as are many enzymes used in metabolism. Yet nearly all of your cells have become specialized in composition, structure, and function. This process, **cell differentiation**, happens during the development of all multicelled organisms. It arises when embryonic cells and their descendants activate and suppress a few of the total number of genes in unique, selective ways. Thus, only the precursors of red blood cells use the genes for hemoglobin. Only certain white blood cells use the genes for weapons called antibodies.

As outlined in Figure 15.3, selective control of gene expression is exerted at many levels. Some genes become amplified or undergo rearrangement before being transcribed. Many are shut down, temporarily or permanently, through chemical modification and chromosome packaging. On-off transcription is rare. Usually, transcription rates rise or fall by degrees, depending on changing concentrations of molecular signals, substrates, or products. Post-transcriptional controls govern transcript processing and transport of mature RNAs from the nucleus. Controls govern rates of translation. And controls govern the modification of new polypeptides or the activation, inhibition, and degradation of the protein products.

Many genes specify enzymes. Thus, controls over the transcription, translation, and expression of those genes extend to *all* metabolic reactions in the cell. They indirectly govern the type and number of all molecules that will be synthesized and degraded at a given time. *And so they govern all short-term and long-term aspects of cell structure and function.*

In multicelled organisms, all cells inherit the same genes and use most of them in the same way. But each differentiated cell type also activates and suppresses a small number of genes in a selective way.

Gene controls operating at many different levels underlie short-term housekeeping tasks *and* intricate, long-term patterns of growth and development.

Figure 15.3 Examples of the levels of control over gene expression in eukaryotes. Compared with prokaryotes, the cells of multicelled eukaryotes require more complex control mechanisms. Gene activity changes as the developmental program unfolds, as cells physically contact one another in the developing tissues, and as they start interacting by way of hormones and other molecular signals.

a Controls Related to Transcription. With respect to transcription, we see less of the rapid, all-or-nothing control that is common among bacteria. (The intercellular environment provides more stable operating conditions.) Many eukaryotic genes concerned with housekeeping tasks are under positive controls that promote continuous, low levels of transcription. Transcription rates for many other genes are often adjusted by degrees, rather than being abruptly switched on or off.

Also, even before some genes are transcribed, a portion of their base sequences may become amplified, rearranged, or chemically modified in temporary or reversible ways. These are not mutations; they are programmed events that affect the manner in which particular genes will be expressed (if at all).

(1) *Gene amplification*. Some cells that require enormous numbers of certain molecules temporarily increase the number of the required genes. In response to a molecular signal, multiple rounds of DNA replication sometimes produce hundreds or thousands of gene copies prior to transcription. This happens in immature amphibian eggs and in glandular cells of some insect larvae (page 242).

(2) *DNA rearrangements*. In a few cell types, many base sequences are alternative regions for the same gene. Before transcription, they are snipped out of the DNA molecule. Combinations of these snippets are spliced together as the gene's final base sequence.

This happens when B lymphocytes, a special class of white blood cells, are forming. As shown in Figure 40.9, these cells transcribe and translate the rearranged DNA into staggeringly diverse versions of the protein weapons called antibody molecules.

(3) *Chemical modification*. Numerous histones and other proteins interact with eukaryotic DNA in highly organized fashion. The DNA-protein packaging, as well as chemical modifications to particular sequences, influences gene activity. Usually, only a small fraction of genes are available for transcription. A more dramatic shutdown is called X chromosome inactivation in female mammals, as described on page 240.

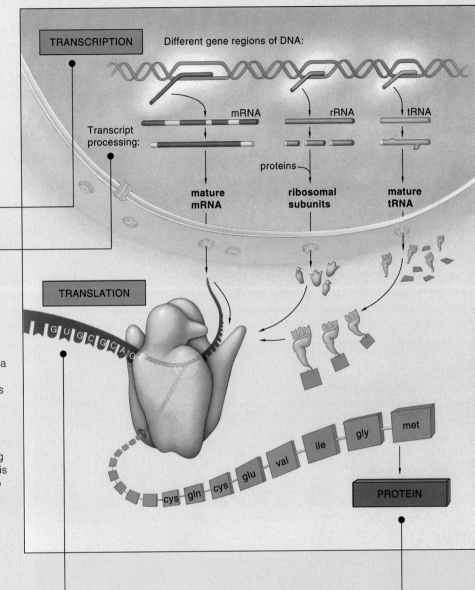

TRANSCRIPTION

Different gene regions of DNA:

mRNA   rRNA   tRNA

Transcript processing:

proteins

mature mRNA   **ribosomal subunits**   **mature tRNA**

TRANSLATION

GUGCGCAG

cys gln cys glu val ile gly met

PROTEIN

b Transcript Processing Controls. As indicated on page 222, pre-mRNA transcripts undergo modifications in the nucleus, as when introns are removed and exons are spliced together in more than one way. Thus, a transcript from a single gene may undergo *alternative splicing*.

Pre-mRNA from a gene that specifies a contractile protein (troponin-T) is like this. Enzymes excise different portions of the mRNA transcript in different cells. Thus, when the exons are spliced together, the final protein-building message is slightly different from one cell to the next. The resulting proteins are all very similar, but each is unique in a certain region of its amino acid sequence. The proteins function in slightly different ways—which may account for subtle variations in the functioning of different types of muscles in the body.

c Controls Over Translation. Controls govern when, how rapidly, and how often an mRNA transcript will be translated. Pages 616, 760, and elsewhere in the book provide elegant examples.

The stability of an mRNA transcript influences the number of protein molecules that can be produced from it. Enzymes destroy the transcripts from the poly-A tail on up (page 222). The tail's length and its attached proteins influence the pace of degradation.

Also, after leaving the nucleus, some kinds of mRNA transcripts are temporarily or permanently inactivated. For example, in unfertilized eggs, many transcripts are chemically inactivated and stored in the cytoplasm. These "masked messengers" will not be available for translation until after fertilization, when many protein molecules will be required for the early cell divisions of the new individual (page 760).

d Controls Following Translation. Before they can become fully functional, many polypeptide chains must pass through the cytomembrane system (page 64). There they undergo modification, as by having specific oligosaccharides or phosphate groups attached to them.

Also, a variety of control mechanisms govern the activation, inhibition, and stability of enzymes and other molecules that control protein synthesis. Allosteric control of tryptophan synthesis, described earlier on page 100, is an example.

15.4 EVIDENCE OF GENE CONTROL

By some estimates, the cells of a multicelled organism rarely use more than 5 to 10 percent of their genes at a given time. One way or another, most of their genes are being repressed. However, *which* genes are being expressed varies, depending on the stage of growth and development that the organism is passing through. The following sections provide you with a few specific examples of the control mechanisms that are operating at different stages.

Transcription in Lampbrush Chromosomes

As they mature, the unfertilized eggs of amphibians grow extremely fast. Growth requires numerous copies of enzymes and other proteins. The germ cells that give rise to these eggs stockpile many RNAs and ribosomes that will be required for rapid protein synthesis. As these germ cells proceed through prophase I of meiosis,

their chromosomes decondense into thousands of loops. The chromosomes look so bristly at this time, they are said to have a "lampbrush" configuration (Figure 15.4). At such times, proteins that structurally organize the DNA loosen their grip, making RNA genes accessible. Among these are the proteins of nucleosomes.

Nucleosomes are the basic unit of organization in eukaryotic chromosomes, as described on page 216. Each consists of a stretch of DNA looped around a core of histone molecules. The diagram in Figure 15.5 is based on micrographs that show the nucleosome packaging being loosened up during transcription.

X Chromosome Inactivation

A mammalian zygote that is destined to develop into a female has inherited two X chromosomes, one from the mother and one from the father. During development, one of the two chromosomes condenses in each cell. In that chromosome, cytosines in the DNA base sequences become heavily methylated, and this may block tran-

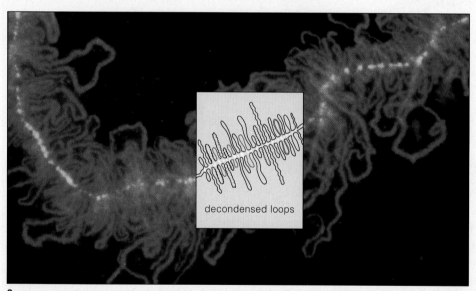

a

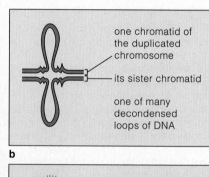

one chromatid of the duplicated chromosome

its sister chromatid

one of many decondensed loops of DNA

b

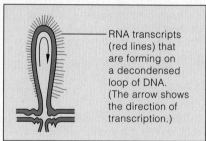

RNA transcripts (red lines) that are forming on a decondensed loop of DNA. (The arrow shows the direction of transcription.)

c

Figure 15.4 Lampbrush chromosome from a germ cell of a newt (*Notophthalmus viridiscens*). During prophase I, gene regions of this duplicated chromosome decondensed into thousands of loops. A *red* fluorescent dye labeled one of the protein components of ribosomal subunits. Ribonucleoproteins are evidence of gene activity. (A *white* fluorescent dye labeled proteins making up the duplicated chromosome's framework.)

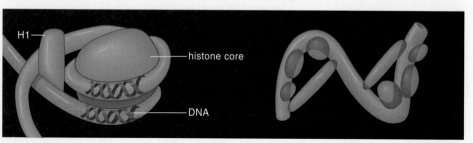

a   b

Figure 15.5 Model of changes in nucleosome structure during gene transcription. (**a**) A nucleosome consists of a portion of the DNA double helix, looped twice around a core of histone proteins. (**b**) During the transcription of specific gene regions, the tight DNA-histone packing loosens up.

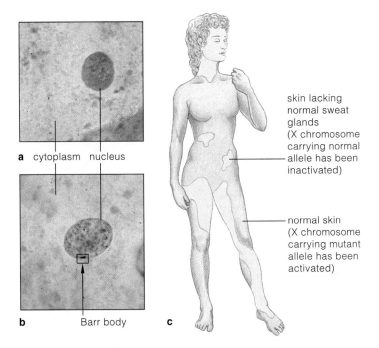

a cytoplasm nucleus

b Barr body

c

skin lacking normal sweat glands (X chromosome carrying normal allele has been inactivated)

normal skin (X chromosome carrying mutant allele has been activated)

Figure 15.6 Comparison of the interphase nucleus of somatic cells from a human male and female. (**a**) In the male's cell, the X chromosome is not condensed. (**b**) In the female's cell, a Barr body (one inactivated X chromosome) is visible as a dark-staining spot on the inside of the nuclear envelope. (**c**) Anhidrotic ectodermal dysplasia. A mosaic pattern of gene expression, arising from random X chromosome inactivation, leads to patches of skin with and without normal sweat glands.

scription of most of its genes. Condensation is a programmed event, but the outcome is random. *Either chromosome may be inactivated.* The condensed X chromosome is visible as a dark spot in the interphase nucleus (Figure 15.6). It is called a **Barr body** after its discoverer, Murray Barr.

When a maternal X chromosome is inactivated in a cell, the same chromosome becomes inactivated in all of the cell's descendants—and in tissue regions that they produce. The same thing happens following inactivation of the paternal X chromosome in a cell. By adulthood, every adult female is a "mosaic" for inactivated X chromosomes. She has patches of tissues in which maternal genes are being expressed—and patches of tissue in which paternal genes are being expressed. Because any pair of alleles on the two X chromosomes may or may not be identical, the tissue patches may or may not have the same characteristics.

The mosaic tissue effect arising from random X chromosome inactivation is called **Lyonization** (after its discoverer, Mary Lyon). We see its effects in human females who are heterozygous for a recessive allele that prevents sweat glands from forming. The allele resides on the X chromosome. In these females, the X chromosome bearing the normal dominant allele has been condensed—and genes on the one bearing the mutated allele are being transcribed in patches of skin that have no sweat glands (Figure 15.6c). This condition is called *anhidrotic ectodermal dysplasia.*

The mosaic effect is wonderfully apparent in female calico cats, which are heterozygous for black and yellow coat-color alleles on the X chromosome. Coat color in a given body region depends on which X chromosome's genes are being transcribed (Figure 15.7).

Figure 15.7 Why is this female calico cat "calico"? One X chromosome in her cells carries a dominant allele for the brownish-black pigment melanin. The allele on her other X chromosome specifies yellow fur. At an early stage of the cat's embryonic development, one of the two X chromosomes was inactivated at random in each cell that had formed by then. In all descendants of those cells, the same chromosome also became inactivated, leaving only one functional allele for the coat-color trait. We see patches of different colors, depending on which allele was inactivated in cells that formed a given tissue region. (The white patches result from a gene interaction involving the "spotting gene," which blocks melanin synthesis entirely.)

15.5 EXAMPLES OF SIGNALING MECHANISMS

A dizzying variety of signals influence gene activity. The following examples from animals and plants will give you an idea of their action at the molecular level.

Hormonal Signals

Hormones are major signaling molecules that can stimulate or inhibit gene activity in target cells. Any cell with receptors for a given hormone is a target. Animal cells secrete hormones into interstitial fluid. Most hormones are picked up by the bloodstream and distributed to cells some distance away. Plant hormones do not travel far from cells that secrete them.

Certain hormones bind to membrane receptors at the surface of target cells. Others enter target cells, bind with regulatory proteins, and so help initiate transcription. In many cases, a hormone molecule must first touch bases, so to speak, with an enhancer. **Enhancers** are base sequences that serve as binding sites for suitable activator proteins. They may or may not be adjacent to a promoter in the same DNA molecule. When some distance separates the two, a loop forms in the DNA and brings them together. RNA polymerase binds avidly to this looped complex.

Consider the effect of ecdysone, a hormone with key roles in the life cycle of many insects. Larvae of these insects grow rapidly and feed continuously on plant or animal tissues. The larvae require copious amounts of saliva to prepare food for digestion. In their salivary gland cells, the DNA has been replicated repeatedly. The copies of the DNA molecules have remained together in parallel array, forming what is called the "polytene chromosome." In response to ecdysone, multiple copies of genes in the DNA are rapidly transcribed. The gene regions affected by this hormonal signal puff out during transcription (Figure 15.8). Afterward, translation of the mRNA transcripts produces the protein components of saliva.

In vertebrates, some hormones have widespread effects on gene expression in many cell types. For instance, the pituitary gland secretes somatotropin (growth hormone). This hormonal signal stimulates synthesis of the proteins required for cell division and, ultimately, the body's growth. Most cells have receptors for somatotropin.

Other vertebrate hormones signal only certain cells at certain times. Prolactin, a secretion from the pituitary gland, is like this. Beginning a few days after a female mammal gives birth, prolactin can be detected in the bloodstream. Prolactin activates genes in certain cells of mammary glands. Those genes alone have responsibil-

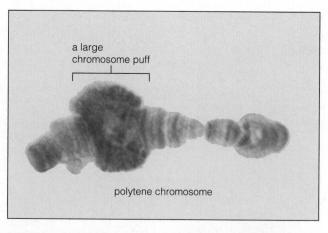

Figure 15.8 Visible evidence of transcription in a polytene chromosome from a midge larva. (A midge is a type of insect.) Staining techniques reveal banding patterns in these chromosomes, in which the DNA has been replicated repeatedly.

Ecdysone, a hormone, serves as a regulatory protein that helps promote transcription of the amplified genes in these chromosomes. Where genes are being transcribed, the chromosome loosens and puffs out. The puffs become large and diffuse when transcription is most intense.

ity for milk production. Liver cells and heart cells have those genes also, but they have no means of responding to signals from prolactin.

Explaining hormonal control of gene activity is like explaining a full symphony orchestra to someone who has never seen one or heard it perform. Many separate parts must be defined before their interactions can be understood! We will return to this topic, starting with Chapter 37 on the endocrine system. As you will see in Chapters 44 and 45, some of the most elegant examples of hormonal controls are drawn from studies of animal reproduction and development.

Sunlight as a Signal

Plant a few seeds from a corn or bean plant in a pot that contains moist, nutrient-rich soil. Let the seeds germinate—but keep them in total darkness. After eight days have passed, they will have developed into seedlings that are spindly and pale, owing to the absence of chlorophyll (Figure 15.9). Next, expose the seedlings to a single burst of dim light from a flashlight. Within ten minutes, they will start converting stockpiled molecules to active forms of chlorophyll, the light-trapping pigment molecules necessary for photosynthesis!

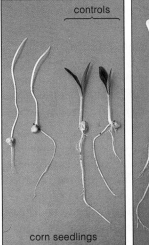

corn seedlings   bean seedlings

Figure 15.9 Effect of the absence of light on corn and bean seedlings. The two seedlings to the right in each photograph served as the control group; they were grown in a greenhouse. The two seedlings positioned next to them were grown in total darkness for eight days. The dark-grown plants could not convert their stockpiled precursors of chlorophyll molecules to active form, and they never did green up.

Phytochrome, a blue-green pigment, is a signaling molecule that helps plants adapt to changes in light conditions. This molecule exists in active and inactive forms. As described on page 538, it becomes inactive at sunset, at night, or even in shade, where far-red wavelengths predominate. The molecule becomes activated at sunrise, when red wavelengths dominate the sky. As days alternate with nights, as days grow shorter and then longer with the changing seasons, the quantity of incoming red or far-red wavelengths varies. By controlling phytochrome activity, these variations control transcription of certain genes at certain times of day and year. The genes specify enzymes and other proteins with roles in germination, stem elongation, branching, leaf expansion, and formation of flowers, fruits, and seeds.

Phytochrome control of gene transcription may work along the lines of the model shown in Figure 15.10. Experiments conducted by Elaine Tobin and her coworkers at the University of California, Los Angeles, provide evidence in favor of such a model. Working with dark-grown seedlings of duckweed (*Lemna*), they discovered a marked increase in the number of certain mRNA transcripts following a one-minute exposure to red light. Exposure enhanced transcription of the genes for proteins that bind the chlorophylls and for the enzyme that mediates carbon fixation. All of these molecules are required for the development and greening of chloroplasts.

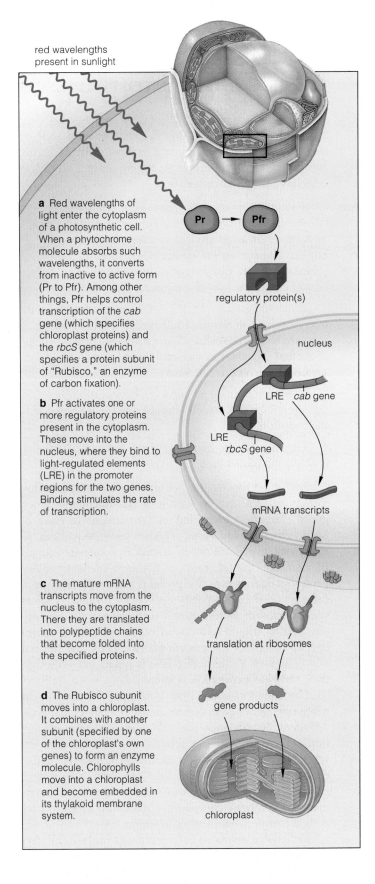

red wavelengths present in sunlight

a Red wavelengths of light enter the cytoplasm of a photosynthetic cell. When a phytochrome molecule absorbs such wavelengths, it converts from inactive to active form (Pr to Pfr). Among other things, Pfr helps control transcription of the *cab* gene (which specifies chloroplast proteins) and the *rbcS* gene (which specifies a protein subunit of "Rubisco," an enzyme of carbon fixation).

b Pfr activates one or more regulatory proteins present in the cytoplasm. These move into the nucleus, where they bind to light-regulated elements (LRE) in the promoter regions for the two genes. Binding stimulates the rate of transcription.

c The mature mRNA transcripts move from the nucleus to the cytoplasm. There they are translated into polypeptide chains that become folded into the specified proteins.

d The Rubisco subunit moves into a chloroplast. It combines with another subunit (specified by one of the chloroplast's own genes) to form an enzyme molecule. Chlorophylls move into a chloroplast and become embedded in its thylakoid membrane system.

Pr → Pfr

regulatory protein(s)

nucleus

LRE   *cab* gene

LRE
rbcS gene

mRNA transcripts

translation at ribosomes

gene products

chloroplast

Figure 15.10 A proposed model for the mechanism by which phytochrome helps control gene transcription in plants. Light of red wavelengths converts the phytochrome molecule from inactive form (here designated Pr) to active form (Pfr). In this form, the phytochrome can serve as a regulator of transcription.

15.6 GENES IMPLICATED IN CANCER

Having touched on the kinds of gene controls that exist in eukaryotes, let's conclude the chapter with a closer look at what may be happening to those controls during cancerous transformations. At the minimum, all cancer cells have these characteristics:

1. *Profound changes in the plasma membrane and cytoplasm.* Membrane permeability is amplified. Some membrane proteins are lost or altered; new ones appear. The cytoskeleton becomes disorganized, shrinks, or both. Enzyme activity shifts, as in an amplified reliance on glycolysis.

2. *Abnormal growth and division.* Controls that prevent overcrowding in tissues are lost. Cell populations increase to high densities. New proteins trigger abnormal increases in the small blood vessels that service the growing cell mass.

3. *Weakened capacity for adhesion.* Recognition proteins are altered or lost, so cells cannot remain anchored in the proper tissue.

4. *Lethality.* Unless cancer cells are eradicated, they will kill the individual.

Any gene having the potential to induce the transformations leading to cancer is an **oncogene**. Such genes were first identified in retroviruses (a class of RNA viruses). However, certain gene sequences in many organisms are nearly identical to oncogenes, yet they rarely trigger cancer! These sequences, called **proto-oncogenes**, specify proteins required for normal cell function. Some specify regulatory proteins. Others specify growth factors (transcriptional signals sent by one cell to trigger growth in other cells). Still others specify receptors for growth factors. Thus, the normal expression of proto-oncogenes is vital, even though their *abnormal* expression is lethal.

The transformation may begin with mutations in proto-oncogenes or in certain control elements that govern them. For instance, this happens with certain insertions of viral DNA into cellular DNA. It can happen when **carcinogens** (cancer-inducing mutagens) cause changes in the DNA. Ultraviolet radiation, x-rays, and gamma rays are common carcinogens. So are many natural and synthetic compounds, including asbestos and certain components of tobacco smoke.

Yet cancer seems to be a multistep process, involving more than one oncogene. Thus, in 1994, researchers identified three such genes that contribute to nearly all colon cancers. About 1 of every 200 individuals carries these oncogenes. As a result of this discovery, it may become possible to diagnose carriers early and observe them closely for cancer development.

SUMMARY

1. Cells exert control over gene expression—that is, which gene products appear, at what times, and in what amounts. At any time, the control mechanisms that are operating depend on the type of cell and its functions, on prevailing chemical conditions, and on outside signals that can change the cell's activities.

2. Regulatory proteins, enzymes, and hormones are common control elements. They interact with one another, with DNA and RNA, and with gene products (polypeptide chains, the building blocks of proteins).

   a. As an example, in prokaryotes, control is exerted at promoters (DNA binding sites that signal the start of a gene), operators (binding sites for regulatory proteins that can inhibit transcription), and enhancers (binding sites for activator proteins that can promote transcription).

   b. Different kinds of controls operate before, during, and after gene transcription and translation.

3. Two common types of control systems operate in cells to block or enhance transcription.

   a. With negative control systems, a regulatory protein binds at a specific DNA sequence and prevents one or more genes from being transcribed; its effect is reversed by the action of a regulatory protein molecule. Repressor proteins in prokaryotes are examples.

   b. With positive control systems, a regulatory protein binds to DNA and promotes the initiation of transcription. Activator proteins, which bind at enhancer sites in eukaryotic DNA, are examples.

4. Most prokaryotes (bacteria) do not live long and do not require great numbers of genes to grow, divide, and reproduce. Commonly, systems involving a few control elements enhance transcription rates only when gene products are required. This is true of operons, which are groupings of related genes and their control elements.

5. Eukaryotes require complex gene controls. Besides changing rapidly in response to short-term shifts in the surroundings, gene activity must change over the long term, during growth and development. Then, cells multiply, physically contact one another in tissues, and chemically interact, as by hormonal signals.

6. All cells in a multicelled eukaryote inherit the same genes, but different cell types activate and suppress some fraction of the genes in different ways.

7. Selective gene control brings about cell differentiation. In other words, cell lineages become specialized in appearance, composition, and function. Later chapters describe the phenotypic consequences of selected gene control.

Review Questions

1. Define these terms: promoter, operator, repressor protein, and activator protein. *236, 237*

2. Cells depend on controls over which gene products are synthesized, at what times, at what rates, and in what amounts. Describe one type of control over transcription in *E. coli*, a type of prokaryote. Then list five general kinds of gene controls involved in eukaryotes. *236–239*

3. Define cell differentiation. How does it arise? *238*

4. A plant, fungus, or animal is composed of diverse cell types. How might this diversity arise, given that the body cells in each organism inherit the same set of genetic instructions? *238*

5. What are the characteristics of cancer cells? Explain the difference between a benign tumor and one that is malignant. *244, 234*

Self-Quiz *(Answers in Appendix IV)*

1. The expression of a given gene depends on _____ .
   a. cell type and functions   c. environmental signals
   b. changing chemical   d. all of the above
      conditions

2. Regulatory proteins are part of control systems that interact with _____ to govern gene expression.
   a. DNA   c. gene products
   b. RNA   d. all of the above

3. A negative control system that governs the lactose operon in *E. coli* operates during _____ .
   a. replication   c. translation
   b. transcription   d. post-translation

4. A base sequence that signals the start of a gene is called a _____ .
   a. promoter   c. enhancer
   b. operator   d. activator protein

5. An operon is a control system that most typically governs _____ .
   a. a bacterial gene   c. genes of all types
   b. a eukaryotic gene   d. DNA replication

6. Prokaryotic cells rely heavily on _____ controls over transcription.
   a. slow, continuous   c. slow, on-off
   b. rapid, continuous   d. rapid, on-off

7. Cell differentiation _____ .
   a. occurs in all multicelled organisms
   b. requires different genes in different cells
   c. involves selective gene expression
   d. both a and c
   e. all of the above

8. In eukaryotic cells, controls over gene expression include _____ .
   a. amplifying genes   d. processing RNA transcripts
   b. rearranging DNA   e. all of the above
   c. chemically modifying
      DNA

9. Lampbrush chromosomes are evidence of _____ .
   a. mutated genes   c. intense transcription
   b. heavily methylated   d. both b and c
      DNA

10. Inactivation of one of the two X chromosomes in cells of mammalian females appears to involve regions of _____ .
    a. mutated genes   c. intense transcription
    b. heavily methylated   d. both a and b
       DNA

11. In eukaryotic cells, various types of hormones interact with _____ .
    a. membrane receptors   c. enhancers
    b. regulatory proteins   d. all of the above

12. Match the terms with their most suitable descriptions.
    _____ phytochrome   a. helps plants adapt to changes
    _____ enhancer   in light conditions
    _____ Barr body   b. a gene having the potential to
    _____ proto-oncogene   induce cancerous transformation
    _____ Lyonization   c. normal expression is vital,
    _____ oncogene   abnormal expression is lethal
                             d. condensed X chromosome in
                                an interphase nucleus
                             e. mosaic tissue effect arising
                                from random X chromosome
                                inactivation
                             f. base sequence that serves as
                                binding site for activator protein

Selected Key Terms

activator protein *237*   oncogene *244*
Barr body *241*   operator *236*
cancer *235*   operon *236*
carcinogen *244*   phytochrome *243*
cell differentiation *238*   positive control system *236*
cooperative regulation *236*   promoter *236*
enhancer *242*   proto-oncogene *244*
hormone *242*   regulatory protein *236*
Lyonization *241*   repressor *236*
metastasis *234*   tumor *234*
negative control system *236*

Readings

Feldman, M., and L. Eisenbach. November 1988. "What Makes a Tumor Cell Metastatic?" *Scientific American* 259(5):60–85.

Kupchella, C. 1987. *Dimensions of Cancer.* Belmont, California: Wadsworth.

Murray, A., and M. Kirschner. March 1991. "What Controls the Cell Cycle." *Scientific American* 264(3):56–63.

Ptashne, M. January 1989. "How Gene Activators Work." *Scientific American* 260(1):41–47.

Weintraub, H. January 1990. "Antisense RNA and DNA." *Scientific American* 262(1):40–46.

16 RECOMBINANT DNA AND GENETIC ENGINEERING

Make Way for Designer Genes

In 1990, when she was four years old, Ashanthi DeSilva received a historic genetic reprieve. Ashanthi was born without defenses against viruses, bacteria, and other agents of disease. She has no immune system. Of her forty-six chromosomes, one bears a defective gene that normally would specify adenosine deaminase (ADA), an enzyme.

Without the enzyme, Ashanthi's cells cannot properly break down excess amounts of a nucleotide (AMP), and a reaction product accumulates that is toxic to lymphoblasts in the bone marrow. Lymphoblasts give rise to white blood cells—the immune system's army. With too few of those cells (or none at all), the outcome is a *severe combined immune deficiency* (SCID), and it leads to a devastating set of disorders. In this particular case, symptoms included dangerous infections of the ears and lungs, high fevers, severe diarrhea, and an inability to gain weight. Ashanthi was so vulnerable to germs, her parents would not let her attend school.

Bone marrow transplants help some individuals with SCID when the donated lymphoblasts go on to produce functional white blood cells. ADA injections help others. But these are not permanent cures.

Figure 16.1 White blood cells on patrol inside a blood vessel. Ashanthi DeSilva and other individuals have drastically reduced numbers of these infection-fighting cells—or none at all. They are candidates for gene therapy, one of the beneficial applications of recombinant DNA research.

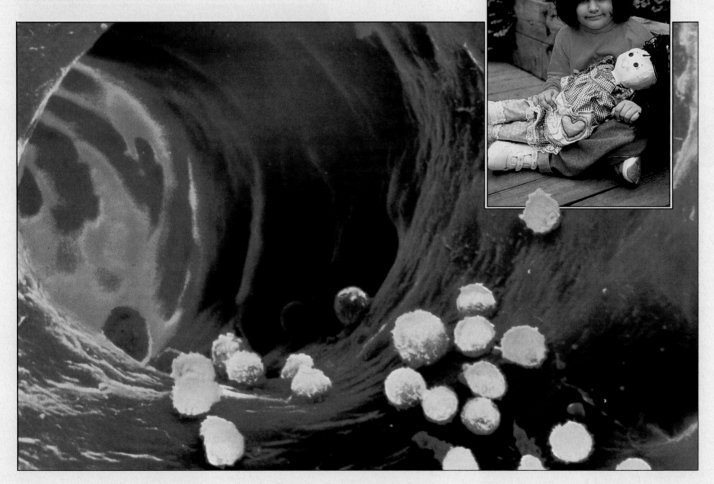

Given the options, the girl's parents consented to the first federally approved gene therapy test for humans. Using recombinant DNA methods of the sort described in this chapter, medical researchers had already identified and isolated the ADA gene, and they were producing quantities of it in the laboratory. They also were able to splice those genes into the genetic material of a harmless type of virus that could serve roughly the same function as a hypodermic needle. That is, by infecting targeted cells, the harmless virus could deliver copies of the "good" gene to them.

The researchers harvested some of Ashanthi's white blood cells, cultured them in petri dishes, then exposed them to the modified virus. The ADA gene became incorporated in some cells—which started to synthesize ADA. Later, the researchers inserted about a billion copies of the genetically modified cells into the girl's bloodstream.

The gene therapy worked. At first Ashanthi received additional infusions of fortified cells every month. Now she receives them once a year. And she is attending school. Four more ADA-deficient patients also have started treatment—and they are doing well at this writing.

Some medical researchers are now employing lymphoblasts. They retrieved small amounts of lymphoblasts from blood in the umbilical cord (which is discarded following childbirth). They exposed these stem cells to viruses that could deliver copies of the ADA gene into them. Afterward, they exposed the cells to factors that stimulated growth and division, then reinserted the cells into the newborn patients. It is too soon to tell whether these patients have permanent stem cells that are producing functional copies of the ADA gene. If this turns out to be the case, gene therapy will have given them a continuous, lifelong supply of the crucial enzyme—and of functional disease fighters (Figure 16.1).

As this example suggests, recombinant DNA technology has staggering potential for medicine. It has equally staggering potential for agriculture and industry. It does not come without risks. With this chapter, we consider some basic aspects of the new technology, and we address ecological, social, and ethical questions related to its application.

KEY CONCEPTS

1. Genetic experiments have been occurring in nature for billions of years as a result of gene mutations, crossing over and recombination, and other events. Humans are now purposefully bringing about genetic changes by way of recombinant DNA technology. Such enterprises are called genetic engineering.

2. With this technology, researchers isolate, cut, and splice together gene regions from different species, then greatly amplify the number of copies of the genes that interest them. The genes, and in some cases their protein products, are produced in quantities that are large enough for research and for practical applications.

3. Three activities are at the heart of recombinant DNA technology. First, procedures based on specific enzymes are used to cut DNA molecules into fragments. Second, the fragments are inserted into cloning tools, such as plasmids. Third, fragments containing the genes of interest are identified, then copied rapidly and repeatedly.

4. Genetic engineering involves isolating, modifying, and inserting particular genes back into the same organism or into a different one. The goal is to modify traits influenced by those genes in beneficial ways. Human gene therapy, which focuses on controlling or curing genetic disorders, is an example.

5. The new technology raises social, legal, ecological, and ethical questions regarding its benefits and risks.

16.1 RECOMBINATION IN NATURE— AND IN THE LABORATORY

For more than 3 billion years, nature has been conducting genetic experiments by way of mutation, then by crossing over between chromosomes and other events. Genetic messages have changed countless times, and this is the source of life's diversity.

Figure 16.2 Plasmids (*blue* arrows) released from a ruptured bacterial cell (*Escherichia coli*).

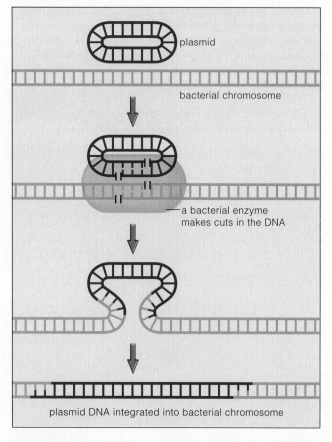

plasmid

bacterial chromosome

a bacterial enzyme makes cuts in the DNA

plasmid DNA integrated into bacterial chromosome

Figure 16.3 Plasmid integration into a bacterial chromosome.

For many thousands of years, we humans have been changing genetically based traits of species. Through artificial selection, we produced modern crop plants and new breeds of cattle, birds, dogs, and cats from wild ancestral stocks. We developed meatier turkeys, sweeter oranges, seedless watermelons, flamboyant ornamental roses, and other wonderfully useful or novel plants. We produced the tangelo (tangerine × grapefruit) and the mule (donkey × horse).

Currently, researchers employ **recombinant DNA technology** to analyze genetic changes. With this technology, they cut and splice together DNA from different species, then insert the modified molecules into bacteria or other types of cells that engage in rapid replications and cell divisions. The cells copy the foreign DNA right along with their own. In short order, huge bacterial populations can produce useful quantities of recombinant DNA molecules. The new technology also is the basis of **genetic engineering**. Genes are being isolated, modified, and inserted back into the same organism or into a different one.

Believe it or not, this astonishing technology had its origins in the innards of bacteria. Bacterial cells have a single chromosome, a circular DNA molecule with all the genes needed for growth and reproduction. Many bacteria also have **plasmids**, which are small, circular molecules of "extra" DNA having a few genes (Figure 16.2). Replication enzymes can copy plasmid DNA, just as they copy the chromosomal DNA.

Many bacteria can transfer plasmid genes to a bacterial neighbor (page 350). Transferred plasmids may even get integrated into a recipient's chromosome, forming a recombinant DNA molecule (Figure 16.3). Specific enzymes mediate the recombination events. A bacterial enzyme recognizes a short nucleotide sequence present in both the plasmid and the chromosome. It cuts the molecules at that sequence, then another enzyme splices the cut ends together. In nature, viruses as well as bacteria dabble in gene transfers and recombinations. So do most eukaryotic organisms.

As we turn now to some key features of the new technology, keep these concepts in mind:

---

1. With recombinant DNA technology, DNA from different species is cut and spliced together, then the recombinant molecules are amplified (by way of copying mechanisms) to produce useful quantities of the genes of interest.

2. The new technology is being used for basic research. It also is being used for genetic engineering—the deliberate modification of genes, followed by their insertion into the same individual or a different one.

16.2 PRODUCING RESTRICTION FRAGMENTS

Bacteria are equipped with many restriction enzymes. These enzymes cut apart foreign DNA that enters the bacterial cell, often during viral infection. Each type of restriction enzyme makes a cut wherever a specific, very short nucleotide sequence occurs in the DNA. Cuts at two identical sequences in the same molecule produce a fragment. Because some types of enzymes make staggered cuts, the fragments may have single-stranded portions at both ends. These are sometimes referred to as sticky ends:

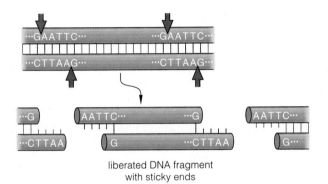

liberated DNA fragment
with sticky ends

By "sticky," we mean that the short, single-stranded ends of a DNA fragment can base-pair with any other DNA molecule that also has been cut by the same restriction enzyme.

Suppose you use the same restriction enzyme to cut plasmids and the DNA from a human cell. When you mix the cut-up molecules together, they base-pair at the cut sites. Then you add **DNA ligase** to the mixture. This enzyme seals the sugar-phosphate backbone of DNA at the cut sites, just as it does during DNA replication. In this way, you create "recombinant plasmids," which have pieces of DNA from another organism inserted into them.

You now have a **DNA library**—a collection of DNA fragments, produced by restriction enzymes, that have been incorporated into plasmids (Figure 16.4).

Recombination is made possible by restriction enzymes that make specific cuts in DNA molecules and by DNA ligases that seal the cut ends.

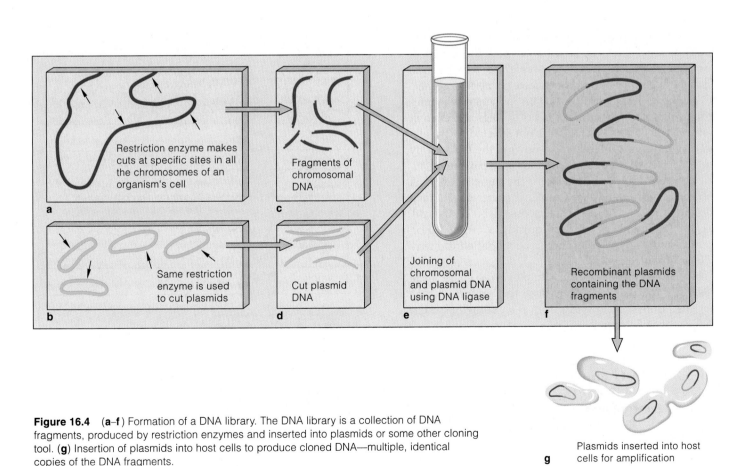

Figure 16.4 (**a–f**) Formation of a DNA library. The DNA library is a collection of DNA fragments, produced by restriction enzymes and inserted into plasmids or some other cloning tool. (**g**) Insertion of plasmids into host cells to produce cloned DNA—multiple, identical copies of the DNA fragments.

a
Restriction enzyme makes cuts at specific sites in all the chromosomes of an organism's cell

b
Same restriction enzyme is used to cut plasmids

c
Fragments of chromosomal DNA

d
Cut plasmid DNA

e
Joining of chromosomal and plasmid DNA using DNA ligase

f
Recombinant plasmids containing the DNA fragments

g
Plasmids inserted into host cells for amplification

16.3 WORKING WITH DNA FRAGMENTS

Amplification Procedures

A DNA library is almost vanishingly small. It must be amplified—copied again and again—into useful amounts. One way to do this is to employ "factories" of bacteria, yeasts, or some other kind of cell that reproduces rapidly and can take up plasmids. In short order, a growing population of such cells can amplify a DNA library. Their repeated replications and divisions yield cloned DNA that has been inserted into plasmids. The "cloned" part of the name simply refers to the multiple, identical copies of DNA fragments.

The **polymerase chain reaction**, or **PCR**, is an alternative method of amplifying fragments of DNA. The reactions proceed in test tubes, not in microbial factories. First, researchers identify short nucleotide sequences located just before and just after a region of DNA from an organism that interests them. Then they synthesize primers (nucleotide sequences that will base-pair with the ones in the DNA). They mix the primers with enzyme molecules, free nucleotides, and all the DNA from one of the organism's cells. Then they subject the mixture to precise temperature cycles.

During the cycles, the two strands of all the DNA molecules unwind from each other. Then, as Figure 16.5 indicates, the short sequences become positioned at the specific target sites, in accordance with the exposed, complementary bases. **DNA polymerase**, a replication enzyme, recognizes the synthesized primers as signals to get busy. This particular DNA polymerase comes from a bacterium that thrives in hot springs and even in hot water heaters. The enzyme is not affected by the elevated temperatures required to unwind the DNA or by the lower temperatures required for base-pairing.

With each round of reactions, the number of DNA molecules doubles. For example, if there are 10 such molecules in the test tube, there soon will be 20, then 40, 80, 160, 320, and so on. Very quickly, a target region from a single DNA molecule can be amplified to many *billions* of molecules.

Thus PCR can amplify samples with very little DNA. As you will see shortly, such tiny samples can be obtained from some fossils—even from a single hair left at the scene of a crime.

Sorting Out and Sequencing Specific DNA Fragments

Gel Electrophoresis of DNA When restriction enzymes cut DNA, the resulting fragments are not all the same length. Researchers can employ gel electrophoresis to separate the fragments from one another

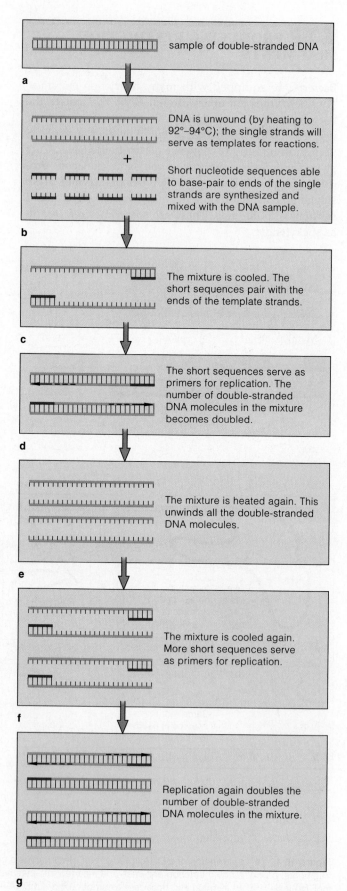

Figure 16.5 The polymerase chain reaction (PCR).

Figure 16.6 One method of sequencing DNA, as initially developed by Frederick Sanger. The method can be used to determine the nucleotide sequence of specific DNA fragments, as this example illustrates.

a Single-stranded DNA fragments are added to a solution in four different test tubes:

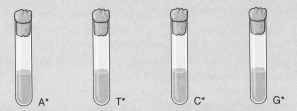

All four tubes contain DNA polymerases, short nucleotide sequences that can serve as primers for replication, and the nucleotide subunits of DNA (that is, A, T, C, and G). One of the four kinds of nucleotides in each tube is modified and radioactively labeled. *Which* one is different in each tube. We can designate these as A*, T*, C*, and G*. Let's follow what happens in the tube with the A* subunits.

b As expected, the DNA polymerase recognizes a primer that has become attached to a fragment, which it uses as a template strand. The enzyme assembles a complementary strand according to base-pairing rules (A only to T, and C only to G). Sooner or later, the enzyme picks up an A* subunit for pairing with a T on the template strand. The modified nucleotide is a chemical roadblock—it prevents the enzyme from adding more nucleotides to the growing complementary strand. In time, the tube contains radioactively labeled strands of different lengths, as dictated by the location of each A* in the sequence:

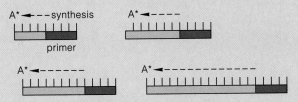

The same thing happens in the other three tubes, with strand lengths dictated by the location of T*, C*, and G*.

c DNA from each of the four tubes is placed in four parallel lanes in the same gel. Then it can be subjected to electrophoresis. The resulting nucleotide sequence can be read off the resulting bands in the gel. Look at the numbers running down the side of this diagram:

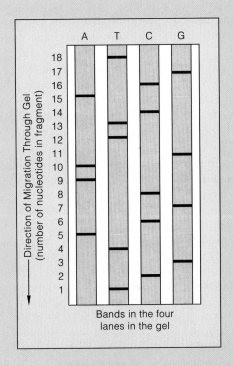

Bands in the four lanes in the gel

Start with "l" and read across the four lanes (A, T, C, G). As you can see, T is closest to the start of the nucleotide sequence; it has migrated farthest through the gel. At "2," the next nucleotide is C, and so on. The entire sequence, read from the first nucleotide to the last, is

T C G T A C G C A A G T T C A C G T

Now, using the rules of base-pairing, you can deduce the sequence of the DNA fragment that served as the template.

according to their length. (As you read on page 220, this procedure also can be used to separate a sample of protein molecules according to their overall electric charge and size.)

A sample of restriction fragments is placed near one end of a slab of gel that is mounted on a glass plate and bathed in an appropriate solution. When an electric current passes through the gel, the fragments respond to it (because of their charged phosphate groups). The extent to which fragments of a given length migrate through the gel depends only on their size—larger ones cannot move as fast through it.

DNA Sequencing Once DNA fragments have been sorted out according to length, researchers can work out the nucleotide sequence of each type. The Sanger method of DNA sequencing will give you a sense of one of the ways this can be done (Figure 16.6).

Restriction fragments can be rapidly amplified by PCR and by large populations of bacterial or yeast cells.

Restriction fragments can be sorted out according to length, and their nucleotide sequences can be determined.

16.4 RFLP ANALYSIS

Imagine that a researcher has used restriction enzymes to break some DNA from one of your cells and now is subjecting the DNA to gel electrophoresis. As the fragments migrate through a slab of gel in response to an electric current, the large ones can't move through it as fast as small ones do. They separate from one another, forming bands in the gel. Now suppose she does the same to DNA samples from other people. When she compares the resulting banding patterns, she finds small variations among them. Variations in the banding patterns of DNA fragments from different individuals have a name. They are called **restriction fragment length polymorphisms**, or **RFLPs** for short. They arise because some base sequences in the DNA vary in molecular form from one person to the next. The molecular differences shift the number and location of sites where restriction enzymes make their cuts.

RFLP analysis is a wonderful new procedure for basic research. As the *Focus* essay suggests, it also has a few rather startling applications in society at large.

16.5 MODIFIED HOST CELLS

Use of DNA Probes

When you mix DNA with living cells, how can you find out which ones take up the DNA and contain a gene of interest? You can use **DNA probes**—short DNA sequences synthesized from radioactively labeled nucleotides (page 21). Part of the probe must be able to base-pair with part of the gene. Base-pairing between nucleotide sequences from different sources is called **nucleic acid hybridization**.

The first step is to "select" cells that have taken up the recombinant plasmids. As it happens, most plasmids contain genes that make their bacterial owners resistant to antibiotics. You put the prospective host cells on a culture medium that has the antibiotic added to it. The antibiotic prevents growth of all cells *except* the ones housing plasmids (because the plasmids carry the antibiotic-resistance genes).

The next step is to locate the particular cells that contain the recombinant plasmids with the gene being

---

Focus on Science

RIFF-lips and DNA Fingerprints

The DNA molecules of any two people are alike in most respects. Yet they also differ in some base sequences, and this affects the lengths of restriction fragments cut from the DNA. Comparisons of DNA electrophoresis patterns, such as the one in Figure *a*, reveal RFLPs (pronounced RIFF-lips), or "restriction fragment length polymorphisms" among individuals. In fact, each individual of any sexually reproducing species has a **DNA fingerprint**, a unique array of RFLPs, inherited from each parent in a Mendelian pattern.

RFLP analysis has uses in basic research, such as the human genome project. This is an effort to establish the nucleotide sequence of human DNA, with its 3.2 billion base pairs. As another example, evolutionary biologists have analyzed RFLPs to decipher DNA from mummies, from fossilized insects and plants, even from mammoths and humans that were preserved many thousands of years ago in glacial ice. Enterprising investigators used RFLP analysis to confirm that bones exhumed from a shallow pit in Siberia belonged to five members of the Russian imperial family, all shot to death in 1918.

RFLP analysis has medical applications. For example, researchers have identified unique restriction sites within and near several mutated genes, including those responsi-

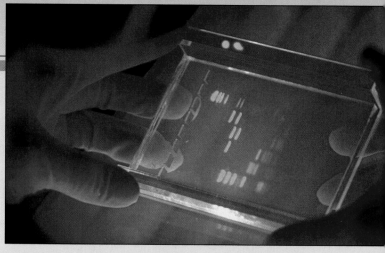

a Running from left to right, columns of separated DNA fragments (labeled with a dye that fluoresces *pink*). The banding pattern reflects differences in their lengths.

ble for sickle-cell anemia, cystic fibrosis, and other genetic disorders. The sites are used to determine whether an individual carries the mutated gene. Thus RFLP analysis can be used for prenatal diagnosis (page 204).

As another example, paternity and maternity cases can be resolved by carefully comparing the child's DNA fingerprint with the disputed parent's. As a final example, murderers can be identified if they lose even a few drops of blood at the scene of the crime or if drops of the victim's blood stain their clothing. A bloodstain may provide enough DNA to identify the perpetrator. Similarly, a rapist can be identified from semen recovered from the victim.

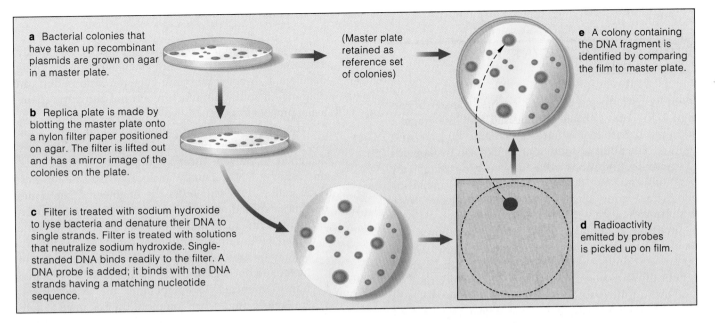

a Bacterial colonies that have taken up recombinant plasmids are grown on agar in a master plate.

b Replica plate is made by blotting the master plate onto a nylon filter paper positioned on agar. The filter is lifted out and has a mirror image of the colonies on the plate.

c Filter is treated with sodium hydroxide to lyse bacteria and denature their DNA to single strands. Filter is treated with solutions that neutralize sodium hydroxide. Single-stranded DNA binds readily to the filter. A DNA probe is added; it binds with the DNA strands having a matching nucleotide sequence.

(Master plate retained as reference set of colonies)

e A colony containing the DNA fragment is identified by comparing the film to master plate.

d Radioactivity emitted by probes is picked up on film.

Figure 16.7 Use of a DNA probe to identify a colony of bacterial cells that have taken up plasmids containing a specific DNA fragment.

studied. As the cells divide, they form colonies. Suppose the colonies are on agar (a gel-like substance) in a petri dish. You blot the agar against a nylon filter. Some cells stick to the filter in locations that mirror the locations of the original colonies (Figure 16.7). You use solutions to rupture the cells, fix the released DNA onto the filter, and make the double-stranded DNA molecules unwind. Then you add the DNA probes. The probes hybridize only with the gene region having the proper base sequence. Probe-hybridized DNA emits radioactivity and allows you to tag the colonies that harbor the gene of interest.

Use of cDNA

We study genes because we want to learn about or use their protein products. Even if a host bacterial cell takes up a gene, however, the protein product may not materialize. For example, recall that human genes contain noncoding sequences (introns). The mRNA transcripts of these genes cannot be used until the introns are snipped out and coding regions (exons) are spliced together into mature form. Bacterial enzymes don't recognize the splice signals, so bacterial host cells cannot always properly translate human DNA.

Sometimes this problem can be circumvented by using **cDNA**, which has been "copied" from a mature mRNA transcript for the desired gene from which the introns have been removed. By a backwards process called **reverse transcription**, a matching DNA strand is assembled on mRNA. The result is a hybrid DNA-RNA molecule (Figure 16.8). Enzymes remove the RNA, then

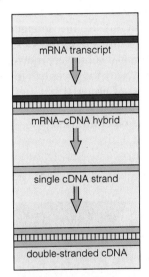

a An mRNA transcript of a desired gene is used as a template for assembling a DNA strand. An enzyme (reverse transcriptase) does the assembling.

b An mRNA-DNA hybrid molecule results.

c Enzyme action removes the mRNA and assembles a second strand of DNA on the remaining DNA strand.

d The result is double-stranded cDNA, "copied" from an mRNA template.

mRNA transcript

mRNA–cDNA hybrid

single cDNA strand

double-stranded cDNA

Figure 16.8 Formation of cDNA from an mRNA transcript.

they synthesize a complementary DNA strand to make double-stranded DNA.

Double-stranded cDNA can be further modified by the addition of a bacterial promoter and other signals that are required for transcription and translation. Then the modified cDNA can be inserted into a plasmid for amplification. The recombinant plasmids can be inserted into bacteria, which may then use the cDNA directions for making a protein.

Host cells that take up modified genes may be identified by procedures involving DNA probes.

Gene expression does not automatically follow the successful uptake of a modified gene by a host cell. Genes must be suitably engineered first.

16.6 GENETIC ENGINEERING OF BACTERIA

Many years have passed since the first transfer of foreign DNA into a bacterial plasmid, yet the transfer started a debate that is sure to continue into the next century. The point of contention is this: Do the benefits of gene modifications and transfers outweigh the potential dangers? Before coming to any conclusion about this, reflect on the following examples of work with bacteria, then with plants and animals.

Imagine a miniaturized factory that churns out insulin or another protein having medical value. This is an apt description for the huge stainless steel vats of genetically engineered, protein-producing bacteria.

Think of the diabetics who require insulin injections for as long as they live (page 623). Insulin is a protein hormone of the pancreas, and until the 1970s, medical supplies of it had to be obtained from cattle and pigs. Some diabetics developed allergic reactions to the foreign insulin. Also, the demand started outstripping the supplies. Then synthetic genes for human insulin were transferred into *E. coli* cells. (Can you say why the genes had to be synthesized?) This was the start of bacterial factories for human insulin and, later, growth hormone, hemoglobin, serum albumin, interferon, and other proteins.

Figure 16.9 Spraying an experimental strawberry patch in California with "ice-minus" bacteria. The sprayer used elaborate protective gear to meet government regulations in effect then.

At this writing, several other lines of genetically engineered bacteria have been established or are being developed. Among them are bacterial strains that can degrade oil spills, manufacture alcohol and other chemicals, process minerals, or leave crop plants alone.

The strains being used are harmless to begin with. They are grown in confined settings, behind barriers designed to prevent escape. As an added precaution, the foreign DNA usually includes "fail-safe" genes. Such genes are silent *unless* the engineered bacterium becomes exposed to conditions characteristic of the outside environment. Upon exposure, the genes become activated, with lethal results. For example, the foreign DNA may include a *hok* gene with an adjacent promoter of the lactose operon (page 236). Thus, if the engineered bacterium manages to escape into the environment, where lactose sugars are common, the *hok* gene is activated. The protein specified by the gene destroys membrane function and so destroys the cell.

Even so, there is concern about possible risks of introducing genetically engineered bacteria into the environment. Consider how Steven Lindow altered a bacterium that can make many crop plants less vulnerable to frost. Proteins at the bacterial surface promote the formation of ice crystals. Lindow excised the ice-forming gene from some cells. He hypothesized that spraying these "ice-minus bacteria" on strawberry plants in an isolated field just before a frost would make the plants more resistant to freezing.

Here was an organism from which a harmful gene had been deleted, yet it triggered a bitter legal debate on the risks of releasing genetically engineered microbes into the environment. The courts finally ruled in favor of allowing the genetically engineered bacteria to be released, and researchers sprayed a small patch of strawberries (Figure 16.9). Nothing bad happened. Since then, rules governing the release of genetically engineered organisms have become less restrictive.

16.7 GENETIC ENGINEERING OF PLANTS

Years ago, Frederick Steward and his coworkers cultured cells of carrot plants and induced the cells to grow into small embryos. Some embryos actually grew into whole plants. Today many plant species, including major crop plants, are regenerated from cultured cells. The culturing methods increase mutation rates, so the cultures are a source of genetic modifications.

Researchers can pinpoint a useful mutation among millions of cells. Suppose a culture medium contains a toxin that is produced by a disease agent. If a few cells have a mutated gene that confers resistance to the toxin, they will end up being the only live cells in the culture.

Figure 16.10 Gene transfers in plants. (**a**) A tumor-inducing plasmid from a common bacterium (*Agrobacterium tumefaciens*) causes crown gall tumors on willow trees and other woody plants. Scientists put this plasmid to work in cultures of plant cells, where it moves desirable genes into the plant chromosomes. (**b**) Evidence of a successful gene transfer: a modified tobacco plant that glows in the dark. A gene from a firefly has become incorporated into its own DNA and is being translated into functional protein, which in this case is the enzyme luciferase.

Figure 16.11 Examples of commercially valuable plants that benefit from genetic engineering. (**a**) Buds of cotton plants are vulnerable to worm attack. (**b**) Buds of a modified cotton plant resist attack. (**c**) A virus damages potato plants, but a modified plant (**d**) resists attack.

Now suppose plants are regenerated from the cells, then hybridized with other varieties. The hybrid plants may end up with the new gene that confers disease resistance.

Today, researchers are successfully inserting genes into cultured plant cells. For example, they have inserted DNA fragments into the "Ti" plasmid from *Agrobacterium tumefaciens*, a bacterium that infects many flowering plants. Some plasmid genes become integrated into the DNA of infected plants, and they induce the formation of crown gall tumors (Figure 16.10*a*). Before the plasmid is introduced into a plant, the tumor-inducing genes are removed from it and desired genes are inserted into it. Then the genetically modified bacterial cells are grown with cultured plant cells. Selected plants that are regenerated from infected cultures contain the foreign genes within their DNA. In some cases, the foreign genes are expressed in the plant tissues, with observable effects.

Vivid evidence of a successful gene transfer came from researchers who used *A. tumefaciens* to deliver a firefly gene into cultured tobacco plant cells. The gene codes for luciferase, an enzyme required for bioluminescence (page 103). Plants regenerated from the infected culture cells have the peculiar ability to glow in the dark (Figure 16.10*b*).

A. tumefaciens only infects the plants called dicots. Wheat, corn, rice, oats, and other major food crop plants are monocots. In some cases, genetic engineers use chemicals or electric shocks to deliver DNA directly into protoplasts (plant cells stripped of their walls). For some species, however, regenerating whole plants from protoplast cultures is not yet possible.

Not long ago, someone came up with the idea to deliver genes into cultured plant cells by shooting them with a pistol. Instead of bullets, blanks are used to drive DNA-coated, microscopic tungsten particles into the cells. Although this "gene gun" might seem analogous to using a battleship cannon to light a match, the shooters are reporting some success.

Despite the obstacles, many improved varieties of crop plants have been developed or are on the horizon. For example, certain cotton plants have been genetically engineered for resistance to worm attacks (Figure 16.11*b*). Such gene insertions are ecologically safer than pesticide applications. They kill only the targeted pest and do not interfere with beneficial insects, including the ladybird beetles that prey on aphids. Also on the horizon are genetically engineered plants that may serve as pharmaceutical factories. Two years ago, genetically engineered tobacco plants that produce human hemoglobin, melanin, and other proteins were planted in a field in North Carolina on a trial basis. A year after the trial was completed, ecologists found no trace of the foreign genes or proteins in the soil or in any plants or animals in the vicinity.

16.8 GENETIC ENGINEERING OF ANIMALS

Supermice and Biotech Barnyards

Mice were the first mammals subjected to genetic engineering experiments. Consider how R. Hammer, R. Palmiter, and R. Brinster managed to correct a hormone deficiency that causes dwarfism in mice. Such mice have trouble producing somatotropin (also called growth hormone). The researchers used a microneedle to inject the gene for rat somatotropin into fertilized mouse eggs, then they implanted the eggs into an adult female. The gene was successfully integrated into the mouse DNA. The baby mice in which the foreign gene was expressed were 1-1/2 times larger than their dwarf littermates. In other experiments, the gene for human somatotropin was transferred into a mouse embryo, where it became integrated into the DNA. A "supermouse" resulted (Figure 16.12).

Today, as part of research into the genetic basis of Alzheimer disease and other genetic disorders, several other human genes are being inserted into mouse embryos. Besides microneedles, microscopic laser beams are being used to open temporary holes in the plasma membrane of cultured cells, although such

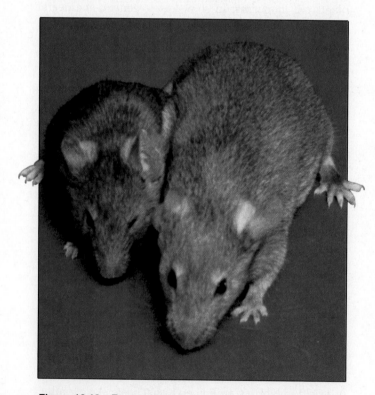

Figure 16.12 Ten-week-old mouse littermates, the one on the left weighing 29 grams, and the one on the right, 44 grams. The larger mouse grew from a fertilized egg into which the gene for human somatotropin (growth hormone) had been inserted.

methods have varying degrees of success. Retroviruses also are used to insert genes into cultured cells, virtually all of which incorporate the foreign genes into their DNA. However, the genetic material of retroviruses often undergoes rearrangments, deletions, and other alterations that render the introduced genes ineffective. There is also the possibility that viral particles can escape and infect other individuals.

Soon, "biotech barnyards" may be competing with bacterial factories as genetically engineered sources of proteins. Farm animals produce the proteins in far greater quantities, at less cost. Consider Herman, a Holstein bull that received the human gene for lactoferrin, a milk protein, when he was just an embryo. His female offspring may mass-produce the protein, which can be used as a supplement for infant formulas. Similarly, goats are providing CFTR protein (used in the treatment of cystic fibrosis), and TPA (which lessens the severity of heart attacks). Sheep are producing alpha-1 antitrypsin, used in the treatment of emphysema. Cattle may soon be producing human collagen, a key component of skin, cartilage, and bone.

Applying the New Technology to Humans

Researchers in laboratories throughout the world are now working their way through the 3.2 billion base pairs that are present in the twenty-three pairs of human chromosomes. This ambitious effort is called the **human genome project**. Some researchers are working on specific chromosomes, others on specific gene regions only. For example, rather than busying themselves with noncoding sequences (introns), J. Venter and Sidney Brenner are isolating mRNAs from brain cells and using them to make cDNA. By sequencing only the cDNAs, they already have identified hundreds of previously unknown genes.

About 99.9 percent of the nucleotide sequence is the same in every human on earth. Thus, once the sequencing project is completed, we will have the ultimate reference book on human biology and genetic disorders. (Or should we say reference *books*—the complete sequence will fill the equivalent of 200 Manhattan telephone directories.)

What will we do with the information? Certainly we will use it in the search for effective treatments and cures for genetic disorders. Of 2,000 or so genes studied so far, 400 already have been linked to genetic disorders. The knowledge opens doors to **gene therapy**, the transfer of one or more normal or modified genes into body cells of an individual to correct a genetic defect or boost resistance to disease. But what about forms of human gene expression that are neither disabling nor life-threatening? Will we tinker with these, also?

Focus on Bioethics

Some Implications of Human Gene Therapy

Recombinant DNA technology and genetic engineering are advancing rapidly. We are only beginning to work our way through their social and ethical implications.

To most of us, human gene therapy to correct genetic abnormalities seems like a socially acceptable goal. Is it also socially acceptable to insert genes into a *normal* human individual (or sperm or egg) to alter or enhance traits? The idea of selecting desirable human traits is called *eugenic engineering*. Yet who decides which forms of a trait are most "desirable"? What if prospective parents could pick the sex of a child by way of genetic engineering? Three-fourths of one survey group said they would choose a boy. So what would be the long-term social implications of a drastic shortage of girls?

Would it be okay to engineer taller or blue-eyed or curlier-haired individuals? If so, would it be okay to engineer "superhuman" offspring with exceptional strength or breathtaking intelligence? Suppose a person of average intelligence moved into a town composed of 800 Einsteins. Would the response go beyond a few mutterings of "There goes the neighborhood"?

Some say that the DNA of any organism must never be altered. Put aside the fact that nature itself alters DNA

much of the time. The concern is that we don't have the wisdom to bring about beneficial changes without causing harm to ourselves or to the environment.

When it comes to manipulating human genes, one is reminded of our human tendency to leap before we look. When it comes to restricting genetic modifications of any sort, one also is reminded of an old saying: "If God had wanted us to fly, he would have given us wings." And yet, something about the human experience gave us the *capacity* to imagine wings of our own making—and that capacity has carried us to the frontiers of space.

Where are we going from here with recombinant DNA technology, this new product of our imagination? To gain perspective on the question, spend some time reading the history of our species. It is a history of survival in the face of all manner of new challenges, threats, bumblings, and sometimes disasters on a grand scale. It is also a story of our connectedness with the environment and with one another.

The questions confronting you today are these: Should we be more cautious, believing that one day the risk takers may go too far? And what do we as a species stand to lose if the risks are *not* taken?

This chapter opened with the first human application of this technology, meant to save the life of a four-year-old girl. It closes with a *Focus* essay that invites you to consider some prospects and problems associated with the application of recombinant DNA technology to our rapidly advancing knowledge of the human genome.

As with any new technology, the potential benefits of recombinant DNA technology and genetic engineering should be weighed carefully against the potential risks.

SUMMARY

1. Gene mutations and other genetic "experiments" have been going on in nature for billions of years. Through artificial selection, humans have been manipulating the genetic character of different species for many thousands of years. Today, recombinant DNA technol-

ogy has enormously expanded our capacity to modify organisms genetically.

2. With recombinant DNA technology, researchers isolate, cut, and splice together gene regions from different species, then greatly amplify the number of copies of those regions into usefully large quantities. Enzymes make the cuts and do the splicing.

a. Researchers use restriction enzymes and DNA ligase to cut and insert DNA into plasmids from bacterial cells. (The plasmids are small, circular DNA molecules that contain extra genes, besides those of the bacterial chromosome.) A DNA library is a collection of DNA fragments, produced by restriction enzymes, and incorporated into plasmids.

b. Restriction fragments may be amplified by a population of rapidly dividing cells (such as bacteria or yeasts). Short DNA fragments may be amplified more rapidly in a test tube by the polymerase chain reaction (PCR).

c. Following amplification, the DNA fragments can be sorted out (according to length), and their nucleotide sequences can be determined.

3. Each individual of a species has a DNA fingerprint: a unique array of RFLPs (restriction fragment length polymorphisms). Molecular variations in the base sequence of their DNA lead to small, identifiable differences in restriction fragments cut from the DNA.

4. Cells that take up foreign genes can be identified by DNA probes, which base-pair with the genes of interest. Base-pairing of nucleotide sequences from different sources is called nucleic acid hybridization.

5. In genetic engineering, genes are isolated, modified, and inserted into the same organism or a different one. In gene therapy, genes are inserted into an individual to correct a genetic defect.

6. Recombinant DNA technology and genetic engineering have enormous potential for research and applications in medicine, agriculture, the home, and industry. As with any new technology, potential benefits must be weighed against potential risks, including ecological and social.

Review Questions

1. In which type of organisms are restriction enzymes produced naturally? What is the function of these enzymes in nature? *249*

2. Recombinant DNA technology involves producing DNA restriction fragments, amplifying DNA, and identifying modified host cells. Briefly describe some examples of how these activities are carried out. *249–250; 252–253*

3. Name two enzymes used in recombinant DNA technology and define their function. *249–250*

4. Besides this chapter's examples, list what you believe might be some potential benefits and risks of genetic engineering. *254–257*

Self-Quiz (Answers in Appendix IV)

1. _____ are small circles of bacterial DNA that are separate from the bacterial chromosome.

2. DNA fragments result when _____ cut DNA molecules at specific sites.
   a. DNA polymerases   c. restriction enzymes
   b. DNA probes   d. RFLPs

3. Recombinant DNA technology involves _____ .
   a. producing DNA fragments   d. all of the above
   b. making DNA libraries   e. a and c
   c. amplifying DNA

4. PCR stands for _____ .
   a. polymerase chain reaction
   b. polyploid chromosome restrictions
   c. polygraphed criminal rating
   d. politically correct research

5. A _____ is a collection of DNA fragments, produced by restriction enzymes and incorporated into plasmids.
   a. DNA clone   c. DNA probe
   b. DNA library   d. gene map

6. A _____ is a collection of multiple, identical copies of DNA fragments.
   a. DNA clone   c. DNA probe
   b. DNA library   d. gene map

7. In reverse transcription, _____ is assembled on _____ .
   a. mRNA; DNA   c. DNA; enzymes
   b. DNA; mRNA   d. DNA; agar

8. _____ is the transfer of normal genes into body cells to correct a genetic defect.
   a. Reverse transcription   c. Gene mutation
   b. Nucleic acid hybridization   d. Gene therapy

9. Tobacco plant leaves that produce hemoglobin are a result of _____ .
   a. gene therapy   c. pressure on tobacco growers
   b. genetic engineering   d. a and b

10. Match the terms appropriately.
   ____ DNA library   a. mutation, crossing over
   ____ plasmid   b. raises social, legal, and
   ____ nature's genetic   ethical questions
        experiments   c. rapid DNA amplification
   ____ polymerase chain   d. cut DNA fragments
        reaction   incorporated into plasmids
   ____ human gene therapy   e. extra bacterial genes

Selected Key Terms

cDNA *253*
DNA fingerprint *252*
DNA library *249*
DNA ligase *249*
DNA polymerase *250*
DNA probe *252*
gene therapy *256*
genetic engineering *248*

human genome project *256*
nucleic acid hybridization *252*
plasmid *248*
polymerase chain reaction (PCR) *250*
recombinant DNA technology *248*
restriction fragment length
   polymorphism (RFLP) *252*
reverse transcription *253*

Readings

Anderson, W. F. 1992. Human Gene Therapy. *Science* 256:808–813.

Gasser, C. S. and R. T. Fraley. 1989. "Genetically Engineering Plants for Crop Improvement." *Science* 244:1293–1299.

Joyce, G. December 1992. "Directed Molecular Evolution." *Scientific American* 267(6):90–97.

Pursel, V. G. et al. 1989. "Genetic Engineering of Livestock." *Science* 244:1281–1288.

Watson, J. D. 1990. "The Human Genome Project: Past, Present, and Future." *Science* 248:44–49.

White, R., and J. Lalouel. February 1988. "Chromosome Mapping with DNA Markers." *Scientific American* 258(2):40–48.

FACING PAGE: *Millions of years ago, a bony fish died, and sediments gradually buried it. Today its fossilized remains are studied as one more piece of the evolutionary puzzle.*

APPENDIX I
A Brief Classification Scheme

The following classification scheme is a composite of several that are used in microbiology, botany, and zoology. The major groupings are more or less agreed upon. There is not always agreement on what to call a given grouping or where it fits in the overall hierarchy. There are several reasons for this.

First, the fossil record varies in its quality and completeness. Therefore, the relationship of one group to others is sometimes open to interpretation. Comparative studies at the molecular level are firming up the picture, but this work is still under way.

Second, since the time of Linnaeus, classification schemes have been based on perceived morphological similarities and differences among organisms. Although some original interpretations are now open to question, we are so used to thinking about organisms in certain ways that reclassification proceeds slowly. Traditionally, for example, birds and reptiles are separate classes (Reptilia and Aves). Yet there are compelling arguments for grouping lizards and snakes as one class, and crocodilians, dinosaurs, and birds as another.

Finally, microbiologists, mycologists, botanists, zoologists, and other researchers have inherited a wealth of literature, based on classification schemes peculiar to their fields. Most see no good reason to give up established terminology and so disrupt access to the past. Until very recently, botanists were using *division*, and zoologists, *phylum* for groupings that are equivalent in the hierarchy. Opinions are polarized with respect to an entire kingdom (the Protista), certain members of which could just as easily be called single-celled plants, fungi, or animals. Indeed, the term protozoan is a holdover from earlier schemes that ranked the amoebas and some other single cells as simple animals.

Given the problems, why do we bother imposing artificial frameworks on the history of life? We do this for the same reason that a writer might decide to break up the history of civilization into several volumes, many chapters, and many more paragraphs. Both efforts are attempts to impart structure to what might otherwise be an overwhelming body of information.

Bear in mind, we include this classification scheme mainly for your reference purposes. It is by no means complete. Numerous existing and extinct organisms of the so-called lesser phyla are not represented. Our strategy is to focus mainly on organisms mentioned in the text. A few examples of organisms also are listed under the entries.

SUPERKINGDOM PROKARYOTA.   Prokaryotes. Single-celled organisms with DNA concentrated in a cytoplasmic region, not in a membrane-bound nucleus.

KINGDOM MONERA.   Bacteria, either single cells or simple associations of cells. Both autotrophs and heterotrophs (refer to Table 22.2). *Bergey's Manual of Systematic Bacteriology*, the authoritative reference in the field, calls this "a time of taxonomic transition." It groups bacteria mainly on the basis of form, physiology, and behavior, not on phylogeny. The scheme presented here does reflect the growing evidence of evolutionary relationships for at least some bacterial groups.

SUBKINGDOM ARCHAEBACTERIA. Methanogens, halophiles, thermophiles. Strict anaerobes, distinct from other bacteria in cell wall, membrane lipids, ribosomes, and RNA sequences. *Methanobacterium, Halobacterium, Sulfolobus.*

SUBKINGDOM EUBACTERIA. Gram-negative and gram-positive forms. Peptidoglycan in cell wall. Photosynthetic autotrophs, chemosynthetic autotrophs, and heterotrophs.

PHYLUM GRACILICUTES.   Typical Gram-negative, thin wall. Autotrophs (photosynthetic and chemosynthetic) and heterotrophs. *Anabaena* and other cyanobacteria. *Escherichia, Pseudomonas, Neisseria, Myxococcus.*

PHYLUM FIRMICUTES.   Typical Gram-positive, thick wall. Heterotrophs. *Bacillus, Staphylococcus, Streptococcus, Clostridium, Actinomycetes.*

PHYLUM TENERICUTES.   Gram-negative, wall absent. Heterotrophs (saprobes, pathogens). *Mycoplasma.*

SUPERKINGDOM EUKARYOTA.   Eukaryotes (single-celled and multicelled organisms. Cells typically have a nucleus (enclosing the DNA) and other membrane-bound organelles.

KINGDOM PROTISTA.   Mostly single-celled eukaryotes, some colonial forms. Diverse autotrophs and heterotrophs. Many lineages apparently related evolutionarily to certain plants, fungi, and possibly animals.

PHYLUM CHYTRIDIOMYCOTA.   Chytrids. Heterotrophs; saprobic decomposers, parasites. *Chytridium.*

PHYLUM OOMYCOTA.   Water molds. Heterotrophs. Decomposers, some parasites. *Saprolegnia, Phytophthora, Plasmopara.*

PHYLUM ACRASIOMYCOTA.   Cellular slime molds. Heterotrophs with amoeboid and spore-bearing stages. *Dictyostelium.*

PHYLUM MYXOMYCOTA.   Plasmodial slime molds. Heterotrophs with amoeboid and spore-bearing stages. *Physarum.*

PHYLUM SARCODINA.   Amoeboid protozoans. Heterotrophs. Soft-bodied; some shelled. Amoebas, foraminiferans, radiolarians, heliozoans. *Amoeba, Entomoeba.*

PHYLUM CILIOPHORA.   Ciliated protozoans. Heterotrophs. Distinctive arrays of cilia, used as motile structures. *Paramecium,* hypotrichs.

PHYLUM MASTIGOPHORA.   Flagellated protozoans. Some free-living, many internal parasites; all with one to several flagella. *Trypanosoma, Trichomonas, Giardia.*

SPOROZOANS.   Parasitic protozoans, many intracellular. "Sporozoans" is the common name with no formal taxonomic status. *Plasmodium, Toxoplasma.*

PHYLUM EUGLENOPHYTA.   Euglenoids. Mostly heterotrophs, some photosynthetic types. Flagellated. *Euglena.*

PHYLUM CHRYSOPHYTA.   Golden algae, yellow-green algae, diatoms. Photosynthetic. Some flagellated, others not. *Mischococcus, Synura, Vaucheria.*

PHYLUM PYRRHOPHYTA.   Dinoflagellates. Photosynthetic, mostly, but some heterotrophs. *Gymnodinium breve.*

PHYLUM RHODOPHYTA.   Red algae. Photosynthetic, nearly all marine, some freshwater. *Porphyra. Bonnemaisonia, Euchema.*

PHYLUM PHAEOPHYTA.   Brown algae. Photosynthetic, nearly all temperate or marine waters. *Macrocystis, Fucus, Sargassum, Ectocarpus, Postelsia.*

PHYLUM CHLOROPHYTA.   Green algae. Photosynthetic. Most freshwater, some marine or terrestrial. *Chlamydomonas, Spirogyra, Ulva, Volvox, Codium, Halimeda.*

KINGDOM FUNGI.   Mostly multicelled eukaryotes. Heterotrophs (mostly saprobes, some parasites). Major decomposers of nearly all communities. Reliance on extracellular digestion of organic matter and absorption of nutrients by individual cells.

PHYLUM ZYGOMYCOTA.   Zygomycetes. All produce nonmotile spores. Bread molds, related forms. *Rhizopus, Philobolus.*

PHYLUM ASCOMYCOTA.   Ascomycetes. Sac fungi. Most yeasts and molds; morels, truffles. *Saccharomycetes, Morchella Neurospora, Sarcoscypha. Claviceps, Ophiostoma.*

PHYLUM BASIDIOMYCOTA.   Basidiomycetes. Club fungi. Mushrooms, shelf fungi, stinkhorns. *Agaricus, Amanita, Puccinia, Ustilago.*

IMPERFECT FUNGI.   Sexual spores absent or undetected. The group has no formal taxonomic status. If better understood, a given species might be grouped with sac fungi or club fungi. *Verticillium, Candida, Microsporum, Histoplasma.*

LICHENS.   Mutualistic interactions between a fungus and a cyanobacterium, green alga, or both. *Usnea, Cladonia.*

KINGDOM PLANTAE.   Nearly all multicelled eukaryotes. Photosynthetic autotrophs, except for a few parasitic types.

PHYLUM RHYNIOPHYTA.   Earliest known vascular plants; extinct. *Cooksonia, Rhynia.*

PHYLUM TRIMEROPHYTA.   Trimerophytes. *Psilophyton.*

PHYLUM PROGYMNOSPERMOPHYTA. Progymnosperms. Ancestral to early seed-bearing plants; extinct. *Archaeopteris.*

PHYLUM CHAROPHYTA.   Stoneworts.

PHYLUM BRYOPHYTA.   Liverworts, hornworts, mosses. *Marchantia, Polytrichum, Sphagnum.*

PHYLUM PSILOPHYTA.   Whisk ferns. *Psilotum.*

PHYLUM LYCOPHYTA.   Lycophytes, club mosses. *Lycopodium, Selaginella.*

PHYLUM SPHENOPHYTA.   Horsetails. *Equisetum.*

PHYLUM PTEROPHYTA.   Ferns.

PHYLUM PTERIDOSPERMOPHYTA.   Seed ferns. Fernlike gymnosperms; extinct.

PHYLUM CYCADOPHYTA.   Cycads. *Zamia.*

PHYLUM GINKGOPHYTA.   Ginkgo. *Ginkgo.*

PHYLUM GNETOPHYTA.   Gnetophytes. *Ephedra, Welwitchia.*

PHYLUM CONIFEROPHYTA.   Conifers.
 Family Pinaceae. Pines, firs, spruces, hemlock, larches, Douglas firs, true cedars. *Pinus.*
 Family Cupressaceae. Junipers, cypresses. *Juniperus.*
 Family Taxodiaceae. Bald cypress, redwoods, Sierra bigtree, dawn redwood. *Sequoia.*
 Family Taxaceae. Yews.

PHYLUM ANTHOPHYTA.   Flowering plants.
 Class Dicotyledonae. Dicotyledons (dicots). Some families of several different orders are listed:
  Family Nymphaeaceae. Water lilies.
  Family Papaveraceae. Poppies.
  Family Brassicaceae. Mustards, cabbages, radishes.
  Family Malvaceae. Mallows, cotton, okra, hibiscus.
  Family Solanaceae. Potatoes, eggplant, petunias.
  Family Salicaceae. Willows, poplars.
  Family Rosaceae. Roses, apples, almonds, strawberries.
  Family Fabaceae. Peas, beans, lupines, mesquite.
  Family Cactaceae. Cacti.
  Family Euphorbiaceae. Spurges, poinsettia.
  Family Cucurbitaceae. Gourds, melons, cucumbers, squashes.
  Family Apiaceae. Parsleys, carrots, poison hemlock.
  Family Aceraceae. Maples.
  Family Asteraceae. Composites. Chrysanthemums, sunflowers, lettuces, dandelions.
 Class Monocotyledonae. Monocotyledons (monocots). Some families of several different orders are listed:
  Family Liliaceae. Lilies, hyacinths, tulips, onions, garlic.
  Family Iridaceae. Irises, gladioli, crocuses.
  Family Orchidaceae. Orchids.
  Family Arecaceae. Date palms, coconut palms.
  Family Cyperaceae. Sedges.
  Family Poaceae. Grasses, bamboos, corn, wheat, sugarcane.
  Family Bromeliaceae. Bromeliads, pineapples, Spanish moss.

KINGDOM ANIMALIA.   Multicelled eukaryotes. Heterotrophs (herbivores, carnivores, omnivores, parasites, detritivores).

PHYLUM PLACOZOA.   Small, organless marine animal. *Trichoplax.*

PHYLUM MESOZOA.   Ciliated, wormlike parasites, about the same level of complexity as *Trichoplax.*

PHYLUM PORIFERA.   Sponges.

PHYLUM CNIDARIA.
 Class Hydrozoa. Hydrozoans. *Hydra, Obelia, Physalia.*
 Class Scyphozoa. Jellyfishes. *Aurelia.*
 Class Anthozoa. Sea anemones, corals. *Telesto.*

PHYLUM CTENOPHORA.   Comb jellies. *Pleurobrachia.*

PHYLUM PLATYHELMINTHES.   Flatworms.
 Class Turbellaria. Triclads (planarians), polyclads. *Dugesia.*
 Class Trematoda. Flukes. *Schistosoma.*
 Class Cestoda. Tapeworms. *Taenia.*

PHYLUM NEMERTEA.   Ribbon worms.

PHYLUM NEMATODA.   Roundworms. *Ascaris, Trichinella.*

PHYLUM ROTIFERA.   Rotifers.

PHYLUM MOLLUSCA.   Mollusks.
 Class Polyplacophora. Chitons.
 Class Gastropoda. Snails (periwinkles, whelks, limpets, abalones, cowries, conches, nudibranchs, tree snails, garden snails), sea slugs, land slugs.
 Class Bivalvia. Clams, mussels, scallops, cockles, oysters, shipworms.
 Class Cephalopoda. Squids, octopuses, cuttlefish, nautiluses. *Loligo.*

PHYLUM BRYOZOA.   Bryozoans (moss animals).

PHYLUM BRACHIOPODA.   Lampshells.

PHYLUM ANNELIDA.   Segmented worms.
 Class Polychaeta. Mostly marine worms.
 Class Oligochaeta. Mostly freshwater and terrestrial worms, but many marine. *Lumbricus* (earthworms).
 Class Hirudinea. Leeches.

PHYLUM TARDIGRADA.   Water bears.

PHYLUM ONYCHOPHORA.   Onychophorans. *Peripatus.*

PHYLUM ARTHROPODA.
 Subphylum Trilobita. Trilobites; extinct.
 Subphylum Chelicerata. Chelicerates. Horseshoe crabs, spiders, scorpions, ticks, mites.
 Subphylum Crustacea. Shrimps, crayfishes, lobsters, crabs, barnacles, copepods, isopods (sowbugs).
 Subphylum Uniramia.
  Superclass Myriapoda. Centipedes, millipedes.
  Superclass Insecta.
   Order Ephemeroptera. Mayflies.
   Order Odonata. Dragonflies, damselflies.

Order Orthoptera. Grasshoppers, crickets, katydids.
Order Dermaptera. Earwigs.
Order Blattodea. Cockroaches.
Order Mantodea. Mantids.
Order Isoptera. Termites.
Order Mallophaga. Biting lice.
Order Anoplura. Sucking lice.
Order Homoptera. Cicadas, aphids, leafhoppers, spittlebugs.
Order Hemiptera. Bugs.
Order Coleoptera. Beetles.
Order Diptera. Flies.
Order Mecoptera. Scorpion flies. *Harpobittacus.*
Order Siphonaptera. Fleas.
Order Lepidoptera. Butterflies, moths.
Order Hymenoptera. Wasps, bees, ants.

PHYLUM ECHINODERMATA.   Echinoderms.
Class Asteroidea. Sea stars.
Class Ophiuroidea. Brittle stars.
Class Echinoidea. Sea urchins, heart urchins, sand dollars.
Class Holothuroidea. Sea cucumbers.
Class Crinoidea. Feather stars, sea lilies.
Class Concentricycloidea. Sea daisies.

PHYLUM HEMICHORDATA.   Acorn worms.

PHYLUM CHORDATA.   Chordates.
Subphylum Urochordata. Tunicates, related forms.
Subphylum Cephalochordata. Lancelets.
Subphylum Vertebrata. Vertebrates.
Class Agnatha. Jawless vertebrates (lampreys, hagfishes).
Class Placodermi. Jawed, heavily armored fishes; extinct.
Class Chondrichthyes. Cartilaginous fishes (sharks, rays, skates, chimaeras).
Class Osteichthyes. Bony fishes.
Subclass Dipnoi. Lungfishes.
Subclass Crossopterygii. Coelacanths, related forms.
Subclass Actinopterygii. Ray-finned fishes.
Order Acipenseriformes. Sturgeons, paddlefishes.
Order Salmoniformes. Salmon, trout.
Order Atheriniformes. Killifishes, guppies.
Order Gasterosteiformes. Seahorses.
Order Perciformes. Perches, wrasses, barracudas, tunas, freshwater bass, mackerels.
Order Lophiiformes. Angler fishes.
Class Amphibia. Mostly tetrapods; embryo enclosed in amnion.
Order Caudata. Salamanders.
Order Anura. Frogs, toads.
Order Apoda. Apodans (caecilians).
Class Reptilia. Skin with scales, embryo enclosed in amnion.
Subclass Anapsida. Turtles, tortoises.
Subclass Lepidosaura. *Sphenodon*, lizards, snakes.
Subclass Archosaura. Dinosaurs (extinct), crocodiles, alligators.
Class Aves. Birds. (In more recent schemes, dinosaurs, crocodilians, and birds are grouped in the same category.)

Order Struthioniformes. Ostriches.
Order Sphenisciformes. Penguins.
Order Procellariiformes. Albatrosses, petrels.
Order Ciconiiformes. Herons, bitterns, storks, flamingoes.
Order Anseriformes. Swans, geese, ducks.
Order Falconiformes. Eagles, hawks, vultures, falcons.
Order Galliformes. Ptarmigan, turkeys, domestic fowl.
Order Columbiformes. Pigeons, doves.
Order Strigiformes. Owls.
Order Apodiformes. Swifts, hummingbirds.
Order Passeriformes. Sparrows, jays, finches, crows, robins, starlings, wrens.
Class Mammalia. Skin with hair; young nourished by milk-secreting glands of adult.
Subclass Prototheria. Egg-laying mammals (duckbilled platypus, spiny anteaters).
Subclass Metatheria. Pouched mammals or marsupials (opossums, kangaroos, wombats).
Subclass Eutheria. Placental mammals.
Order Insectivora. Tree shrews, moles, hedgehogs.
Order Scandentia. Insectivorous tree shrews.
Order Chiroptera. Bats.
Order Primates.
Suborder Strepsirhini (prosimians). Lemurs, lorises.
Suborder Haplorhini (tarsioids and anthropoids).
Infraorder Tarsiiformes. Tarsiers.
Infraorder Platyrrhini (New World monkeys).
Family Cebidae. Spider monkeys, howler monkeys, capuchin.
Infraorder Catarrhini (Old World monkeys and hominoids).
Superfamily Cercopithecoidea. Baboons, macaques, langurs.
Superfamily Hominoidea. Apes and humans.
Family Hylobatidae. Gibbons.
Family Pongidae. Chimpanzees, gorillas, orangutans.
Family Hominidae. Humans and most recent ancestors of humans.
Order Carnivora. Carnivores.
Suborder Feloidea. Cats, civets, mongooses, hyenas.
Suborder Canoidea. Dogs, weasels, skunks, otters, raccoons, pandas, bears.
Order Proboscidea. Elephants; mammoths (extinct).
Order Sirenia. Sea cows (manatees, dugongs).
Order Perissodactyla. Odd-toed ungulates (horses, tapirs, rhinos).
Order Artiodactyla. Even-toed ungulates (camels, deer, bison, sheep, goats, antelopes, giraffes).
Order Edentata. Anteaters, tree sloths, armadillos.
Order Tubulidentata. African aardvarks.
Order Cetacea. Whales, porpoises.
Order Rodentia. Most gnawing animals (squirrels, rats, mice, guinea pigs, porcupines).

APPENDIX II
Units of Measure

Metric-English Conversions

Length

English		Metric
inch	=	2.54 centimeters
foot	=	0.30 meter
yard	=	0.91 meter
mile (5,280 feet)	=	1.61 kilometer

To convert	multiply by	to obtain
inches	2.54	centimeters
feet	30.00	centimeters
centimeters	0.39	inches
millimeters	0.039	inches

Weight

English		Metric
grain	=	64.80 milligrams
ounce	=	28.35 grams
pound	=	453.60 grams
ton (short) (2,000 pounds)	=	0.91 metric ton

To convert	multiply by	to obtain
ounces	28.3	grams
pounds	453.6	grams
pounds	0.45	kilograms
grams	0.035	ounces
kilograms	2.2	pounds

Volume

English		Metric
cubic inch	=	16.39 cubic centimeters
cubic foot	=	0.03 cubic meter
cubic yard	=	0.765 cubic meters
ounce	=	0.03 liter
pint	=	0.47 liter
quart	=	0.95 liter
gallon	=	3.79 liters

To convert	multiply by	to obtain
fluid ounces	30.00	milliliters
quart	0.95	liters
milliliters	0.03	fluid ounces
liters	1.06	quarts

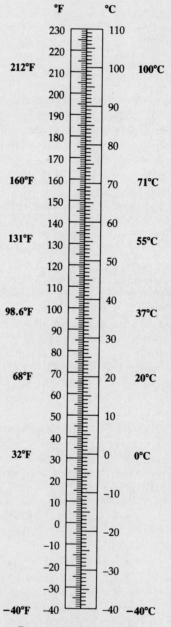

To convert temperature scales:

Fahrenheit to Celsius: $°C = 5/9 (°F - 32)$

Celsius to Fahrenheit: $°F = 9/5 (°C) + 32$

APPENDIX III
Answers to Genetics Problems

Chapter 11

1. a. *AB*
   b. *AB* and *aB*
   c. *Ab* and *ab*
   d. *AB*, *aB*, *Ab*, and *ab*

2. a. *AaBB* will occur in all the offspring.
   b. 25% *AABB*; 25% *AaBB*; 25% *AABb*; 25% *AaBb*
   c. 25% *AaBb*; 25% *Aabb*; 25% *aaBb*; 25% *aabb*
   d. | 1/16 | *AABB* | (6.25%) |
      |------|--------|---------|
      | 1/8 | *AaBB* | (12.5%) |
      | 1/16 | *aaBB* | (6.25%) |
      | 1/8 | *AABb* | (12.5%) |
      | 1/4 | *AaBb* | (25%) |
      | 1/8 | *aaBb* | (12.5%) |
      | 1/16 | *AAbb* | (6.25%) |
      | 1/8 | *Aabb* | (12.5%) |
      | 1/16 | *aabb* | (6.25%) |

3. Yellow is recessive. Because the first-generation plants must be heterozygous and have a green phenotype, green must be dominant over the recessive yellow.

4. a. Mother must be heterozygous for both genes; father is homozygous recessive for both genes. The first child is also homozygous recessive for both genes.
   b. The probability that the second child will not be able to roll the tongue and will have detached earlobes is 1/4 (25%).

5. a. *ABC*
   b. *ABc* and *aBc*
   c. *ABC*, *aBC*, *ABc*, and *aBc*
   d. *ABC*, *aBC*, *AbC*, *abC*, *ABc*, *aBc*, *Abc*, and *abc*

6. The first-generation plants must all be double heterozygotes. When these plants are self-pollinated 1/4 (25%) of the second-generation plants will be doubly heterozygous.

7. The most direct way to accomplish this would be to allow a true-breeding mouse having yellow fur to mate with a true-breeding mouse having brown fur. Such true-breeding strains could be obtained by repeated inbreeding (mating of related individuals; for example, a male and a female of the same litter) of yellow and brown strains. In this way, it should be possible to obtain homozygous yellow and homozygous brown mice.

   When true-breeding yellow and true-breeding brown mice are crossed, the progeny should all be heterozygous. If the progeny phenotype is either yellow or brown, then the dominance is simple or complete, and the phenotype reflects the dominant allele. If the phenotype is intermediate between yellow and brown, there is incomplete dominance. If the phenotype shows both yellow and brown, there is codominance.

8. a. The mother must be heterozygous ($I^A i$). The man having type B blood could have fathered the child if he were also heterozygous ($I^B i$).

   b. If the man is heterozygous, then he *could be* the father. However, because any other type B heterozygous male also could be the father, one cannot say that this particular man absolutely must be. Actually, any male who could contribute an O allele (*i*) could have fathered the child. This would include males with type O blood (*ii*) or type A blood who are heterozygous ($I^A i$).

9. a. F_1 genotypes and phenotypes; 100% *Bb Cc*, brown progeny. F_2 phenotypes: 9/16 brown + 3/16 tan + 4/16 albino.

   F_2 genotypes $\begin{cases} \text{1/16 } BBCC + \text{2/16 } BBCc + \text{2/16} \\ BbCC + \text{4/16 } BbCc; \text{(9/16 brown)} \\ \text{1/16 } bbCC + \text{2/16 } bbCc; \text{(3/16 tan)} \\ \text{1/16 } BBcc + \text{2/16 } Bbcc + \text{1/16 } bbcc; \\ \text{(4/16 albino)} \end{cases}$

   b. Backcross phenotypes: 1/4 brown + 1/4 tan + 2/4 albino.

   Backcross genotypes $\begin{cases} \text{1/4 } BbCc; \text{(1/4 brown)} \\ \text{1/4 } bbCc; \text{(1/4 tan)} \\ \text{1/4 } Bbcc + \text{1/4 } bbcc; \text{(1/2 albino)} \end{cases}$

10. If the young man selects a white guinea pig to mate with his black guinea pig, he knows the genotype of at least one of the parents—the white one is *ww*. The cross is: *ww* × *W*? If a number of matings occur and a white offspring is obtained, the young man knows that the unknown genotype of the black guinea pig is hybrid, *Ww*. It would be impossible for a pure black, *WW*, to have a white offspring. This problem demonstrates the 1:1 ratio associated with Mendel's testcross.

11. a. *RR* × *Rr* yields 1/2 red flowers and 1/2 pink flowers.
    b. *RR* × *rr* yields all pink flowers.
    c. *Rr* × *Rr* yields 1/4 red flowers, 1/2 pink flowers, and 1/4 white flowers.
    d. *Rr* × *rr* yields 1/2 pink flowers and 1/2 white flowers.

12. Both parents have sickle-cell trait.
    a. The expected ratio is that 1/4 of their offspring will have sickle-cell disease ($Hb^S Hb^S$).
    b. Yes, the expected ratio is that 1/4 of their offspring will be homozygous normal ($Hb^A Hb^A$).
    c. The expected ratio is that 1/2 of their offspring will have sickle-cell trait ($Hb^A Hb^S$).

13. a. Parents that are *Aa* × *Aa* can produce both albino (*aa*) and normal (*AA* or *Aa*) children.
    b. Parents that are *aa* × *aa* can only produce albino (*aa*) children.
    c. The normal woman must be *Aa* and the albino man must be *aa*; this mating can produce both albino and normal children.

14. 9/16 walnut (pea-rose); 3/16 rose; 3/16 pea; 1/16 single

15. a. All offspring from this mating will be medium, *AaBb*.
    b. 1 white; 4 light; 6 medium; 4 dark; 1 black

Chapter 12

1. a. Males inherit their X chromosome from their mothers.
   b. A male can produce two types of gametes with respect to an X-linked gene. One type will lack this gene and possess a Y chromosome. The other will have an X chromosome and the linked gene.
   c. A female homozygous for an X-linked gene will produce just one type of gamete containing an X chromosome with the gene.
   d. A female heterozygous for an X-linked gene will produce two types of gametes. One will contain an X chromosome with the dominant allele, and the other type will contain an X chromosome with the recessive allele.

2. a. Because this gene is only carried on Y chromosomes, females would not be expected to have hairy pinnae because they normally do not have Y chromosomes.
   b. Because sons always inherit a Y chromosome from their fathers and because daughters never do, a man having hairy pinnae will always transmit this trait to his sons and never to his daughters.

3. A 0% crossover frequency means that 50% of the gametes will be *AB* and 50% will be *ab*.

4. The first-generation females must be heterozygous for both genes. The 42 red-eyed, vestigial-winged and the 30 purple-eyed, long-winged progeny represent recombinant gametes from these females. Because the first-generation females must have produced 600 gametes to give these 600 progeny, and because 42 + 30 of these were recombinant, the percentage of recombinant gametes is 72/600, or 12%, which implies that 12 map units separate the two genes.

5. The rare vestigial-winged flies could be explained by a deletion of the dominant allele from one of the chromosomes, due to the action of the X-rays. Alternatively, the radiation may have induced a mutation in the dominant allele.

6. If the longer-than-normal chromosome 14 represented the translocation of most of chromosome 21 to the end of a normal chromosome 14, then this individual would be afflicted with Down syndrome due to the presence of this attached chromosome 21 as well as the two normal chromosomes 21. The total chromosome number, however, would be 46.

7. Using *c* as the symbol for color blindness and *C* for normal vision, then the cross can be diagrammed as follows:

C(Y) female × *Cc* male.

In mugwumps, a son receives one sex-linked allele from each of his parents, but a daughter inherits her unpaired sex-linked allele solely from her father. In this cross, half of the sons will be *CC* and half *Cc*, but none will be color blind. Of the daughters, half will be *C*(Y) and half *c*(Y). There is a 50% chance that a daughter will be color blind. Note: this answer is backward from the way it would be in humans; but it is correct not only for mugwumps but also for all birds and Lepidoptera (moths and butterflies), as well as a few other forms.

8. To produce a female who will suffer from childhood muscular dystrophy, not only must the mother be a carrier of the disease, but the father must have it. Few, if any, such males who survive to adulthood are capable of having children.

APPENDIX IV
Answers to Self-Quizzes

Chapter 1
1. cell
2. Metabolism
3. Homeostasis
4. adaptive
5. mutations
6. d
7. d
8. d
9. c

Chapter 2
1. electrons
2. d
3. electrons
4. Isotopes
5. c
6. b
7. d
8. b
9. b
10. d
11. c, e, a, b, d

Chapter 3
1. carbon
2. a
3. carbohydrates, lipids, proteins, nucleic acids
4. c
5. c
6. d
7. b
8. c, e, b, d, a

Chapter 4
1. a
2. c
3. c
4. c
5. d
6. d
7. b
8. d
9. h, g, a, b, d, e, i, f, c

Chapter 5
1. c
2. a
3. c
4. a
5. a
6. a
7. centrifuge
8. d
9. d
10. b
11. f, d, e, a, b, c

Chapter 6
1. metabolism
2. thermodynamics
3. c
4. a
5. c
6. d
7. d
8. c
9. c, d, a, b

Chapter 7
1. carbon
2. carbon dioxide, sunlight
3. d
4. c
5. b
6. b and d
7. c
8. c
9. c, d, e, b, a

Chapter 8
1. ATP
2. pyruvate
3. d
4. c
5. d
6. c
7. b
8. b
9. b, c, a, d

Chapter 9
1. b
2. a
3. b
4. b
5. a
6. c
7. a
8. d
9. b
10. d, b, c, a

Chapter 10
1. d
2. d
3. a
4. b
5. d
6. c
7. c
8. d
9. d, a, c, b

Chapter 11
1. a
2. b
3. a
4. c
5. b
6. a
7. d
8. b, d, a, c

Chapter 12
1. c
2. e
3. e
4. c
5. e
6. c
7. d
8. d
9. d
10. c, e, d, b, a

Chapter 13
1. c
2. d
3. d
4. c
5. a
6. a
7. d
8. d, e, b, c, a

Chapter 14
1. three
2. e
3. b
4. c
5. c
6. a
7. c
8. b
9. a
10. c
11. c
12. a
13. e, c, d, b, a

Chapter 15
1. d
2. d
3. b
4. a
5. a
6. d
7. d
8. e
9. c
10. b
11. d
12. a, f, d, c, e, b

Chapter 16
1. Plasmids
2. c
3. d
4. a
5. b
6. a
7. b
8. d
9. b
10. d, e, a, c, b

APPENDIX V
A Closer Look at Some Major Metabolic Pathways

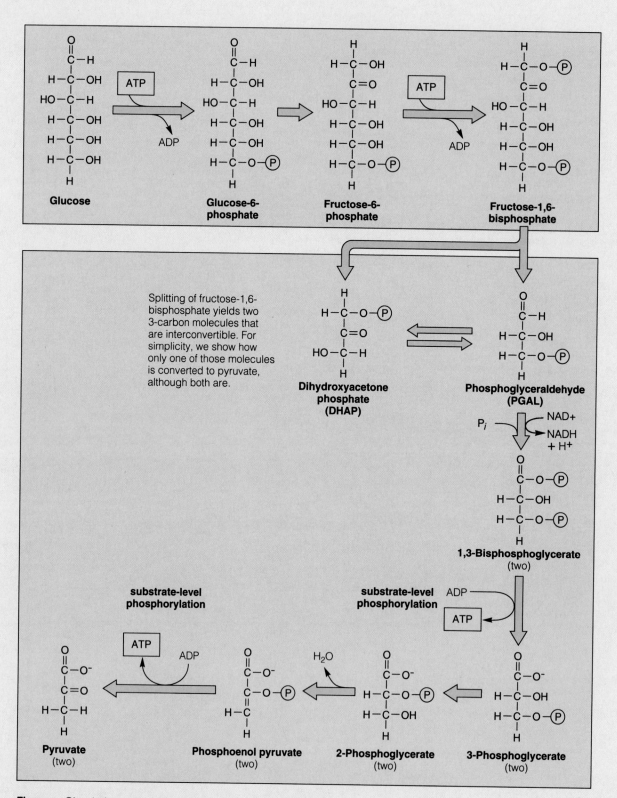

ENERGY-REQUIRING STEPS OF GLYCOLYSIS

(two ATP invested)

ENERGY-RELEASING STEPS OF GLYCOLYSIS

(four ATP produced)

Glucose

Glucose-6-phosphate

Fructose-6-phosphate

Fructose-1,6-bisphosphate

Splitting of fructose-1,6-bisphosphate yields two 3-carbon molecules that are interconvertible. For simplicity, we show how only one of those molecules is converted to pyruvate, although both are.

Dihydroxyacetone phosphate (DHAP)

Phosphoglyceraldehyde (PGAL)

1,3-Bisphosphoglycerate (two)

3-Phosphoglycerate (two)

2-Phosphoglycerate (two)

Phosphoenol pyruvate (two)

Pyruvate (two)

substrate-level phosphorylation

substrate-level phosphorylation

Figure a Glycolysis, ending with two 3-carbon pyruvate molecules for each 6-carbon glucose entering the reactions. The *net* energy yield is two ATP molecules (two invested, four produced).

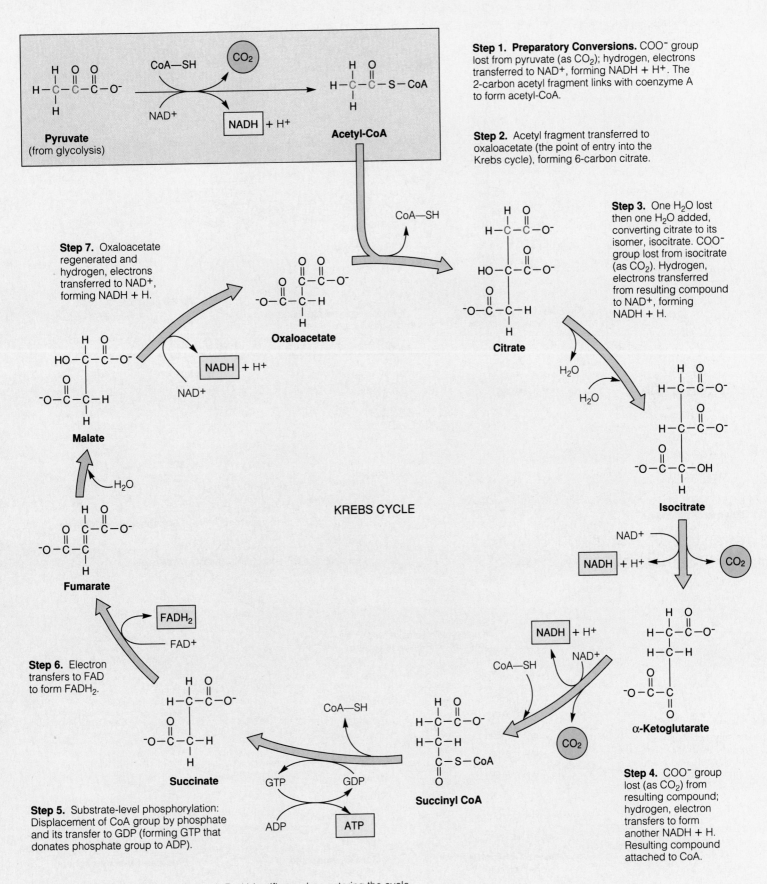

Step 1. Preparatory Conversions. COO^- group lost from pyruvate (as CO_2); hydrogen, electrons transferred to NAD^+, forming $NADH + H^+$. The 2-carbon acetyl fragment links with coenzyme A to form acetyl-CoA.

Step 2. Acetyl fragment transferred to oxaloacetate (the point of entry into the Krebs cycle), forming 6-carbon citrate.

Step 3. One H_2O lost then one H_2O added, converting citrate to its isomer, isocitrate. COO^- group lost from isocitrate (as CO_2). Hydrogen, electrons transferred from resulting compound to NAD^+, forming $NADH + H$.

Step 7. Oxaloacetate regenerated and hydrogen, electrons transferred to NAD^+, forming $NADH + H$.

Step 6. Electron transfers to FAD to form $FADH_2$.

Step 5. Substrate-level phosphorylation: Displacement of CoA group by phosphate and its transfer to GDP (forming GTP that donates phosphate group to ADP).

Step 4. COO^- group lost (as CO_2) from resulting compound; hydrogen, electron transfers to form another $NADH + H$. Resulting compound attached to CoA.

KREBS CYCLE

Pyruvate (from glycolysis)

CoA—SH

Acetyl-CoA

Oxaloacetate

Citrate

Isocitrate

α-Ketoglutarate

Succinyl CoA

Succinate

Fumarate

Malate

Figure b Krebs cycle (citric acid cycle). Red identifies carbon entering the cycle by way of acetyl-CoA.

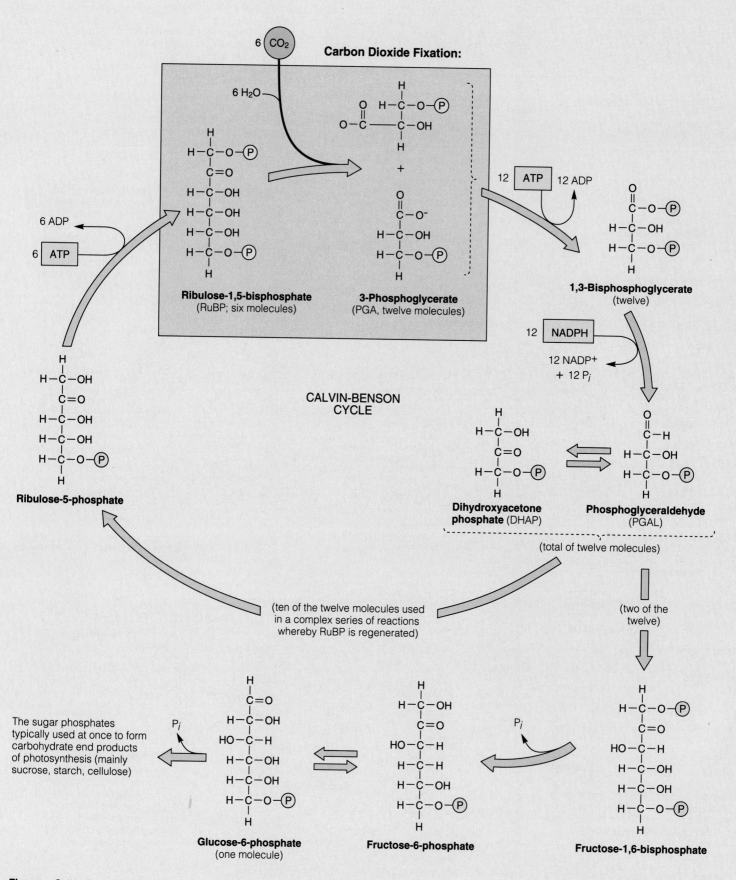

Figure c Calvin-Benson cycle of the light-independent reactions of photosynthesis.

APPENDIX VI
Periodic Table of the Elements

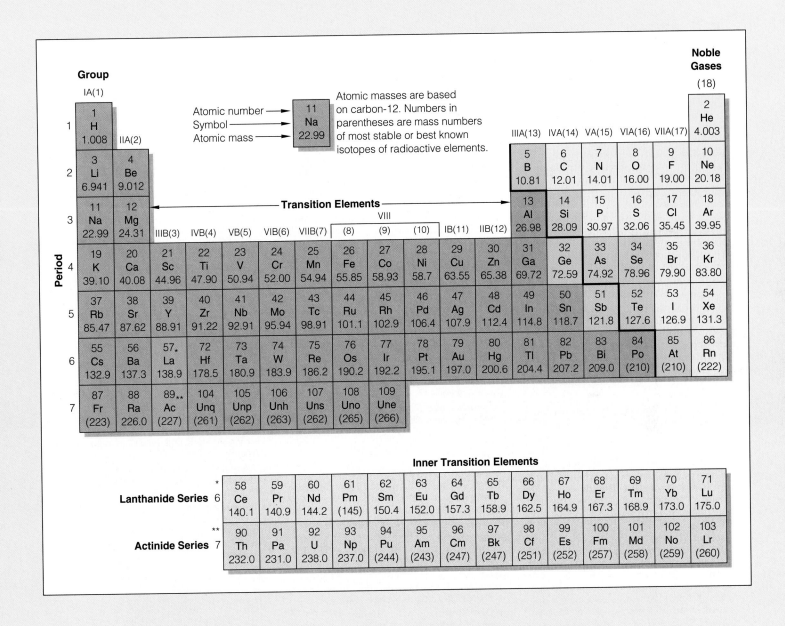

A GLOSSARY OF BIOLOGICAL TERMS

ABO blood typing Method of characterizing an individual's blood according to whether one or both of two protein markers, A and B, are present at the surface of red blood cells. The O signifies that neither marker is present.

abortion Spontaneous or induced expulsion of the embryo or fetus from the uterus.

abscisic acid (ab-SISS-ik) Plant hormone that promotes stomatal closure, bud dormancy, and seed dormancy.

abscission (ab-SIH-zhun) [L. *abscindere*, to cut off] The dropping of leaves, flowers, fruits, or other plant parts due to hormonal action.

absorption Of complex animals, the movement of nutrients, fluid, and ions across the gut lining and into the internal environment.

accessory pigment A light-trapping pigment that contributes to photosynthesis by extending the range of usable wavelengths beyond those absorbed by the chlorophylls.

acid [L. *acidus*, sour] A substance that releases hydrogen ions (H$^+$) in water.

acid rain The falling to earth of snow or rain that contains sulfur and nitrogen oxides. Also called wet acid deposition (as opposed to dry acid deposition, or the falling to earth of airborne particles of sulfur and nitrogen oxides).

acoelomate (ay-SEE-luh-mate) Type of animal that has no fluid-filled cavity between the gut and body wall.

acoustical signal Sounds that are used as a communication signal.

actin (AK-tin) A globular contractile protein. In muscle cells, actin interacts with another protein, myosin, to bring about contraction.

action potential An abrupt, brief reversal in the steady voltage difference across the plasma membrane (that is, the resting membrane potential) of a neuron and some other cells.

activation energy The minimum amount of collision energy required to bring reactant molecules to an activated condition (the transition state) at which a reaction will proceed spontaneously. Enzymes enhance reaction rates by lowering the activation energy (they put substrates on a precise collision course).

active site A crevice on the surface of an enzyme molecule where a specific reaction is catalyzed.

active transport The pumping of one or more specific solutes through a transport protein that spans the lipid bilayer of a cell membrane. Most often, the solute is transported against its concentration gradient. The protein is activated by an energy boost, as from ATP.

adaptation [L. *adaptare*, to fit] In evolutionary biology, the process of becoming adapted (or more adapted) to a given set of environmental conditions. Of sensory neurons, a decrease in the frequency of action potentials (or their cessation) even when a stimulus is maintained at constant strength.

adaptive behavior A behavior that promotes the propagation of an individual's genes and that tends to increase in frequency in a population over time.

adaptive radiation A burst of speciation events, with lineages branching away from one another as they partition the existing environment or invade new ones.

adaptive trait Any aspect of form, function, or behavior that helps an organism survive and reproduce under a given set of environmental conditions.

adaptive zone A way of life, such as "catching insects in the air at night." A lineage must have physical, ecological, and evolutionary access to an adaptive zone to become a successful occupant of it.

adenine (AH-de-neen) A purine; a nitrogen-containing base found in nucleotides.

adenosine diphosphate (ah-DEN-uh-seen die-FOSS-fate) ADP, a molecule involved in cellular energy transfers; typically formed by hydrolysis of ATP.

adenosine phosphates Any of several relatively small molecules, some of which function as chemical messengers within and between cells, and others that function as energy carriers.

adenosine triphosphate *See* ATP.

ADH Antidiuretic hormone. Produced by the hypothalamus and released by the posterior pituitary, it stimulates reabsorption in the kidneys and so reduces urine volume.

adipose tissue A type of connective tissue having an abundance of fat-storing cells and blood vessels for transporting fats.

ADP Adenosine diphosphate. A nucleotide coenzyme that accepts unbound phosphate or a phosphate group to become ATP.

ADP/ATP cycle In cells, a mechanism of ATP renewal. When ATP donates a phosphate group to other molecules (and so energizes them), it reverts to ADP, then forms again by phosphorylation of ADP.

adrenal cortex (ah-DREE-nul) Outer portion of the adrenal gland; its hormones have roles in metabolism, inflammation, maintaining extracellular fluid volume, and other functions.

adrenal medulla Inner region of the adrenal gland; its hormones help control blood circulation and carbohydrate metabolism.

aerobic respiration (air-OH-bik) [Gk. *aer*, air, + *bios*, life] The main energy-releasing metabolic pathway of ATP formation, in which oxygen is the final acceptor of electrons stripped from glucose or some other organic compound. The pathway proceeds from glycolysis through the Krebs cycle and electron transport phosphorylation. A typical net yield is 36 ATP for each glucose molecule.

age structure Of a population, the number of individuals in each of several or many age categories.

agglutination (ah-glue-tin-AY-shun) Clumping together of foreign cells that have invaded the body (as pathogens or in tissue grafts or transplants). Clumping is induced by cross-linking between antibody molecules that have already latched onto antigen at the surface of the foreign cells.

aging A range of processes, including the breakdown of cell structure and function, by which the body gradually deteriorates. All organisms showing extensive cell differentiation undergo aging.

AIDS Acquired immunodeficiency syndrome. A set of chronic disorders following infection by the human immunodeficiency virus (HIV), which destroys key cells of the immune system.

alcoholic fermentation Anaerobic pathway of ATP formation in which pyruvate from glycolysis is broken down to acetaldehyde, which accepts electrons from NADH to become ethanol, and NAD$^+$ is regenerated. Its net yield is two ATP.

aldosterone (al-DOSS-tuh-rohn) Hormone secreted by the adrenal cortex that helps regulate sodium reabsorption.

allantois (ah-LAN-twahz) [Gk. *allas*, sausage] Of vertebrates, one of four extraembryonic membranes that form during embryonic development. It functions in respiration and storage of metabolic wastes in reptiles, birds, and some mammals. In humans, it functions in early blood formation and development of the urinary bladder.

allele (uh-LEEL) For a given location on a chromosome, one of two or more slightly different molecular forms of a gene that code for different versions of the same trait.

allele frequency Of a given gene locus, the relative abundances of each kind of allele carried by the individuals of a population.

allergy An immune response made against a normally harmless substance.

allopatric speciation [Gk. *allos*, different, + *patria*, native land] Speciation that follows geographic isolation of populations of the same species.

allosteric control (AL-oh-STARE-ik) Control over a metabolic reaction or pathway that operates through the binding of a specific substance at a control site on a specific enzyme.

alpine tundra A type of biome that exists at high elevations in mountains throughout the world.

altruistic behavior (al-true-ISS-tik) Self-sacrificing behavior; the individual behaves in a way that helps others but decreases its own chances of reproductive success.

alveolar sac (al-VEE-uh-lar) Any of the pouch-like clusters of alveoli in the lungs; the major sites of gas exchange.

alveolus (ahl-VEE-uh-lus), plural **alveoli** [L. *alveus*, small cavity] Any of the many cup-shaped, thin-walled outpouchings of respiratory bronchioles. A site where oxygen diffuses from air in the lungs to the blood, and carbon dioxide diffuses from blood to the lungs.

amino acid (uh-MEE-no) A small organic molecule having a hydrogen atom, an amino group, an acid group, and an R group covalently bonded to a central carbon atom. The subunit of polypeptide chains, which represent the primary structure of proteins.

ammonification (uh-moan-ih-fih-KAY-shun) Together with decomposition, a process by which certain bacteria and fungi break down nitrogen-containing wastes and remains of other organisms.

amnion (AM-nee-on) Of land vertebrates, one of four extraembryonic membranes. It becomes a fluid-filled sac in which the embryo (and fetus) can grow, move freely, and be protected from sudden temperature shifts and impacts.

amniote egg A type of egg, often with a leathery or calcified shell, that contains extraembryonic membranes, including the amnion. An adaptation that figured in the vertebrate invasion of land.

amphibian A type of vertebrate somewhere between fishes and reptiles in body plan and reproductive mode; salamanders, frogs and toads, and caecilians are the existing groups.

anaerobic pathway (an-uh-ROW-bik) [Gk. *an*, without, + *aer*, air] Metabolic pathway in which a substance other than oxygen serves as the final acceptor of electrons that have been stripped from substrates.

analogous structures Body parts, once different in separate lineages, that were put to comparable uses in similar environments and that came to resemble one another in form and function. They are evidence of morphological convergence.

anaphase (AN-uh-faze) The stage at which microtubules of a spindle apparatus separate sister chromatids of each chromosome and move them to opposite spindle poles. During anaphase I of *meiosis*, the two members of each pair of homologous chromosomes separate. During anaphase II, sister chromatids of each chromosome separate.

aneuploidy (AN-yoo-ploy-dee) A change in the chromosome number following inheritance of one extra or one less chromosome.

angiosperm (AN-gee-oh-spurm) [Gk. *angeion*, vessel, and *spermia*, seed] A flowering plant.

animal A heterotroph that eats or absorbs nutrients from other organisms; is multicelled, usually with tissues arranged in organs and organ systems; is usually motile during at least part of the life cycle; and goes through a period of embryonic development.

annelid The type of invertebrate classified as a segmented worm; an oligochaete (such as an earthworm), leech, or polychaete.

annual plant A flowering plant that completes its life cycle in one growing season.

anther [Gk. *anthos*, flower] In flowering plants, the pollen-bearing part of the male reproductive structure (stamen).

antibiotic A normal metabolic product of certain microorganisms that kills or inhibits the growth of other microorganisms.

antibody [Gk. *anti*, against] Any of a variety of Y-shaped receptor molecules with binding sites for specific antigens. Only B cells produce antibodies, then position them at their surface or secrete them.

anticodon In a tRNA molecule, a sequence of three nucleotide bases that can pair with an mRNA codon.

antigen (AN-tih-jen) [Gk. *anti*, against, + *genos*, race, kind] Any molecular configuration that is recognized as foreign to the body and that triggers an immune response. Most antibodies are protein molecules at the surface of infectious agents or tumor cells.

antigen-presenting cell A macrophage or some other white blood cell that engulfs and digests antigen, then displays it with certain MHC molecules at its surface. Recognition of antigen-MHC complexes by T and B cells triggers an immune response.

aorta (ay-OR-tah) [Gk. *airein*, to lift, heave] Main artery of systemic circulation; carries oxygenated blood away from the heart to all body regions except the lungs.

apical dominance The inhibitory influence of a terminal bud on the growth of lateral buds.

apical meristem (AY-pih-kul MARE-ih-stem) [L. *apex*, top, + Gk. *meristos*, divisible] Of most plants, a mass of self-perpetuating cells responsible for primary growth (elongation) at root and shoot tips.

appendicular skeleton (ap-en-DIK-yoo-lahr) In vertebrates, bones of the limbs, hips, and shoulders.

appendix A slender projection from the cup-shaped pouch at the start of the colon.

archaebacteria One of three great prokaryotic lineages that arose early in the history of life; now represented by methanogens, halophiles, and thermophiles.

arctic tundra A type of biome that lies between the polar ice cap and boreal forests of North America, Europe, and Asia.

arteriole (ar-TEER-ee-ole) Any of the blood vessels between arteries and capillaries. They are control points where the volume of blood delivered to different body regions can be adjusted.

artery Any of the large-diameter blood vessels that conduct oxygen-poor blood to the lungs and oxygen-enriched blood to all body tissues. Their thick, muscular wall allows them to smooth out pulsations in blood pressure caused by heart contractions.

arthropod An invertebrate having a hardened exoskeleton, specialized segments, and jointed appendages. Spiders, crabs, and insects are examples.

asexual reproduction Mode of reproduction by which offspring arise from a single parent, and inherit the genes of that parent only.

atmosphere A region of gases, airborne particles, and water vapor enveloping the earth; 80 percent of its mass is distributed within seventeen miles of the earth's surface.

atmospheric cycle A biogeochemical cycle in which the atmosphere is the largest reservoir of an element. The carbon and nitrogen cycles are examples.

atom The smallest unit of matter that is unique to a particular element.

atomic number The number of protons in the nucleus of each atom of an element; it differs for each element.

ATP Adenosine triphosphate (ah-DEN-uh-seen try-FOSS-fate). A nucleotide composed of adenine, ribose, and three phosphate groups. As the main energy carrier in cells, it directly or indirectly delivers energy to or picks up energy from nearly all metabolic pathways.

atrium (AYE-tree-um) Of the human heart, one of two chambers that receive blood. *Compare* ventricle.

australopith (OHSS-trah-low-pith) [L. *australis*, southern, + Gk. *pithekos*, ape] Any of the earliest known species of hominids, that is, the first species of the evolutionary branch leading to humans.

autoimmune response Misdirected immune response in which lymphocytes mount an attack against normal body cells.

autonomic nervous system (auto-NOM-ik) Those nerves leading from the central nervous system to the smooth muscle, cardiac muscle, and glands of internal organs and structures, that is, to the visceral portion of the body.

autosomal dominant inheritance Condition arising from the presence of a dominant allele on an autosome (not a sex chromosome). The allele is always expressed to some extent, even in heterozygotes.

autosomal recessive inheritance Condition arising from a recessive allele on an autosome (not a sex chromosome). Only recessive homozygotes show the resulting phenotype.

autosome Any of the chromosomes that are of the same number and kind in both males and females of the species.

autotroph (AH-toe-trofe) [Gk. *autos*, self, + *trophos*, feeder] An organism able to build its own large organic molecules by using carbon dioxide and energy from the physical environment. Photosynthetic autotrophs use sunlight energy; chemosynthetic autotrophs extract energy from chemical reactions involving inorganic substances. *Compare* heterotroph.

auxin (AWK-sin) Any of a class of growth-regulating hormones in plants; auxins promote stem elongation as one effect.

axial skeleton (AX-ee-uhl) In vertebrates, the skull, backbone, ribs, and breastbone (sternum).

axon Of a neuron, a long, cylindrical extension from the cell body, with finely branched endings. Action potentials move rapidly, without alteration, along an axon; their arrival at axon endings may trigger the release of neurotransmitter molecules that influence an adjacent cell.

B lymphocyte, or **B cell** The only white blood cell that produces antibodies, then positions them at the cell surface or secretes them as weapons in immune responses.

bacterial conjugation The transfer of plasmid DNA from one bacterial cell to another.

bacterial flagellum Of many bacterial cells, a whiplike motile structure that does not contain a core of microtubules.

bacteriophage (bak-TEER-ee-oh-fahj) [Gk. *baktērion*, small staff, rod, + *phagein*, to eat] Category of viruses that infect bacterial cells.

balanced polymorphism The maintenance of two or more forms of a trait in fairly stable proportions over generations.

Barr body In the cells of female mammals, a condensed X chromosome that was inactivated during early embryonic development.

basal body A centriole which, after having given rise to the microtubules of a flagellum or cilium, remains attached to its base in the cytoplasm.

base A substance that releases OH⁻ in water.

base pair A pair of hydrogen-bonded nucleotide bases in two strands of nucleic acids. In a DNA double helix, adenine pairs with thymine, and guanine with cytosine. When an mRNA strand forms on a DNA strand during transcription, uracil (U) pairs with the DNA's adenine.

base sequence The particular order in which one nucleotide base follows the next in a strand of DNA or RNA. The order differs to some extent for each kind of organism.

basophil Fast-acting white blood cells that secrete histamine and other substances during inflammation.

behavior, animal A response to external and internal stimuli, following integration of sensory, neural, endocrine, and effector components. Because behavior has a genetic basis, it is subject to natural selection and commonly can be modified through experience.

benthic province All of the sediments and rocky formations of the ocean bottom; begins with the continental shelf and extends down through deep-sea trenches.

biennial A flowering plant that lives through two growing seasons.

bilateral symmetry Body plan in which the left and right halves of the animal are mirror-images of each other.

binary fission Of bacteria, a mode of asexual reproduction in which the parent cell replicates its single chromosome, then divides into two genetically identical daughter cells.

biogeochemical cycle The movement of an element such as carbon or nitrogen from the environment to organisms, then back to the environment.

biogeographic realm [Gk. *bios*, life, + *geographein*, to describe the surface of the earth] Any of six major land regions, each having distinguishing types of plants and animals and generally retaining its identity because of climate and geographic barriers to gene flow.

biological clocks Internal time-measuring mechanisms that have roles in adjusting an organism's daily activities, seasonal activities, or both in response to environmental cues.

biological magnification The increasing concentration of a nondegradable or slowly degradable substance in body tissues as it is passed along food chains.

biological systematics Branch of biology that assesses patterns of diversity based on information from taxonomy, phylogenetic reconstruction, and classification.

bioluminescence A flashing of light that emanates from an organism when excited electrons of luciferins, or highly fluorescent substances, return to a lower energy level.

biomass The combined weight of all the organisms at a particular trophic (feeding) level in an ecosystem.

biome A broad, vegetational subdivision of a biogeographic realm shaped by climate, topography, and composition of regional soils.

biosphere [Gk. *bios*, life, + *sphaira*, globe] All regions of the earth's waters, crust, and atmosphere in which organisms live.

biosynthetic pathway A metabolic pathway in which small molecules are assembled into lipids, proteins, and other large organic molecules.

biotic potential Of a population, the maximum rate of increase per individual under ideal conditions.

bipedalism A habitual standing and walking on two feet, as by ostriches and humans.

bird A type of vertebrate, the only one having feathers, with strong resemblances and evolutionary connections to reptiles.

blastocyst (BLASS-tuh-sist) [Gk. *blastos*, sprout, + *kystis*, pouch] In mammalian development, a blastula stage consisting of a hollow ball of surface cells and an inner cell mass.

blastula (BLASS-chew-lah) An embryonic stage consisting of a ball of cells produced by cleavage.

blood A fluid connective tissue composed of water, solutes, and formed elements (blood cells and platelets); it carries substances to and from cells and helps maintain an internal environment that is favorable for cell activities.

blood pressure Fluid pressure, generated by heart contractions, that keeps blood circulating.

blood-brain barrier Set of mechanisms that helps control which blood-borne substances reach neurons in the brain.

bone The mineral-hardened connective tissue of bones.

bones In vertebrate skeletons, organs that function in movement and locomotion, protection of other organs, mineral storage, and (in some bones) blood cell production.

bottleneck An extreme case of genetic drift. A catastrophic decline in population size leads to a random shift in the allele frequencies among survivors. Because the population must rebuild from so few individuals, severely limited genetic variation may be an outcome.

Bowman's capsule The cup-shaped portion of a nephron that receives water and solutes filtered from blood.

brain Of most nervous systems, the most complex integrating center; it receives, processes, and integrates sensory input and issues coordinated commands for response.

brainstem The vertebrate midbrain, pons, and medulla oblongata, the core of which contains the reticular formation that helps govern activity of the nervous system as a whole.

bronchiole Of most vertebrates, a component of the finely branched bronchial tree inside each lung.

bronchus, plural **bronchi** (BRONG-CUSS, BRONG-kee) [Gk. *bronchos*, windpipe] Tubelike branchings of the trachea that lead into the lungs of most vertebrates.

brown alga A type of aquatic plant, found in nearly all marine habitats, that has an abundance of xanthophyll pigments.

bryophyte A nonvascular land plant that requires free water to complete its life cycle.

bud An undeveloped shoot of mostly meristematic tissue; often covered and protected by scales (modified leaves).

buffer A substance that can combine with hydrogen ions, release them, or both. Buffers help stabilize the pH of blood and other fluids.

bulk Of human digestion, a volume of fiber and other undigested material that absorption processes in the colon cannot decrease.

bulk flow In response to a pressure gradient, a movement of more than one kind of molecule in the same direction in the same medium (as in blood, sap, or air).

C4 pathway Of many plants, a pathway of photosynthesis in which carbon dioxide is fixed twice, in two different cell types. Carbon dioxide accumulates in the leaf and helps counter photorespiration. The first compound formed is the 4-carbon oxaloacetate.

Calvin-Benson cycle Cyclic reactions that are the "synthesis" part of the light-independent reactions of photosynthesis. In land plants, RuBP, or some other compound to which carbon has been affixed, undergoes rearrangements that lead to formation of a sugar phosphate and to regeneration of the RuBP. The cycle runs on ATP and NADPH from light-dependent reactions.

CAM plant A plant that conserves water by opening stomata only at night, when it fixes carbon dioxide by way of a C4 pathway.

cambium (KAM-bee-um), plural **cambia** In vascular plants, one of two types of meristems that are responsible for secondary growth (increases in stem and root diameter). Vascular cambium gives rise to secondary xylem and phloem; cork cambium gives rise to periderm.

camouflage An outcome of an organism's form, patterning, color, or behavior that helps it blend with its surroundings and escape detection.

cancer A type of malignant tumor, the cells of which show profound abnormalities in the plasma membrane and cytoplasm, abnormal growth and division, and weakened capacity for adhesion within the parent tissue (leading to metastasis). Unless eradicated, cancer is lethal.

canine A pointed tooth that functions in piercing food.

capillary [L. *capillus*, hair] A thin-walled blood vessel that functions in the exchange of gases and other substances between blood and interstitial fluid.

capillary bed A diffusion zone, consisting of numerous capillaries, where substances are exchanged between blood and interstitial fluid.

carbohydrate [L. *carbo*, charcoal, + *hydro*, water] A simple sugar or large molecule composed of sugar units. All cells use carbohydrates as structural materials, energy stores, and transportable forms of energy. The three classes of carbohydrates include: monosaccharides, oligosaccharides, or polysaccharides.

carbon cycle A biogeochemical cycle in which carbon moves from its largest reservoir in the atmosphere, through oceans and organisms, then back to the atmosphere.

carbon dioxide fixation First step of the light-independent reactions of photosynthesis. Carbon (from carbon dioxide) becomes affixed to a carbon compound (such as RuBP) that can enter the Calvin-Benson cycle.

carcinogen (kar-SIN-uh-jen) An environmental agent or substance, such as ultraviolet radiation, that can trigger cancer.

cardiac cycle (KAR-dee-ak) [Gk. *kardia*, heart, + *kyklos*, circle] The sequence of muscle contraction and relaxation constituting one heartbeat.

cardiac pacemaker Sinoatrial (SA) node; the basis of the normal rate of heartbeat. The self-excitatory cardiac muscle cells that spontaneously generate rhythmic waves of excitation over the heart chambers.

cardiovascular system Of most animals, an organ system that is composed of blood, one or more hearts, and blood vessels and that functions in the rapid transport of substances to and from cells.

carnivore [L. *caro, carnis*, flesh, + *vovare*, to devour] An animal that eats other animals; a type of heterotroph.

carotenoids (kare-OTT-en-oyds) Light-sensitive, accessory pigments that transfer absorbed energy to chlorophylls. They absorb violet and blue wavelengths but transmit red, orange, and yellow.

carpel (KAR-pul) The female reproductive part of a flower; sometimes called a pistil. The lower portion of a single carpel (or of a structure composed of two or more carpels) is an ovary, where eggs develop and are fertilized and where seeds mature. The upper portion has a stigma (a pollen-capturing surface tissue) and often a style (a slender extension of the ovary wall).

carrier protein Type of transport protein that binds specific substances and changes shape in ways that shunt the substances across a plasma membrane. Some carrier proteins function passively, others require an energy input.

carrying capacity The maximum number of individuals in a population (or species) that can be sustained indefinitely by a given environment.

cartilage A type of connective tissue with solid yet pliable intercellular material that resists compression.

Casparian strip In the exodermis and endodermis of roots, a waxy band that acts as an impermeable barrier between the walls of abutting cells.

cDNA Any DNA molecule copied from a mature mRNA transcript by way of reverse transcription.

cell [L. *cella*, small room] The smallest living unit; an organized unit that can survive and reproduce on its own, given DNA instructions and suitable environmental conditions, including appropriate sources of energy and raw materials.

cell count The number of cells of a given type in a microliter of blood.

cell cycle Events during which a cell increases in mass, roughly doubles its number of cytoplasmic components, duplicates its DNA, then undergoes nuclear and cytoplasmic division. It extends from the time a new cell is produced until it completes its own division.

cell differentiation The developmental process in which different cell types activate and suppress a fraction of their genes in different ways and so become specialized in composition, structure, and function. Regulatory proteins, enzymes, hormonal signals, and control sites built into DNA interact to bring about this selective gene expression.

cell junction Of multicelled organisms, a point of contact that physically links two cells or that provides functional links between their cytoplasm.

cell plate Of a plant cell undergoing cytoplasmic division, a disklike structure that forms from remnants of the spindle; it becomes a crosswall that partitions the cytoplasm.

cell theory A theory in biology, the key points of which are that (1) all organisms are composed of one or more cells, (2) the cell is the smallest unit that still retains a capacity for independent life, and (3) all cells arise from preexisting cells.

cell wall A rigid or semirigid wall outside the plasma membrane that supports a cell and imparts shape to it; a cellular feature of plants, fungi, protistans, and most bacteria.

central nervous system The brain and spinal cord of vertebrates.

central vacuole Of mature, living plant cells, a fluid-filled organelle that stores amino acids, sugars, ions, and toxic wastes. Its enlargement during growth causes the increases in surface area that improve nutrient uptake.

centriole (SEN-tree-ohl) A cylinder of triplet microtubules that gives rise to the microtubules of cilia and flagella.

centromere (SEN-troh-meer) [Gk. *kentron*, center, + *meros*, a part] A small, constricted region of a chromosome having attachment sites for microtubules that help move the chromosome during nuclear division.

cephalization (sef-ah-lah-ZAY-shun) [Gk. *kephalikos*, head] During the evolution of bilateral animals, the concentration of sensory structures and nerve cells in the head.

cerebellum (ser-ah-BELL-um) [L. diminutive of *cerebrum*, brain] Hindbrain region with reflex centers for maintaining posture and refining limb movements.

cerebral cortex Thin surface layer of the cerebral hemispheres. Some regions of the cortex receive sensory input, others integrate information and coordinate appropriate motor responses.

cerebrospinal fluid Clear extracellular fluid that surrounds and cushions the brain and spinal cord.

cerebrum (suh-REE-bruhm) Part of the vertebrate forebrain that originally integrated olfactory input and selected motor responses to it. In mammals, it evolved into the most complex integrating center.

channel protein Type of transport protein that serves as a pore through which ions or other water-soluble substances move across the plasma membrane. Some channels remain open, while others are gated and open and close in controlled ways.

chemical bond A union between the electron structures of two or more atoms or ions.

chemical synapse (SIN-aps) [Gk. *synapsis*, union] A small gap, the synaptic cleft, that separates two neurons (or a neuron and a muscle cell or gland cell) and that is bridged by neurotransmitter molecules released from the presynaptic neuron.

chemiosmotic theory (kim-ee-OZ-MOT-ik) Theory that an electrochemical gradient across a cell membrane drives ATP formation. Metabolic reactions cause hydrogen ions (H^+) to accumulate in a compartment formed by the membrane. The combined force of the resulting concentration and electric gradients propels hydrogen ions down the gradient, through channel proteins. Through enzyme action at these proteins, ADP and inorganic phosphate combine to form ATP.

chemoreceptor (KEE-moe-ree-sep-tur) Sensory receptor that detects chemical energy (ions or molecules) dissolved in the surrounding fluid.

chemosynthetic autotroph (KEE-moe-sin-THET-ik) One of a few kinds of bacteria able to synthesize its own organic molecules using carbon dioxide as the carbon source and certain inorganic substances (such as sulfur) as the energy source.

chlorofluorocarbon (KLORE-oh-FLOOR-oh-car-bun), or **CFC** One of a variety of odorless, invisible compounds of chlorine, fluorine, and carbon, widely used in commercial products, that are contributing to the destruction of the ozone layer above the earth's surface.

chlorophylls (KLOR-uh-fills) [Gk. *chloros*, green, + *phyllon*, leaf] Light-sensitive pigment molecules that absorb violet-to-blue and red wavelengths but that transmit green. Certain chlorophylls donate the electrons required for photosynthesis.

chloroplast (KLOR-uh-plast) An organelle that specializes in photosynthesis in plants and certain protistans.

chordate An animal having a notochord, a dorsal hollow nerve cord, a pharynx, and gill slits in the pharynx wall for at least part of its life cycle.

chorion (CORE-ee-on) Of placental mammals, one of four extraembryonic membranes; it becomes a major component of the placenta. Absorptive structures (villi) that develop at its surface are crucial for the transfer of substances between the embryo and mother.

chromatid Of a duplicated eukaryotic chromosome, one of two DNA molecules and its associated proteins. One chromatid remains attached to its "sister" chromatid at the centromere until they are separated from each other during a nuclear division; then each is a separate chromosome.

chromosome (CROW-moe-some) [Gk. *chroma*, color, + *soma*, body] Of eukaryotes, a DNA molecule with many associated proteins. A bacterial chromosome does not have a comparable profusion of proteins associated with the DNA.

chromosome number Of eukaryotic species, the number of each type of chromosome in all cells except dividing germ cells or gametes.

chytrid A type of single-celled fungus of muddy and aquatic habitats.

ciliated protozoan One of four major groups of protozoans.

cilium (SILL-ee-um), plural **cilia** [L. *cilium*, eyelid] Of eukaryotic cells, a short, hairlike projection that contains a regular array of microtubules. Cilia serve as motile structures, help create currents of fluids, or are part of sensory structures. They typically are more profuse than flagella.

circadian rhythm (ser-KAYD-ee-un) [L. *circa*, about, + *dies*, day] Of many organisms, a cycle of physiological events that is completed every 24 hours or so, even when environmental conditions remain constant.

circulatory system Of multicelled animals, an organ system consisting of a muscular pump (heart, most often), blood vessels, and blood; the system transports materials to and from cells and often helps stabilize body temperature and pH.

cladistics An approach to biological systematics in which organisms are grouped according to similarities that are derived from a common ancestor.

cladogram Branching diagram that represents patterns of relative relationships between organisms based on discrete morphological, physiological, and behavioral traits that vary among taxa being studied.

classification system A way of organizing and retrieving information about species.

cleavage Stage of animal development when mitotic cell divisions convert a zygote to a ball of cells, the blastula.

cleavage furrow Of an animal cell undergoing cytoplasmic division, a shallow, ringlike depression that forms at the cell surface as contractile microfilaments pull the plasma membrane inward. It defines where the cytoplasm will be cut in two.

climate Prevailing weather conditions for an ecosystem, including temperature, humidity, wind speed, cloud cover, and rainfall.

climax community Following primary or secondary succession, the array of species that remains more or less steady under prevailing conditions.

clonal selection theory Theory that lymphocytes activated by a specific antigen rapidly multiply and differentiate into huge subpopulations of cells, all having the parent cell's specificity against that antigen.

cloned DNA Multiple, identical copies of DNA fragments that have been inserted into plasmids or some other cloning vector.

club fungus One of a highly diverse group of multicelled fungi, the reproductive structures of which have microscopic, club-shaped cells that produce and bear spores.

cnidarian A radially symmetrical invertebrate, usually marine, that has tissues (not organs), and nematocysts. Two body forms (medusae and polyps) are common. Jellyfishes, corals, and sea anemones are examples.

coal A nonrenewable source of energy that formed more than 280 million years ago from submerged, undecayed plant remains that were buried in sediments, compressed, then compacted further by heat and pressure.

codominance Condition in which a pair of nonidentical alleles are both expressed even though they specify two different phenotypes.

codon One of a series of base triplets in an mRNA molecule, most of which code for a sequence of amino acids of a specific polypeptide chain. (Of sixty-four codons, sixty-one specify different amino acids and three of these also serve as start signals for translation; one other serves only as a stop signal for translation.)

coelum (SEE-lum) [Gk. *koilos*, hollow] Of many animals, a type of body cavity located between the gut and body wall and having a distinctive lining (peritoneum).

coenzyme A type of nucleotide that transfers hydrogen atoms and electrons from one reaction site to another. NAD$^+$ is an example.

coevolution The joint evolution of two or more closely interacting species; when one species evolves, the change affects selection pressures operating between the two species, so the other also evolves.

cofactor A metal ion or coenzyme; it helps catalyze a reaction or serves briefly as an agent that transfers electrons, atoms, or functional groups from one substrate to another.

cohesion Condition in which molecular bonds resist rupturing when under tension.

cohesion theory of water transport Theory that water moves up through vascular plants due to hydrogen bonding among water molecules confined inside the xylem pipelines. The collective cohesive strength of those bonds allows water to be pulled up as columns in response to transpiration (evaporation from leaves).

collenchyma One of the simple tissues of flowering plants; lends flexible support to primary tissues, such as those of lengthening stems.

colon (CO-lun) The large intestine.

commensalism [L. *com*, together, + *mensa*, table] Two-species interaction in which one species benefits significantly while the other is neither helped nor harmed to any notable extent.

communication signal Of social animals, an action or cue sent by one member of a species (the signaler) that can change the behavior of another member (the signal receiver).

community The populations of all species occupying a habitat; also applied to groups of organisms with similar life-styles in a habitat (such as the bird community).

companion cell A specialized parenchyma cell that helps load dissolved organic compounds into the conducting cells of the phloem.

comparative morphology [Gk. *morph*, form] Anatomical comparisons of major lineages.

competitive exclusion Theory that populations of two species competing for a limited resource cannot coexist indefinitely in the same habitat; the population better adapted to exploit the resource will enjoy a competitive (hence reproductive) edge and will eventually exclude the other population from the habitat.

complement system A set of about twenty proteins circulating in blood plasma with roles in nonspecific defenses and in immune responses. Some induce lysis of pathogens, others promote inflammation, and others stimulate phagocytes to engulf pathogens.

compound A substance in which the relative proportions of two or more elements never vary. Organic compounds have a backbone of carbon atoms arranged as a chain or ring structure. The simpler, inorganic compounds do not have comparable backbones.

concentration gradient A difference in the number of molecules (or ions) of a substance between two adjacent regions, as in a volume of fluid.

condensation reaction Enzyme-mediated reaction leading to the covalent linkage of small molecules and, often, the formation of water as a by-product.

cone cell In the vertebrate eye, a type of photoreceptor that responds to intense light and contributes to sharp daytime vision and color perception.

conifer A type of plant belonging to the dominant group of gymnosperms; mostly evergreen, woody trees and shrubs with pollen- and seed-bearing cones.

connective tissue proper A category of animal tissues, all having mostly the same components but in different proportions. These tissues contain fibroblasts and other cells, the secretions of which form fibers (of collagen and elastin) and a ground substance (of modified polysaccharides).

consumers [L. *consumere*, to take completely] Of ecosystems, heterotrophic organisms that obtain energy and raw materials by feeding on the tissues of other organisms. Herbivores, carnivores, omnivores, and parasites are examples.

continuous variation A more or less continuous range of small differences in a given trait among all the individuals of a population.

contractile vacuole (kun-TRAK-till VAK-you-ohl) [L. *contractus*, to draw together] In some protistans, a membranous chamber that takes up excess water in the cell body, then contracts, expelling the water outside the cell through a pore.

control group In a scientific experiment, a group used to evaluate possible side effects of a test involving an experimental group. Ideally, the control group should differ from the experimental group only with respect to the variable being studied.

convergence, morphological *See* morphological convergence.

cork cambium A type of lateral meristem that produces a tough, corky replacement for epidermis on parts of woody plants showing extensive secondary growth.

corpus callosum (CORE-pus ka-LOW-sum) A band of axons (200 million in humans) that functionally link two cerebral hemispheres.

corpus luteum (CORE-pus LOO-tee-um) A glandular structure; it develops from cells of a ruptured ovarian follicle and secretes progesterone and some estrogen, both of which maintain the lining of the uterus (endometrium).

cortex [L. *cortex*, bark] In general, a rindlike layer; the kidney cortex is an example. In vascular plants, ground tissue that makes up most of the primary plant body, supports plant parts, and stores food.

cotyledon A seed leaf, which develops as part of a plant embryo; cotyledons provide nourishment for the germinating seedling.

courtship display Social behavior by which individuals assess and respond to sexual overtures of potential partners.

covalent bond (koe-VAY-lunt) [L. *con*, together, + *valere*, to be strong] A sharing of one or more electrons between atoms or groups of atoms. When electrons are shared equally, the bond is nonpolar. When electrons are shared unequally, the bond is polar—slightly positive at one end and slightly negative at the other.

cross-bridge formation Of muscle cells, the interaction between actin and myosin filaments that is the basis of contraction.

crossing over During prophase I of meiosis, an interaction between a pair of homologous chromosomes. Their nonsister chromatids break at the same place along their length and exchange corresponding segments at the break points. Crossing over breaks up old combinations of alleles and puts new ones together in chromosomes.

culture The sum total of behavior patterns of a social group, passed between generations by learning and by symbolic behavior, especially language.

cuticle (KEW-tih-kull) A body covering. Of land plants, a covering of waxes and lipid-rich cutin deposited on the outer surface of epidermal cell walls. Of annelids, a thin, flexible surface coat. Of arthropods, a hardened yet lightweight covering with protein and chitin components that functions as an external skeleton.

cycad A type of gymnosperm of the tropics and subtropics; slow growing, with massive, cone-shaped structures that bear ovules or pollen.

cyclic AMP (SIK-lik) Cyclic adenosine monophosphate. A nucleotide that has roles in intercellular communication, as when it serves as a second messenger (a cytoplasmic mediator of a cell's response to signaling molecules).

cyclic pathway of ATP formation Photosynthetic pathway in which excited electrons move from a photosystem to an electron transport system, and back to the photosystem. The electron flow contributes to the formation of ATP from ADP and inorganic phosphate.

cyst Of some microorganisms, a walled, resting structure that forms during the life cycle.

cytochrome (SIGH-toe-krome) [Gk. *kytos*, hollow vessel, + *chrōma*, color] Iron-containing protein molecule; a component of electron transport systems used in photosynthesis and aerobic respiration.

cytokinesis (SIGH-toe-kih-NEE-sis) [Gk. *kinesis*, motion] Cytoplasmic division; the splitting of a parental cell into daughter cells.

cytokinin (SIGH-tow-KY-nin) Any of the class of plant hormones that stimulate cell division, promote leaf expansion, and retard leaf aging.

cytomembrane system [Gk. *kytos*, hollow vessel] Organelles, functioning as a system to modify, package, and distribute newly formed proteins and lipids. Endoplasmic reticulum, Golgi bodies, lysosomes, and a variety of vesicles are its components.

cytoplasm (SIGH-toe-plaz-um) [Gk. *plassein*, to mold] All cellular parts, particles, and semifluid substances enclosed by the plasma membrane except for the region of DNA (which in eukaryotes, is the nucleus).

cytosine (SIGH-toe-seen) A pyrimidine; one of the nitrogen-containing bases in nucleotides.

cytoskeleton Of eukaryotic cells, an internal "skeleton." Its microtubules and other components structurally support the cell, organize and move its internal components. The cytoskeleton also helps free-living cells move through their environment.

cytotoxic T cell A T lymphocyte that eliminates infected body cells or tumor cells with a single hit of toxins and perforins.

decomposers [L. *de-*, down, away, + *companere*, to put together] Of ecosystems, heterotrophs that obtain energy by chemically breaking down the remains, products, or wastes of other organisms. Their activities help cycle nutrients back to producers. Certain fungi and bacteria are examples.

deforestation The removal of all trees from a large tract of land, such as the Amazon Basin or the Pacific Northwest.

degradative pathway A metabolic pathway by which molecules are broken down in stepwise reactions that lead to products of lower energy.

deletion A change in a chromosome's structure after one of its regions is lost as a result of irradiation, viral attack, chemical action, or some other factor.

demographic transition model Model of human population growth in which changes in the growth pattern correspond to different stages of economic development. These are a preindustrial stage, when birth and death rates are both high, a transitional stage, an industrial stage, and a postindustrial stage, when the death rate exceeds the birth rate.

denaturation (deh-NAY-chur-AY-shun) Of any molecule, the loss of three-dimensional shape following disruption of hydrogen bonds and other weak bonds.

dendrite (DEN-drite) [Gk. *dendron*, tree] A short, slender extension from the cell body of a neuron.

denitrification (DEE-nite-rih-fih-KAY-shun) The conversion of nitrate or nitrite by certain bacteria to gaseous nitrogen (N_2) and a small amount of nitrous oxide (N_2O).

density-dependent controls Factors such as predation, parasitism, disease, and competition for resources, which limit population growth by reducing the birth rate, increasing the rates of death and dispersal, or all of these.

density-independent controls Factors such as storms or floods that increase a population's death rate more or less independently of its density.

dentition (den-TIH-shun) The type, size, and number of an animal's teeth.

dermal tissue system Of vascular plants, the tissues that cover and protect the plant surfaces.

dermis The layer of skin underlying the epidermis, consisting mostly of dense connective tissue.

desert A type of biome that exists where the potential for evaporation greatly exceeds rainfall and vegetation cover is limited.

desertification (dez-urt-ih-fih-KAY-shun) The conversion of grasslands, rain-fed cropland, or irrigated cropland to desertlike conditions, with a drop in agricultural productivity of 10 percent or more.

detrital food web Of most ecosystems, the flow of energy mainly from plants through detritivores and decomposers.

detritivores (dih-TRY-tih-vorez) [L. *detritus*; after *deterere*, to wear down] Of ecosystems, heterotrophs that consume dead or decomposing particles of organic matter. Earthworms, crabs, and nematodes are examples.

deuterostome (DUE-ter-oh-stome) [Gk. *deuteros*, second, + *stoma*, mouth] Any of the bilateral animals, including echinoderms and chordates, in which the first indentation in the early embryo develops into the anus.

diaphragm (DIE-uh-fram) [Gk. *diaphragma*, to partition] Muscular partition between the thoracic and abdominal cavities, the contraction and relaxation of which contribute to breathing. Also, a contraceptive device used temporarily to prevent sperm from entering the uterus during sexual intercourse.

dicot (DIE-kot) [Gk. *di*, two, + *kotylēdōn*, cup-shaped vessel] Short for dicotyledon; class of flowering plants characterized generally by seeds having embryos with two cotyledons (seed leaves), net-veined leaves, and floral parts arranged in fours, fives, or multiples of these.

differentiation See cell differentiation.

diffusion Net movement of like molecules (or ions) down their concentration gradient. In the absence of other forces, molecular motion and random collisions cause their net outward movement from one region into a neighboring region where they are less concentrated (because collisions are more frequent where the molecules are most crowded together).

digestive system An internal tube or cavity from which ingested food is absorbed into the internal environment; often divided into regions specialized for food transport, processing, and storage.

dihybrid cross An experimental cross in which offspring inherit two gene pairs, each consisting of two nonidentical alleles.

dinoflagellate A photosynthetic or heterotrophic protistan, often flagellated, that is a component of plankton.

diploid number (DIP-loyd) For many sexually reproducing species, the chromosome number of somatic cells and of germ cells prior to meiosis. Such cells have two chromosomes of each type (that is, pairs of homologous chromosomes). *Compare* haploid number.

directional selection Of a population, a shift in allele frequencies in a steady, consistent direction in response to a new environment or to a directional change in the old one. The outcome is that forms of traits at one end of the range of phenotypic variation become more common than the intermediate forms.

disaccharide (die-SAK-uh-ride) [Gk. *di*, two, + *sakcharon*, sugar] A type of simple carbohydrate, of the class called oligosaccharides; two monosaccharides covalently bonded.

disruptive selection Of a population, a shift in allele frequencies to forms of traits at both ends of a range of phenotypic variation and away from intermediate forms.

distal tubule The tubular section of a nephron most distant from the glomerulus; a major site of water and sodium reabsorption.

divergence Accumulation of differences in allele frequencies between populations that have become reproductively isolated from one another.

divergence, morphological See morphological divergence.

diversity, organismic Sum total of variations in form, function, and behavior that have accumulated in different lineages. Those variations generally are adaptive to prevailing conditions or were adaptive to conditions that existed in the past.

DNA Deoxyribonucleic acid (dee-OX-ee-RYE-bow-new-CLAY-ik). For all cells (and many viruses), the molecule of inheritance. A category of nucleic acids, each usually consisting of two nucleotide strands twisted together helically and held together by hydrogen bonds. The nucleotide sequence encodes the instructions for assembling proteins, and, ultimately, new individuals of a particular species.

DNA-DNA hybridization See nucleic acid hybridization.

DNA fingerprint Of each individual, a unique array of RFLPs, resulting from the DNA sequences inherited (in a Mendelian pattern) from each parent.

DNA library A collection of DNA fragments produced by restriction enzymes and incorporated into plasmids.

DNA ligase (LYE-gaze) Enzyme that seals together the new base-pairings during DNA replication; also used by recombinant DNA technologists to seal base-pairings between DNA fragments and cut plasmid DNA.

DNA polymerase (poe-LIM-uh-raze) Enzyme that assembles a new strand on a parent DNA strand during replication; also takes part in DNA repair.

DNA probe A short DNA sequence that has been assembled from radioactively labeled nucleotides and that can base-pair with part of a gene under investigation.

DNA repair Following an alteration in the base sequence of a DNA strand, a process that restores the original sequence, as carried out by DNA polymerases, DNA ligases, and other enzymes.

DNA replication Of cells, the process by which the hereditary material is duplicated for distribution to daughter nuclei. An example is the duplication of eukaryotic chromosomes during interphase, prior to mitosis.

dominance hierarchy Form of social organization in which some members of the group have adopted a subordinate status to others.

dominant allele In a diploid cell, an allele that masks the expression of its partner on the homologous chromosome.

dormancy [L. *dormire*, to sleep] Of plants, the temporary, hormone-mediated cessation of growth under conditions that might appear to be quite suitable for growth.

double fertilization Of flowering plants only, the fusion of one sperm nucleus with the egg nucleus (to produce a zygote), *and* fusion of a second sperm nucleus with the two nuclei of the endosperm mother cell, which gives rise to triploid (3*n*) nutritive tissue.

doubling time The length of time it takes for a population to double in size.

drug addiction Chemical dependence on a drug, following habituation and tolerance of

it; the drug takes on an "essential" biochemical role in the body.

dry shrubland A type of biome that exists where annual rainfall is less than 25 to 60 centimeters and where short, woody, multi-branched shrubs predominate; chaparral is an example.

dry woodland A type of biome that exists when annual rainfall is about 40 to 100 centimeters; there may be tall trees, but these do not form a dense canopy.

dryopith A type of hominoid, one of the first to appear during the Miocene about the time of the divergences that led to gorillas, chimpanzees, and humans.

duplication A change in a chromosome's structure resulting in the repeated appearance of the same gene sequence.

early *Homo* A type of early hominid that may have been the maker of stone tools that date from about 2.5 million years ago.

echinoderm A type of invertebrate that has calcified spines, needles, or plates on the body wall. It is radially symmetrical but with some bilateral features. Sea stars and sea urchins are examples.

ecology [Gk. *oikos*, home, + *logos*, reason] Study of the interactions of organisms with one another and with their physical and chemical environment.

ecosystem [Gk. *oikos*, home] An array of organisms and their physical environment, all of which interact through a flow of energy and a cycling of materials.

ecosystem modeling Analytical method of predicting unforeseen effects of disturbances to an ecosystem, based on computer programs and models.

ectoderm [Gk. *ecto*, outside, + *derma*, skin] Of animal embryos, the outermost primary tissue layer (germ layer) that gives rise to the outer layer of the integument and to tissues of the nervous system.

effector Of homeostatic systems, a muscle (or gland) that responds to signals from an integrator (such as the brain) by producing movement (or chemical change) that helps adjust the body to changing conditions.

effector cell Of the differentiated subpopulations of lymphocytes that form during an immune response, the type of cell that engages and destroys the antigen-bearing agent that triggered the response.

egg A type of mature female gamete; also called an ovum.

El Niño A recurring, massive displacement to the east of warm surface waters of the western equatorial Pacific, which in turn displaces the cooler waters of the Humboldt Current off the coast of Peru.

electron Negatively charged unit of matter, with both particulate and wavelike properties, that occupies one of the orbitals around the atomic nucleus. Atoms can gain, lose, or share electrons with other atoms.

electron transport phosphorylation (FOSS-for-ih-LAY-shun) Final stage of aerobic respiration, in which ATP forms after hydrogen ions and electrons (from the Krebs cycle) are sent through a transport system that gives up the electrons to oxygen.

electron transport system An organized array of enzymes and cofactors, bound in a cell membrane, that accept and donate electrons in sequence. When such systems operate, hydrogen ions (H$^+$) flow across the membrane, and the flow drives ATP formation and other reactions.

element Any substance that cannot be decomposed into substances with different properties.

embryo (EM-bree-oh) [Gk. *en*, in, + probably *bryein*, to swell] Of animals generally, the stage formed by way of cleavage, gastrulation, and other early developmental events. Of seed plants, the young sporophyte, from the first cell divisions after fertilization until germination.

embryo sac The female gametophyte of flowering plants.

emulsification Of chyme in the small intestine, a suspension of droplets of fat coated with bile salts.

end product A substance present at the end of a metabolic pathway.

endangered species A species poised at the brink of extinction, owing to the extremely small size and severely limited genetic diversity of its remaining populations.

endergonic reaction (en-dur-GONE-ik) Chemical reaction showing a net gain in energy.

endocrine gland Ductless gland that secretes hormones into interstitial fluid, after which they are distributed by way of the bloodstream.

endocrine system System of cells, tissues, and organs that is functionally linked to the nervous system and that exerts control by way of its hormones and other chemical secretions.

endocytosis (EN-doe-sigh-TOE-sis) Movement of a substance into cells; the substance becomes enclosed by a patch of plasma membrane that sinks into the cytoplasm, then forms a vesicle around it. Phagocytic cells also engulf pathogens or prey in this manner.

endoderm [Gk. *endon*, within, + *derma*, skin] Of animal embryos, the inner primary tissue layer, or germ layer, that gives rise to the inner lining of the gut and organs derived from it.

endodermis A sheetlike wrapping of single cells around the vascular cylinder of a root; it functions in controlling the uptake of water and dissolved nutrients. An impermeable barrier (Casparian strip) prevents water from passing between the walls of abutting endodermal cells.

endometrium (EN-doh-MEET-ree-um) [Gk. *metrios*, of the womb] Inner lining of the uterus, consisting of connective tissues, glands, and blood vessels.

endoplasmic reticulum or **ER** (EN-doe-PLAZ-mik reh-TIK-yoo-lum) An organelle that begins at the nucleus and curves through the cytoplasm. In rough ER (which has many ribosomes on its cytoplasmic side), many new polypeptide chains acquire specialized side chains. In many cells, smooth ER (with no attached ribosomes) is the main site of lipid synthesis.

endoskeleton [Gk. *endon*, within, + *sklēros*, hard, stiff] In chordates, the internal framework of bone, cartilage, or both. Together with skeletal muscle, supports and protects other body parts, helps maintain posture, and moves the body.

endosperm (EN-doe-sperm) Nutritive tissue that surrounds and serves as food for a flowering plant embryo and, later, for the germinating seedling.

endospore Of certain bacteria, a resistant body that forms around DNA and some cytoplasm; it germinates and gives rise to new bacterial cells when conditions become favorable.

endosymbiosis A permanent, mutually beneficial interdependency between two species, one of which resides permanently inside the other's body.

energy The capacity to do work.

energy carrier A molecule that delivers energy from one metabolic reaction site to another. ATP is the most widely travelled of these; it readily donates energy to nearly all metabolic reactions.

energy flow pyramid A pyramid-shaped representation of an ecosystem's trophic structure, illustrating the energy losses at each transfer to a different trophic level.

entropy (EN-trow-pee) A measure of the degree of disorder in a system (how much energy has become so disorganized and dispersed, usually as heat, that it is no longer readily available to do work).

enzyme (EN-zime) One of a class of proteins that greatly speed up (catalyze) reactions between specific substances, usually at their functional groups. The substances that each type of enzyme acts upon are called its substrates.

eosinophil Fast-acting, phagocytic white blood cell that takes part in inflammation but not in immune responses.

epidermis The outermost tissue layer of a multicelled plant or animal.

epiglottis A flaplike structure at the start of the larynx, the position of which directs the movement of air into the trachea or food into the esophagus.

epistasis (eh-PISS-tih-sis) A type of gene interaction, whereby two alleles of a gene influence the expression of alleles of a different gene.

epithelium (EP-ih-THEE-lee-um) An animal tissue consisting of one or more layers of adhering cells that covers the body's external surfaces and lines its internal cavities and tubes. Epithelium has one free surface; the opposite surface rests on a basement membrane between it and an underlying connective tissue. Epidermis or skin is an example.

equilibrium, dynamic [Gk. *aequus*, equal, + *libra*, balance] The point at which a chemical reaction runs forward as fast as it runs in reverse; thus the concentrations of reactant molecules and product molecules show no net change.

erythrocyte (eh-RITH-row-site) [Gk. *erythros*, red, + *kytos*, vessel] Red blood cell.

esophagus (ee-SOF-uh-gus) Tubular portion of a digestive system that receives swallowed food and leads to the stomach.

essential amino acid Any of eight amino acids that certain animals cannot synthesize for themselves and must obtain from food.

essential fatty acid Any of the fatty acids that certain animals cannot synthesize for themselves and must obtain from food.

estrogen (ESS-trow-jen) A sex hormone that helps oocytes mature, induces changes in the uterine lining during the menstrual cycle and pregnancy, and maintains secondary sexual traits; also influences bodily growth and development.

estrus (ESS-truss) [Gk. *oistrus*, frenzy] For mammals generally, the cyclic period of a female's sexual receptivity to the male.

estuary (EST-you-ehr-ee) A partly enclosed coastal region where seawater mixes with freshwater from rivers, streams, and runoff from the surrounding land.

ethylene (ETH-il-een) Plant hormone that stimulates fruit ripening and triggers abscission.

eubacteria The subkingdom of all bacterial species except the archaebacteria; one of the three great prokaryotic lineages that arose early in the history of life.

euglenoid A type of flagellated protistan, most of which are photosynthesizers in stagnant or freshwater ponds.

eukaryotic cell (yoo-CARRY-oh-tic) [Gk. *eu*, good, + *karyon*, kernel] A type of cell that has a "true nucleus" and other distinguishing membrane-bound organelles. *Compare* prokaryotic cell.

eutrophication Nutrient enrichment of a body of water, such as a lake, that typically results in reduced transparency and a phytoplankton-dominated community.

evaporation [L. *e*-, out, + *vapor*, steam] Conversion of a substance from the liquid to the gaseous state; some or all of its molecules leave in the form of vapor.

evolution, biological [L. *evolutio*, act of unrolling] Change within a line of descent over time. A population is evolving when some forms of a trait are becoming more or less common, relative to the other kinds of traits. The shifts are evidence of changes in the relative abundances of alleles for that trait, as brought about by mutation, natural selection, genetic drift, and gene flow.

evolutionary tree A treelike diagram in which branches represent separate lines of descent from a common ancestor.

excitatory postsynaptic potential or **EPSP** One of two competing signals at an input zone of a neuron; a graded potential that brings the neuron's plasma membrane closer to threshold.

excretion Any of several processes by which excess water, excess or harmful solutes, or waste materials leave the body by way of the urinary system or certain glands.

exergonic reaction (EX-ur-GONE-ik) A chemical reaction that shows a net loss in energy.

exocrine gland (EK-suh-krin) [Gk. *es*, out of, + *krinein*, to separate] Glandular structure that secretes products, usually through ducts or tubes, to a free epithelial surface.

exocytosis (EK-so-sigh-TOE-sis) Movement of a substance out of a cell by means of a transport vesicle, the membrane of which fuses with the plasma membrane, so that the vesicle's contents are released outside.

exodermis Layer of cells just inside the root epidermis of most flowering plants; helps control the uptake of water and solutes.

exon Of eukaryotic cells, any of the nucleotide sequences of a pre-mRNA molecule that are spliced together to form the mature mRNA transcript and are ultimately translated into protein.

exoskeleton [Gk. *exo*, out, + *sklēros*, hard, stiff] An external skeleton, as in arthropods.

experiment A test in which some phenomenon in the natural world is manipulated in controlled ways to gain insight into its function, structure, operation, or behavior.

exploitation competition Interaction in which both species have equal access to a required resource but differ in how fast or efficiently they exploit it.

exponential growth (EX-po-NEN-shul) Pattern of population growth in which greater and greater numbers of individuals are produced during the successive doubling times; the pattern that emerges when the per capita birth rate remains even slightly above the per capita death rate, putting aside the effects of immigration and emigration.

extinction, background A steady rate of species turnover that characterizes lineages through most of their histories.

extinction, mass An abrupt increase in the rate at which major taxa disappear, with several taxa being affected simultaneously.

extracellular fluid In animals generally, all the fluid not inside cells; includes plasma (the liquid portion of blood) and interstitial fluid (which occupies the spaces between cells and tissues).

extracellular matrix A material, largely secreted, that helps hold many animal tissues together in certain shapes; it consists of fibrous proteins and other components in a ground substance.

FAD Flavin adenine dinucleotide, a nucleotide coenzyme. When delivering electrons and unbound protons (H$^+$) from one reaction to another, it is abbreviated FADH$_2$.

fall overturn The vertical mixing of a body of water in autumn. Its upper layer cools, increases in density, and sinks; dissolved oxygen moves down and nutrients from bottom sediments are brought to the surface.

family pedigree A chart of genetic relationships of the individuals in a family through successive generations.

fat A lipid with a glycerol head and one, two, or three fatty acid tails. The tails of saturated fats have only single bonds between carbon atoms and hydrogen atoms attached to all other bonding sites. Tails of unsaturated fats additionally have one or more double bonds between certain carbon atoms.

fatty acid A long, flexible hydrocarbon chain with a —COOH group at one end.

feedback inhibition Of cells, a control mechanism by which the production (or secretion) of a substance triggers a change in some activity that in turn shuts down further production of the substance.

fermentation [L. *fermentum*, yeast] A type of anaerobic pathway of ATP formation, it starts with glycolysis, ends when electrons are transferred back to one of the breakdown products or intermediates, and regenerates the NAD$^+$ required for the reaction. Its net yield is two ATP per glucose molecule degraded.

fern One of the seedless vascular plants, mostly of wet, humid habitats; requires free water to complete its life cycle.

fertilization [L. *fertilis*, to carry, to bear] Fusion of a sperm nucleus with the nucleus of an egg, which thereupon becomes a zygote.

fever A body temperature higher than a set point that is preestablished in the brain region governing temperature.

fibrous root system Of most monocots, all the lateral branchings of adventitious roots, which arose earlier from the young stem.

filtration Of urine formation, the process by which blood pressure forces water and solutes out of glomerular capillaries and into the cupped portion of a nephron wall (Bowman's capsule).

fin Of fishes generally, an appendage that helps propel, stabilize, and guide the body through water.

first law of thermodynamics [Gk. *therme*, heat, + *dynamikos*, powerful] Law stating that the total amount of energy in the universe remains constant. Energy cannot be created and existing energy cannot be destroyed. It can only be converted from one form to another.

fish An aquatic animal of the most ancient vertebrate lineage; jawless fishes (such as lampreys and hagfishes), cartilaginous fishes (such as sharks), and bony fishes (such as coelacanths and salmon) are the three existing groups.

fixed action pattern An instinctive response that is triggered by a well-defined, simple stimulus and that is performed in its entirety once it has begun.

flagellated protozoan A member of one of four major groups of protozoans, many of which cause serious diseases.

flagellum (fluh-JELL-um), plural **flagella** [L. whip] Tail-like motile structure of many free-living eukaryotic cells; it has a distinctive 9 + 2 array of microtubules.

flatworm A type of invertebrate having bilateral symmetry, a flattened body, and a saclike gut; a turbellarian, fluke, or tapeworm.

flower The reproductive structure that distinguishes angiosperms from other seed plants and often attracts pollinators.

fluid mosaic model Model of membrane structure in which proteins are embedded in a lipid bilayer or attached to one of its surfaces. The lipid molecules give the membrane its basic structure, impermeability to water-soluble molecules, and (through packing variations and movements) fluidity. Proteins carry out most membrane functions, such as transport, enzyme action, and reception of signals or substances.

follicle (FOLL-ih-kul) In a mammalian ovary, a primary oocyte (immature egg) together with the surrounding layer of cells.

food chain A straight-line sequence of who eats whom in an ecosystem.

food web A network of cross-connecting, interlinked food chains, encompassing primary producers and an array of consumers, detritivores, and decomposers.

forebrain Brain region that includes the cerebrum and cerebral cortex, the olfactory lobes, and the hypothalamus.

forest A type of biome where tall trees grow together closely enough to form a fairly continuous canopy over a broad region.

fossil Recognizable evidence of an organism that lived in the distant past. Most fossils are skeletons, shells, leaves, seeds, and tracks that were buried in rock layers before they decomposed.

fossil fuel Coal, petroleum, or natural gas; a nonrenewable source of energy formed in sediments by the compression of carbon-containing plant remains over hundreds of millions of years.

founder effect An extreme case of genetic drift. By chance, a few individuals that leave a population and establish a new one carry fewer (or more) alleles for certain traits. Increased variation between the two populations is one outcome. Limited genetic variability in the new population is another.

free radical A highly reactive, unbound molecular fragment with the wrong number of electrons.

fruit [L. after *frui*, to enjoy] Of flowering plants, the expanded and ripened ovary of one or more carpels, sometimes with accessory structures incorporated.

FSH Follicle-stimulating hormone. Produced and secreted by the anterior lobe of the pituitary gland, this hormone has roles in the reproductive functions of both males and females.

functional group An atom or group of atoms that is covalently bonded to the carbon backbone of an organic compound and that influences its behavior.

Fungi The kingdom of fungi.

fungus A eukaryotic heterotroph that uses extracellular digestion and absorption; it secretes enzymes able to break down an external food source into molecules small enough to be absorbed by its cells. Saprobic types feed on nonliving organic matter; parasitic types feed on living organisms. Fungi as a group are major decomposers.

gall bladder Organ of the digestive system that stores bile secreted from the liver.

gamete (GAM-eet) [Gk. *gametēs*, husband, and *gametē*, wife] A haploid cell that functions in sexual reproduction. Sperm and eggs are examples.

gamete formation Generally, the formation of gametes by way of meiosis. Of animals, the first stage of development, in which sperm or eggs form and mature within reproductive tissues of parents.

gametophyte (gam-EET-oh-fite) [Gk. *phyton*, plant] The haploid, multicelled, gamete-producing phase in the life cycle of most plants.

ganglion (GANG-lee-un), plural **ganglia** [Gk. *ganglion*, a swelling] A distinct clustering of cell bodies of neurons in regions other than the brain or spinal cord.

gastrulation (gas-tru-LAY-shun) Of animals, the stage of embryonic development in which cells become arranged into two or three primary tissue layers (germ layers); in humans, the layers are an inner endoderm, an intermediate mesoderm, and a surface ectoderm.

gene [short for German *pangan*, after Gk. *pan*, all + *genes*, to be born] A unit of information about a heritable trait that is passed on from parents to offspring. Each gene has a specific location on a chromosome.

gene flow A microevolutionary process; a physical movement of alleles out of a population as individuals leave (emigrate) or enter (immigrate), the outcome being changes in allele frequencies.

gene frequency More precisely, allele frequency: the relative abundances of all the different alleles for a trait that are carried by the individuals of a population.

gene locus A given gene's particular location on a chromosome.

gene mutation [L. *mutatus*, a change] Change in DNA due to the deletion, addition, or substitution of one to several bases in the nucleotide sequence.

gene pair In diploid cells, the two alleles at a given locus on a pair of homologous chromosomes.

gene pool Sum total of all genotypes in a population. More accurately, allele pool.

gene therapy Generally, the transfer of one or more normal genes into the body cells of an organism in order to correct a genetic defect.

genetic code [After L. *genesis*, to be born] The correspondence between nucleotide triplets in DNA (then in mRNA) and specific sequences of amino acids in the resulting polypeptide chains; the basic language of protein synthesis.

genetic disorder An inherited condition that results in mild to severe medical problems.

genetic drift A microevolutionary process; a change in allele frequencies over the generations due to chance events alone.

genetic engineering Altering the information content of DNA through use of recombinant DNA technology.

genetic equilibrium Hypothetical state of a population in which allele frequencies for a trait remain stable through the generations; a reference point for measuring rates of evolutionary change.

genetic recombination Presence of a new combination of alleles in a DNA molecule compared to the parental genotype; the result of processes such as crossing over at meiosis, chromosome rearrangements, gene mutation, and recombinant DNA technology.

genome All the DNA in a haploid number of chromosomes of a given species.

genotype (JEEN-oh-type) Genetic constitution of an individual. Can mean a single gene pair or the sum total of the individual's genes. *Compare* phenotype.

genus, plural **genera** (JEEN-US, JEN-er-ah) [L. *genus*, race, origin] A taxon into which all species exhibiting certain phenotypic similarities and evolutionary relationship are grouped.

geologic time scale A time scale for earth history, the subdivisions of which have been refined by radioisotope dating work.

germ cell Of animals, one of a cell lineage set aside for sexual reproduction; germ cells give rise to gametes. *Compare* somatic cell.

germ layer Of animal embryos, one of two or three primary tissue layers that form during gastrulation and that gives rise to certain tissues of the adult body. *Compare* ectoderm; endoderm; mesoderm.

germination (jur-min-AY-shun) Generally, the resumption of growth following a rest stage; of seed plants, the time at which an embryo sporophyte breaks through its seed coat and resumes growth.

gibberellin (JIB-er-ELL-un) Any of a class of plant hormones that promote stem elongation.

gill A respiratory organ, typically with a moist, thin vascularized layer of epidermis that functions in gas exchange.

ginkgo A type of gymnosperm with fan-shaped leaves and fleshy coated seeds; now represented by a single species of deciduous trees.

gland A secretory cell or multicelled structure derived from epithelium and often connected to it.

glomerular capillaries The set of blood capillaries inside Bowman's capsule of the nephron.

glomerulus (glow-MARE-you-luss) [L. *glomus*, ball] The first portion of the nephron, where water and solutes are filtered from blood.

glucagon (GLUE-kuh-gone) Hormone that stimulates conversion of glycogen and amino acids to glucose; secreted by alpha cells of the pancreas when the flow of glucose decreases.

glyceride (GLISS-er-eyed) One of the molecules, commonly called fats and oils, that has one, two, or three fatty acid tails attached to a glycerol backbone. They are the body's most abundant lipids and its richest source of energy.

glycerol (GLISS-er-oh) [Gk. *glykys*, sweet, + L. *oleum*, oil] A three-carbon molecule with three hydroxyl groups attached; together with fatty acids, a component of fats and oils.

glycogen (GLY-kuh-jen) In animals, a storage polysaccharide that is a main food reserve; can be readily broken down into glucose subunits.

glycolysis (gly-CALL-ih-sis) [Gk. *glykys*, sweet, + *lysis*, loosening or breaking apart] Initial reactions of both aerobic and anaerobic pathways by which glucose (or some other organic compound) is partially broken down to pyruvate, with a net yield of two ATP. Gly-

colysis proceeds in the cytoplasm of all cells, and oxygen has no role in it.

gnetophyte A type of gymnosperm limited to deserts and tropics.

Golgi body (GOHL-gee) Organelle in which newly synthesized polypeptide chains as well as lipids are modified and packaged in vesicles for export or for transport to specific locations within the cytoplasm.

gonad (GO-nad) Primary reproductive organ in which gametes are produced.

graded potential Of neurons, a local signal that slightly changes the voltage difference across a small patch of the plasma membrane. Such signals vary in magnitude, depending on the stimulus. With prolonged or intense stimulation, they may spread to a trigger zone of the membrane and initiate an action potential.

granum, plural **grana** Within many chloroplasts, any of the stacks of flattened, membranous compartments with chlorophyll and other light-trapping pigments and reaction sites for ATP formation.

grassland A type of biome with flat or rolling land, 25 to 100 centimeters of annual rainfall, warm summers, and often grazing and periodic fires that regenerate the dominant species.

gravitropism (GRAV-ih-TROPE-izm) [L. *gravis*, heavy, + Gk. *trepein*, to turn] The tendency of a plant to grow directionally in response to the earth's gravitational force.

gray matter Of vertebrates, the dendrites, neuron cell bodies, and neuroglial cells of the spinal cord and cerebral cortex.

grazing food web Of most ecosystems, the flow of energy from plants to herbivores, then through an array of carnivores.

green alga One of a group or division of aquatic plants with an abundance of chlorophylls a and b; early members of its lineage may have given rise to the bryophytes and vascular plants.

green revolution In developing countries, the use of improved crop varieties, modern agricultural practices (including massive inputs of fertilizers and pesticides), and equipment to increase crop yields.

greenhouse effect Warming of the lower atmosphere due to the presence of greenhouse gases—carbon dioxide, methane, nitrous oxide, ozone, water vapor, and chlorofluorocarbons.

ground meristem (MARE-ih-stem) [Gk. *meristos*, divisible] Of vascular plants, a primary meristem that produces the ground tissue system, hence the bulk of the plant body.

ground substance Of certain animal tissues, the intercellular material made up of cell secretions and other noncellular components.

ground tissue system Tissues that make up the bulk of the vascular plant body; parenchyma is the most common of these.

guanine A nitrogen-containing base; present in one of the four nucleotide building blocks of DNA and RNA.

guard cell Either of two adjacent cells having roles in the movement of gases and water vapor across leaf or stem epidermis. An opening (stoma) forms when both cells swell with water and move apart; it closes when they lose water and collapse against each other.

gut A body region where food is digested and absorbed; of complete digestive systems, the gastrointestinal tract (the portions from the stomach onward).

gymnosperm (JIM-noe-sperm) [Gk. *gymnos*, naked, + *sperma*, seed] A plant that bears seeds at exposed surfaces of reproductive structures, such as cone scales. Pine trees are examples.

habitat [L. *habitare*, to live in] The type of place where an organism normally lives, characterized by physical features, chemical features, and the presence of certain other species.

hair cell Type of mechanoreceptor that may give rise to action potentials when bent or tilted.

halophile A type of archaebacterium that lives in extremely salty habitats.

haploid number (HAP-loyd) The chromosome number of a gamete which, as an outcome of meiosis, is only half that of the parent germ cell (it has only one of each pair of homologous chromosomes). *Compare* diploid number.

HCG Human chorionic gonadotropin. A hormone that helps maintain the lining of the uterus during the menstrual cycle and during the first trimester of pregnancy.

heart Muscular pump that keeps blood circulating through the animal body.

helper T cell One of the T lymphocytes; when activated, it produces and secretes interleukins that promote formation of huge populations of effector and memory cells for immune responses.

hemoglobin (HEEM-oh-glow-bin) [Gk. *haima*, blood, + L. *globus*, ball] Iron-containing, oxygen-transporting protein that gives red blood cells their color.

hemostasis (HEE-mow-STAY-sis) [Gk. *haima*, blood, + *stasis*, standing] Stopping of blood loss from a damaged blood vessel through coagulation, blood vessel spasm, platelet plug formation, and other mechanisms.

herbivore [L. *herba*, grass, + *vovare*, to devour] Plant-eating animal.

heterocyst (HET-er-oh-sist) Of some filamentous cyanobacteria, a type of thick-walled, nitrogen-fixing cell that forms when nitrogen is scarce.

heterotroph (HET-er-oh-trofe) [Gk. *heteros*, other, + *trophos*, feeder] Organism that cannot synthesize its own organic compounds and must obtain nourishment by feeding on autotrophs, each other, or organic wastes. Animals, fungi, many protistans, and most bacteria are heterotrophs. *Compare* autotroph.

heterozygous condition (HET-er-oh-ZYE-guss) [Gk. *zygoun*, join together] For a given trait, having nonidentical alleles at a particular locus on a pair of homologous chromosomes.

hindbrain One of the three divisions of the vertebrate brain; the medulla oblongata, cerebellum, and pons; includes reflex centers for respiration, blood circulation, and other basic functions; also coordinates motor responses and many complex reflexes.

histone Any of a class of proteins that are intimately associated with DNA and that are largely responsible for its structural (and possibly functional) organization in eukaryotic chromosomes.

homeostasis (HOE-me-oh-STAY-sis) [Gk. *homo*, same, + *stasis*, standing] Of multicelled organisms, a physiological state in which the physical and chemical conditions of the internal environment are being maintained within tolerable ranges.

homeostatic feedback loop An interaction in which an organ (or structure) stimulates or inhibits the output of another organ, then shuts down or increases this activity when it detects that the output has exceeded or fallen below a set point.

hominid [L. *homo*, man] All species on the evolutionary branch leading to modern humans. *Homo sapiens* is the only living representative.

hominoid Apes, humans, and their recent ancestors.

Homo erectus A hominid lineage that emerged between 1.5 million and 300,000 years ago and that may include the direct ancestors of modern humans.

Homo sapiens The hominid lineage of modern humans that emerged between 300,000 and 200,000 years ago.

homologous chromosome (huh-MOLL-uh-gus) [Gk. *homologia*, correspondence] Of sexually reproducing species, one of a pair of chromosomes that resemble each other in size, shape, and the genes they carry, and that line up with each other at meiosis I. The X and Y chromosomes differ in these respects but still function as homologues.

homologous structures The same body parts, modified in different ways, in different lines of descent from a common ancestor.

homozygous condition (HOE-moe-ZYE-guss) Having two identical alleles at a given locus (on a pair of homologous chromosomes).

homozygous dominant condition Having two dominant alleles at a given locus (on a pair of homologous chromosomes).

homozygous recessive condition Having two recessive alleles at a given gene locus (on a pair of homologous chromosomes).

hormone [Gk. *hormon*, to stir up, set in motion] Any of the signaling molecules secreted from endocrine glands, endocrine cells, and some neurons that the bloodstream distributes to nonadjacent target cells (any cell having receptors for that hormone).

horsetail One of the seedless vascular plants, which require free water to complete the life cycle; only one genus has survived to the present.

human genome project A basic research project in which researchers throughout the world are working together to sequence the estimated 3 billion nucleotides present in the DNA of human chromosomes.

hydrogen bond Type of chemical bond in which an atom of a molecule interacts weakly with a neighboring atom that is already taking part in a polar covalent bond.

hydrogen ion A free (unbound) proton; a hydrogen atom that has lost its electron and so bears a positive charge (H^+).

hydrologic cycle A biogeochemical cycle, driven by solar energy, in which water moves slowly through the atmosphere, on or through surface layers of land masses, to the ocean, and back again.

hydrolysis (high-DRAWL-ih-sis) [L. *hydro*, water, + Gk. *lysis*, loosening or breaking apart] Enzyme-mediated reaction in which covalent bonds break, splitting a molecule into two or more parts, and H^+ and OH^- (derived from a water molecule) become attached to the exposed bonding sites.

hydrophilic substance [Gk. *philos*, loving] A polar substance that is attracted to the polar water molecule and so dissolves easily in water. Sugars are examples.

hydrophobic substance [Gk. *phobos*, dreading] A nonpolar substance that is repelled by the polar water molecule and so does not readily dissolve in water. Oil is an example.

hydrosphere All liquid or frozen water on or near the earth's surface.

hydrothermal vent ecosystem A type of ecosystem that exists near fissures in the ocean floor and is based on chemosynthetic bacteria that use hydrogen sulfide as the energy source.

hypha (HIGH-fuh), plural **hyphae** [Gk. *hyphe*, web] Of fungi, a generally tube-shaped filament with chitin-reinforced walls and, often, reinforcing cross-walls; component of the mycelium.

hypodermis A subcutaneous layer having stored fat that helps insulate the body; although not part of skin, it anchors skin while allowing it some freedom of movement.

hypothalamus [Gk. *hypo*, under, + *thalamos*, inner chamber or possibly *tholos*, rotunda] Of vertebrate forebrains, a brain center that monitors visceral activities (such as salt-water balance, temperature control, and reproduction) and that influences related forms of behavior (as in hunger, thirst, and sex).

hypothesis A possible explanation of a specific phenomenon.

immune response A series of events by which B and T lymphocytes recognize a specific antigen, undergo repeated cell divisions that form huge lymphocyte populations, and differentiate into subpopulations of effector and memory cells. Effector cells engage and destroy antigen-bearing agents. Memory cells enter a resting phase and are activated during subsequent encounters with the same antigen.

immunization Various processes, including vaccination, that promote increased immunity against specific diseases.

immunoglobulins (Ig) Four classes of antibodies, each with binding sites for antigen and binding sites used in specialized tasks. Examples are IgM antibodies (first to be secreted during immune responses) and IgG antibodies (which activate complement proteins and neutralize many toxins).

implantation A process by which a blastocyst adheres to the endometrium and begins to establish connections by which the mother and embryo will exchange substances during pregnancy.

imprinting Category of learning in which an animal that has been exposed to specific key stimuli early in its behavioral development forms an association with the object.

incisor A tooth, shaped like a flat chisel or cone, used in nipping or cutting food.

incomplete dominance Of heterozygotes, the appearance of a version of a trait that is somewhere between the homozygous dominant and recessive conditions.

independent assortment Mendelian principle that each gene pair tends to assort into gametes independently of other gene pairs located on nonhomologous chromosomes.

indirect selection A theory in evolutionary biology that self-sacrificing individuals can indirectly pass on their genes by helping relatives survive and reproduce.

induced-fit model Model of enzyme action whereby a bound substrate induces changes in the shape of the enzyme's active site, resulting in a more precise molecular fit between the enzyme and its substrate.

industrial smog A type of gray-air smog that develops in industrialized regions when winters are cold and wet.

inflammation, acute In response to tissue damage or irritation, fast-acting phagocytes and plasma proteins, including complement proteins, leave the bloodstream, then defend and help repair the tissue. Proceeds during both nonspecific and specific (immune) defense responses.

inheritance The transmission, from parents to offspring, of structural and functional patterns that have a genetic basis and are characteristic of each species.

inhibiting hormone A signaling molecule produced and secreted by the hypothalamus that controls secretions by the anterior lobe of the pituitary gland.

inhibitor A substance that can bind with an enzyme and interfere with its functioning.

inhibitory postsynaptic potential, or **IPSP** Of neurons, one of two competing types of graded potentials at an input zone; tends to drive the resting membrane potential away from threshold.

instinctive behavior A complex, stereotyped response to a particular environmental cue that often is quite simple.

insulin Hormone that lowers the glucose level in blood; it is secreted from beta cells of the pancreas and stimulates cells to take up glucose; also promotes protein and fat synthesis and inhibits protein conversion to glucose.

integration, neural [L. *integrare*, to coordinate] Moment-by-moment summation of all excitatory and inhibitory synapses acting on a neuron; occurs at each level of synapsing in a nervous system.

integrator Of homeostatic systems, a control point where different bits of information are pulled together in the selection of a response. The brain is an example.

integument Of animals, a protective body covering such as skin. Of flowering plants, a protective layer around the developing ovule; when the ovule becomes a seed, its integument(s) harden and thicken into a seed coat.

integumentary exchange (in-teg-you-MEN-tuh-ree) Of some animals, a mode of respiration in which oxygen and carbon dioxide diffuse across a thin, vascularized layer of moist epidermis at the body surface.

interference competition Interaction in which one species may limit another species' access to some resource regardless of whether the resource is abundant or scarce.

interleukin One of a variety of communication signals, secreted by macrophages and by helper T cells, that drive immune responses.

intermediate compound A compound that forms between the start and the end of a metabolic pathway.

intermediate filament A cytoskeletal component that consists of different proteins in different types of animal cells.

interneuron Any of the neurons in the vertebrate brain and spinal cord that integrate information arriving from sensory neurons and that influence other neurons in turn.

internode In vascular plants, the stem region between two successive nodes.

interphase Of cell cycles, the time interval between nuclear divisions in which a cell increases its mass, roughly doubles the number of its cytoplasmic components, and finally duplicates its chromosomes (replicates its DNA). The interval is different for different species.

interspecific competition Two-species interaction in which both species can be harmed due to overlapping niches.

interstitial fluid (IN-ter-STISH-ul) [L. *interstitus*, to stand in the middle of something] In multicelled animals, that portion of the extracellular fluid occupying spaces between cells and tissues.

intertidal zone Generally, the area on a rocky or sandy shoreline that is above the low water mark and below the high water mark; organisms inhabiting it are alternately submerged, then exposed, by tides.

intervertebral disk One of a number of disk-shaped structures containing cartilage that serve as shock absorbers and flex points between bony segments of the vertebral column.

intraspecific competition Interaction among individuals of the same species that are competing for the same resources.

intron A noncoding portion of a newly formed mRNA molecule.

inversion A change in a chromosome's structure after a segment separated from it was then inserted at the same place, but in reverse. The reversal alters the position and order of the chromosome's genes.

invertebrate Animal without a backbone.

ion, negatively charged (EYE-on) An atom or a compound that has gained one or more electrons, and hence has acquired an overall negative charge.

ion, positively charged An atom or a compound that has lost one or more electrons, and hence has acquired an overall positive charge.

ionic bond An association between ions of opposite charge.

isotonic condition Equality in the relative concentrations of solutes in two fluids; for two fluids separated by a cell membrane, there is no net osmotic (water) movement across the membrane.

isotope (EYE-so-tope) For a given element, an atom with the same number of protons as the other atoms but with a different number of neutrons.

J-shaped curve A curve, obtained when population size is plotted against time, that is characteristic of unrestricted, exponential growth.

joint An area of contact or near-contact between bones.

karyotype (CARRY-oh-type) Of eukaryotic individuals (or species), the number of metaphase chromosomes in somatic cells and their defining characteristics.

keratin A tough, water-insoluble protein manufactured by most epidermal cells.

keratinization (care-AT-in-iz-AY-shun) Process by which keratin-producing epidermal cells of skin die and collect at the skin surface as keratinized "bags" that form a barrier against dehydration, bacteria, and many toxic substances.

kidney In vertebrates, one of a pair of organs that filter mineral ions, organic wastes, and other substances from the blood, and help regulate the volume and solute concentrations of extracellular fluid.

kilocalorie 1,000 calories of heat energy, or the amount of energy needed to raise the temperature of 1 kilogram of water by 1°C; the unit of measure for the caloric value of foods.

kinetochore A specialized group of proteins and DNA at the centromere of a chromosome that serves as an attachment point for several spindle microtubules during mitosis or meiosis. Each chromatid of a duplicated chromosome has its own kinetochore.

Krebs cycle Together with a few conversion steps that precede it, the stage of aerobic respiration in which pyruvate is completely broken down to carbon dioxide and water. Coenzymes accept the unbound protons (H^+) and electrons stripped from intermediates during the reactions and deliver them to the next stage.

lactate fermentation Anaerobic pathway of ATP formation in which pyruvate from glycolysis is converted to the three-carbon compound lactate, and NAD^+ (a coenzyme used in the reactions) is regenerated. Its net yield is two ATP.

lactation The production of milk by hormone-primed mammary glands.

lake A body of fresh water having littoral, limnetic, and profundal zones.

lancelet An invertebrate chordate having a body that tapers sharply at both ends, segmented muscles, and a full-length notochord.

large intestine The colon; a region of the gut that receives unabsorbed food residues from the small intestine and concentrates and stores feces until they are expelled from the body.

larva, plural **larvae** Of animals, a sexually immature, free-living stage between the embryo and the adult.

larynx (LARE-inks) A tubular airway that leads to the lungs. In humans, contains vocal cords, where sound waves used in speech are produced.

lateral meristem Of vascular plants, a type of meristem responsible for secondary growth; either vascular cambium or cork cambium.

lateral root Of taproot systems, a lateral branching from the first, primary root.

leaching The movement of soil water, with dissolved nutrients, out of a specified area.

leaf For most vascular plants, a structure having chlorophyll-containing tissue that is the major region of photosynthesis.

learned behavior The use of information gained from specific experiences to vary or change a response to stimuli.

lethal mutation A gene mutation that alters one or more traits in such a way that the individual inevitably dies.

LH Leutinizing hormone. Secreted by the anterior lobe of the pituitary gland, this hormone has roles in the reproductive functions of both males and females.

lichen (LY-kun) A symbiotic association between a fungus and a captive photosynthetic partner such as a green alga.

life cycle A recurring, genetically programmed frame of events in which individuals grow, develop, maintain themselves, and reproduce.

life table A tabulation of age-specific patterns of birth and death for a population.

ligament A strap of dense connective tissue that bridges a joint.

light-dependent reactions First stage of photosynthesis in which the energy of sunlight is trapped and converted to the chemical energy of ATP alone (by the cyclic pathway) or ATP and NADPH (by the noncyclic pathway).

light-independent reactions Second stage of photosynthesis, in which sugar phosphates form with the help of the ATP (and NADPH, in land plants) that were produced during the first stage. The sugar phosphates are used in other reactions by which starch, cellulose, and other end products of photosynthesis are assembled.

lignification Of mature land plants, a process by which lignin is deposited in secondary cell walls. The deposits impart strength and rigidity by anchoring cellulose strands in the walls, stabilize and protect other wall components, and form a waterproof barrier around the cellulose. Probably a key factor in the evolution of vascular plants.

lignin A substance that strengthens and waterproofs cell walls in certain tissues of vascular plants.

limbic system Brain regions that, along with the cerebral cortex, collectively govern emotions.

limiting factor Any essential resource that is in short supply and so limits population growth.

lineage (LIN-ee-age) A line of descent.

linkage The tendency of genes located on the same chromosome to end up in the same gamete. For any two of those genes, the probability that crossing over will disrupt the linkage is proportional to the distance separating them.

lipid A greasy or oily compound of mostly carbon and hydrogen that shows little tendency to dissolve in water, but that dissolves in nonpolar solvents (such as ether). Cells use lipids as energy stores and structural materials, especially in membranes.

lipid bilayer The structural basis of cell membranes, consisting of two layers of mostly phospholipid molecules. Hydrophilic heads force all fatty acid tails of the lipids to become sandwiched between the hydrophilic heads.

liver Glandular organ with roles in storing and interconverting carbohydrates, lipids, and proteins absorbed from the gut, maintaining blood; disposing of nitrogen-containing wastes; and other tasks.

local signaling molecules Secretions from cells in many different tissues that alter chemical conditions in the immediate vicinity where they are secreted, then are swiftly degraded.

locus (LOW-cuss) The specific location of a particular gene on a chromosome.

logistic population growth (low-JIS-tik) Pattern of population growth in which a low-density population slowly increases in size, goes through a rapid growth phase, then levels off once the carrying capacity is reached.

loop of Henle The hairpin-shaped, tubular region of a nephron that functions in reabsorption of water and solutes.

lung An internal respiratory surface in the shape of a cavity or sac.

lycophyte A type of seedless vascular plant of mostly wet or shade habitats; requires free water to complete its life cycle.

lymph (LIMF) [L. *lympha*, water] Tissue fluid that has moved into the vessels of the lymphatic system.

lymph capillary A small-diameter vessel of the lymph vascular system that has no pronounced entrance; tissue fluid moves inward by passing between overlapping endothelial cells at the vessel's tip.

lymph node A lymphoid organ that serves as a battleground of the immune system; each lymph node is packed with organized arrays of macrophages and lymphocytes that cleanse lymph of pathogens before it reaches the blood.

lymph vascular system [L. *lympha*, water, + *vasculum*, a small vessel] The vessels of the lymphatic system, which take up and transport excess tissue fluid and reclaimable solutes as well as fats absorbed from the digestive tract.

lymphatic system An organ system that supplements the circulatory system. Its vessels take up fluid and solutes from interstitial fluid and deliver them to the bloodstream; its lymphoid organs have roles in immunity.

lymphocyte Any of various white blood cells that take part in nonspecific and specific (immune) defense responses.

lymphoid organs The lymph nodes, spleen, thymus, tonsils, adenoids, and other organs with roles in immunity.

lysis [Gk. *lysis*, a loosening] Gross structural disruption of a plasma membrane that leads to cell death.

lysosome (LYE-so-sohm) The main organelle of digestion, with enzymes that can break down polysaccharides, proteins, nucleic acids, and some lipids.

lysozyme An infection-fighting enzyme that digests bacterial cell walls. Present in mucous membranes that line the body's surfaces.

lytic pathway During a viral infection, viral DNA or RNA quickly directs the host cell to produce the components necessary to produce new virus particles, which are released by lysis.

macroevolution The large-scale patterns, trends, and rates of change among groups of species.

macrophage One of the phagocytic white blood cells. It engulfs anything detected as foreign. Some also become antigen-presenting cells that serve as the trigger for immune responses by T and B lymphocytes. *Compare* antigen-presenting cell.

mammal A type of vertebrate; the only animal having offspring that are nourished by milk produced by mammary glands of females.

mass extinction An abrupt rise in extinction rates above the background level; a catastrophic, global event in which major taxa are wiped out simultaneously.

mass number The total number of protons and neutrons in an atom's nucleus. The relative masses of atoms are also called atomic weights.

maternal chromosome One of the chromosomes bearing the alleles that are inherited from a female parent.

mechanoreceptor Sensory cell or cell part that detects mechanical energy associated with changes in pressure, position, or acceleration.

medulla oblongata Part of the vertebrate brainstem with reflex centers for respiration, blood circulation, and other vital functions.

medusa (meh-DOO-sah) [Gk. *Medusa*, one of three sisters in Greek mythology having snake-entwined hair; this image probably evoked by the tentacles and oral arms extending from the medusa] Free-swimming, bell-shaped stage in cnidarian life cycles.

megaspore Of gymnosperms and flowering plants, a haploid spore that forms in the ovary; one of its cellular descendants develops into an egg.

meiosis (my-OH-sis) [Gk. *meioun*, to diminish] Two-stage nuclear division process in which the chromosome number of a germ cell is reduced by half, to the haploid number. (Each daughter nucleus ends up with one of each type of chromosome.) Meiosis is the basis of gamete formation and (in plants) of spore formation. *Compare* mitosis.

meltdown Events which, if unchecked, can blow apart a nuclear power plant, with a release of radioactive material into the environment.

membrane excitability A membrane property of any cell that can produce action potentials in response to appropriate stimulation.

memory The storage and retrieval of information about previous experiences; underlies the capacity for learning.

memory cell One of the subpopulations of cells that form during an immune response and that enters a resting phase, from which it is released during a secondary immune response.

memory lymphocyte Any of the various B or T lymphocytes of the immune system that are formed in response to invasion by a foreign agent and that circulate for some period, available to mount a rapid attack if the same type of invader reappears.

Mendel's theory of independent assortment Stated in modern terms, during meiosis, the gene pairs of homologous chromosomes tend to be stored independently of how gene pairs on other chromosomes are sorted for forthcoming gametes. The theory does not take into account the effects of gene linkage and crossing over.

Mendel's theory of segregation Stated in modern terms, diploid cells have two of each kind of gene (on pairs of homologous chromosomes), and the two segregate during meiosis so that they end up in different gametes.

menopause (MEN-uh-pozz) [L. *mensis*, month, + *pausa*, stop] End of the period of a human female's reproductive potential.

menstrual cycle The cyclic release of oocytes and priming of the endometrium (lining of the uterus) to receive a fertilized egg; the complete cycle averages about 28 days in female humans.

menstruation Periodic sloughing of the blood-enriched lining of the uterus when pregnancy does not occur.

mesoderm (MEH-so-derm) [Gk. *mesos*, middle, + *derm*, skin] In most animal embryos, a primary tissue layer (germ layer) between ectoderm and endoderm. Gives rise to muscle; organs of circulation, reproduction, and excretion; most of the internal skeleton (when present); and connective tissue layers of the gut and body covering.

mesophyll Of vascular plants, a type of parenchyma tissue with photosynthetic cells and an abundance of air spaces.

messenger RNA A linear sequence of ribonucleotides transcribed from DNA and translated into a polypeptide chain; the only type of RNA that carries protein-building instructions.

metabolic pathway One of many orderly sequences of enzyme-mediated reactions by which cells normally maintain, increase, or decrease the concentrations of substances.

Different pathways are linear or circular, and often they interconnect.

metabolism (meh-TAB-oh-lizm) [Gk. *meta*, change] All controlled, enzyme-mediated chemical reactions by which cells acquire and use energy. Through these reactions, cells synthesize, store, break apart, and eliminate substances in ways that contribute to growth, survival, and reproduction.

metamorphosis (met-uh-MOR-foe-sis) [Gk. *meta*, change, + *morphe*, form] Transformation of a larva into an adult form.

metaphase Of mitosis or meiosis II, the stage when each duplicated chromosome has become positioned at the midpoint of the microtubular spindle, with its two sister chromatids attached to microtubules from opposite spindle poles. Of meiosis I, the stage when all pairs of homologous chromosomes are positioned at the spindle's midpoint, with the two members of each pair attached to opposite spindle poles.

metazoan Any multicelled animal.

methanogen A type of archaebacterium that lives in oxygen-free habitats and that produces methane gas as a metabolic by-product.

MHC marker Any of a variety of proteins that are self-markers. Some occur on all body cells of an individual; others are unique to the macrophages and lymphocytes.

micelle formation Formation of a small droplet that consists of bile salts and products of fat digestion (fatty acids and monoglycerides) and that assists in their absorption from the small intestine.

microevolution Changes in allele frequencies brought about by mutation, genetic drift, gene flow, and natural selection.

microfilament [Gk. *mikros*, small, + L. *filum*, thread] In animal cells, one of a variety of cytoskeletal components. Actin and myosin filaments are examples.

microorganism An organism, usually single-celled, that is too small to be observed without the aid of a microscope.

microspore Of gymnosperms and flowering plants, a haploid spore, encased in a sculpted wall, that develops into a pollen grain.

microtubular spindle Of eukaryotic cells, a bipolar structure composed of organized arrays of microtubules that forms during nuclear division and that moves the chromosomes.

microtubule Hollow cylinder of mainly tubulin subunits; a cytoskeletal element with roles in cell shape, motion, and growth and in the structure of cilia and flagella.

microtubule organizing center, or **MTOC** Small mass of proteins and other substances in the cytoplasm; the number, type, and location of MTOCs determine the organization and orientation of microtubules.

microvillus (MY-crow-VILL-us) [L. *villus*, shaggy hair] A slender, cylindrical extension of the animal cell surface that functions in absorption or secretion.

midbrain Of vertebrates, a brain region that evolved as a coordination center for reflex responses to visual and auditory input; together with the pons and medulla oblongata,

part of the brainstem, which includes the reticular formation.

migration Of certain animals, a cyclic movement between two distant regions at times of year corresponding to seasonal change.

mimicry (MIM-ik-ree) Situation in which one species (the mimic) bears deceptive resemblance in color, form, and/or behavior to another species (the model) that enjoys some survival advantage.

mineral An inorganic substance required for the normal functioning of body cells.

mitochondrion (MY-toe-KON-dree-on), plural **mitochondria** Organelle that specializes in ATP formation; it is the site of the second and third stages of aerobic respiration, an oxygen-requiring pathway.

mitosis (my-TOE-sis) [Gk. *mitos*, thread] Type of nuclear division that maintains the parental chromosome number for daughter cells. It is the basis of bodily growth and, in many eukaryotic species, asexual reproduction.

molar One of the cheek teeth; a tooth with a platform having cusps (surface bumps) that help crush, grind, and shear food.

molecular clock With respect to the presumed regular accumulation of neutral mutations in highly conserved genes, a way of calculating the time of origin of one species or lineage relative to others.

molecule A unit of matter in which chemical bonding holds together two or more atoms of the same or different elements.

mollusk A type of invertebrate having a tissue fold (mantle) draped around a soft, fleshy body; snails, clams, and squids are examples.

molting The shedding of hair, feathers, horns, epidermis, or a shell (or some other exoskeleton) in a process of growth or periodic renewal.

Monera The kingdom of bacteria.

moneran A bacterium; a single-celled prokaryote.

monocot (MON-oh-kot) Short for monocotyledon; a flowering plant in which seeds have only one cotyledon, whose floral parts generally occur in threes (or multiples of three), and whose leaves typically are parallel-veined. *Compare* dicot.

monohybrid cross [Gk. *monos*, alone] An experimental cross in which offspring inherit a pair of nonidentical alleles for a single trait being studied, so that they are heterozygous.

monophyletic group A set of independently evolving lineages that share a common evolutionary heritage.

monosaccharide (MON-oh-SAK-ah-ride) [Gk. *monos*, alone, single, + *sakharon*, sugar] The simplest carbohydrate, with only one sugar unit. Glucose is an example.

monosomy Abnormal condition in which one chromosome of diploid cells has no homologue.

morphogenesis (MORE-foe-JEN-ih-sis) [Gk. *morphe*, form, + *genesis*, origin] Processes by which differentiated cells in an embryo become organized into tissues and organs, under genetic controls and environmental influences.

morphological convergence A macroevolutionary pattern of change in which separate lineages adopt similar lifestyles, put comparable body parts to similar uses, and in time resemble one another in structure and function. Analogous structures are evidence of this pattern.

morphological divergence A macroevolutionary pattern of change from a common ancestral form. Homologous structures are evidence of this pattern.

motor neuron A type of neuron; it delivers signals from the brain and spinal cord that can stimulate or inhibit the body's effectors (muscles, glands, or both).

mouth An oral cavity; in human digestion, the site where polysaccharide breakdown begins.

multicelled organism An organism that has differentiated cells arranged into tissues, organs, and often organ systems.

multiple allele system Three or more different molecular forms of the same gene (alleles) that exist in a population.

muscle fatigue A decline in tension of a muscle that has been kept in a state of tetanic contraction as a result of continuous, high-frequency stimulation.

muscle tension A mechanical force, exerted by a contracting muscle, that resists opposing forces such as gravity and the weight of objects being lifted.

muscle tissue Tissue having cells able to contract in response to stimulation, then passively lengthen and so return to their resting stage.

mutagen (MEW-tuh-jen) An environmental agent that can permanently modify the structure of a DNA molecule. Certain viruses and ultraviolet radiation are examples.

mutation [L. *mutatus*, a change, + *-ion*, result or a process or an act] A heritable change in the DNA. Generally, mutations are the source of all the different molecular versions of genes (alleles) and, ultimately, of life's diversity. *See also* lethal mutation; neutral mutation.

mutualism [L. *mutuus*, reciprocal] An interaction between two species that benefits both.

mycelium (my-SEE-lee-um), plural **mycelia** [Gk. *mykes*, fungus, mushroom, + *helos*, callus] A mesh of tiny, branching filaments (hyphae) that is the food-absorbing part of a multicelled fungus.

mycorrhiza (MY-coe-RISE-uh) "Fungus-root;" a symbiotic arrangement between fungal hyphae and the young roots of many vascular plants. The fungus obtains carbohydrates from the plant and in turn releases dissolved mineral ions to the plant roots.

myelin sheath Of many sensory and motor neurons, an axonal sheath that affects how fast action potentials travel; formed from the plasma membranes of Schwann cells that are wrapped repeatedly around the axon and are separated from each other by a small node.

myofibril (MY-oh-FY-brill) One of many thread-like structures inside a muscle cell; each is functionally divided into sarcomeres, the basic units of contraction.

myosin (MY-uh-sin) A type of protein with a head and long tail. In muscle cells, it interacts with actin, another protein, to bring about contraction.

NAD⁺ Nicotinamide adenine dinucleotide; a nucleotide coenzyme. When carrying electrons and unbound protons (H^+) between reaction sites, it is abbreviated NADH.

NADP⁺ Nicotinamide adenine dinucleotide phosphate; a phosphorylated nucleotide coenzyme. When carrying electrons and unbound protons (H^+) between reaction sites, it is abbreviated $NADPH_2$.

nasal cavity Of a respiratory system, the region where air is warmed, moistened, and filtered of airborne particles and dust.

natural selection A microevolutionary process; a difference in survival and reproduction among members of a population that vary in one or more traits.

negative feedback mechanism A homeostatic feedback mechanism in which an activity changes some condition in the internal environment and so triggers a response that reverses the changed condition.

nematocyst (NEM-ad-uh-sist) [Gk. *nema*, thread, + *kystis*, pouch] Of cnidarians only, a stinging capsule that assists in prey capture and possibly protection.

nephridium (neh-FRID-ee-um), plural **nephridia** Of earthworms and some other invertebrates, a system of regulating water and solute levels.

nephron (NEFF-ron) [Gk. *nephros*, kidney] Of the vertebrate kidney, a slender tubule in which water and solutes filtered from blood are selectively reabsorbed and in which urine forms.

nerve Cordlike communication line of the peripheral nervous system, composed of axons of sensory neurons, motor neurons, or both packed within connective tissue. In the brain and spinal cord, similar cord-like bundles are called nerve pathways or tracts.

nerve cord Of many animals, a cordlike communication line consisting of axons of neurons.

nerve impulse *See* action potential.

nerve net Cnidarian nervous system.

nervous system System of neurons oriented relative to one another in precise message-conducting and information-processing pathways.

nervous tissue A type of connective tissue composed of neurons.

net energy Of energy resources available to the human population, the amount of energy that is left over after subtracting the energy used to locate, extract, transport, store, and deliver energy to consumers.

net population growth rate per individual (r) Of population growth equations, a single variable in which birth and death rates, which are assumed to remain constant, are combined.

neuroglial cell (NUR-oh-GLEE-uhl) Of vertebrates, one of the cells that provide structural and metabolic support for neurons and that collectively represent about half the volume of the nervous system.

neuromodulator Type of signaling molecule that influences the effects of transmitter substances by enhancing or reducing membrane responses in target neurons.

neuromuscular junction Chemical synapses between axon terminals of a motor neuron and a muscle cell.

neuron A nerve cell; the basic unit of communication in nervous systems. Neurons collectively sense environmental change, integrate sensory inputs, then activate muscles or glands that initiate or carry out responses.

neurotransmitter Any of the class of signaling molecules that are secreted from neurons, act on immediately adjacent cells, and are then rapidly degraded or recycled.

neutral mutation A gene mutation that has neither harmful nor helpful effects on the individual's ability to survive and reproduce.

neutron Unit of matter, one or more of which occupies the atomic nucleus, that has mass but no electric charge.

neutrophil Fast-acting, phagocytic white blood cell that takes part in inflammatory responses against bacteria.

niche (NITCH) [L. *nidas*, nest] Of a species, the full range of physical and biological conditions under which its members can live and reproduce.

nitrification (nye-trih-fih-KAY-shun) A chemosynthetic process in which certain bacteria strip electrons from ammonia or ammonium present in soil. The end product, nitrite (NO_2^-), is broken down to nitrate (NO_3^-) by different bacteria.

nitrogen cycle Biogeochemical cycle in which the atmosphere is the largest reservoir of nitrogen.

nitrogen fixation Process by which a few kinds of bacteria convert gaseous nitrogen (N_2) to ammonia. This dissolves rapidly in their cytoplasm to form ammonium, which can be used in biosynthetic pathways.

NK cell Natural killer cell, possibly of the lymphocyte lineage, that reconnoiters and kills tumor cells and infected body cells.

nociceptor A receptor, such as a free nerve ending, that detects any stimulus causing tissue damage.

node In vascular plants, a point on a stem where one or more leaves are attached.

noncyclic pathway of ATP formation (non-SIK-lik) [L. *non*, not, + Gk. *kylos*, circle] Photosynthetic pathway in which excited electrons derived from water molecules flow through two photosystems and two transport chains, and ATP and NADPH form.

nondisjunction Failure of one or more chromosomes to separate properly during mitosis or meiosis.

nonsteroid hormone A type of water-soluble hormone, such as a protein hormone, that cannot cross the lipid bilayer of a target cell. These hormones enter the cell by receptor-mediated endocytosis, or they bind to receptors that activate membrane proteins or second messengers within the cell.

notochord (KNOW-toe-kord) Of chordates, a rod of stiffened tissue (not cartilage or bone) that serves as a supporting structure for the body.

nuclear envelope A double membrane (two lipid bilayers and associated proteins) that is the outermost portion of a cell nucleus.

nucleic acid (new-CLAY-ik) A long, single- or double-stranded chain of four different kinds of nucleotides joined one after the other at their phosphate groups. They differ in which nucleotide base follows the next in sequence. DNA and RNA are examples.

nucleic acid hybridization The base-pairing of nucleotide sequences from different sources, as used in genetics, genetic engineering, and studies of evolutionary relationship based on similarities and differences in the DNA or RNA of different species.

nucleoid Of bacteria, a region in which DNA is physically organized apart from other cytoplasmic components.

nucleolus (new-KLEE-oh-lus) [L. *nucleolus*, a little kernel] Within the nucleus of a nondividing cell, a site where the protein and RNA subunits of ribosomes are assembled.

nucleosome (NEW-klee-oh-sohm) Of eukaryotic chromosomes, one of many organizational units, each consisting of a small stretch of DNA looped twice around a "spool" of histone molecules, which another histone molecule stabilizes.

nucleotide (NEW-klee-oh-tide) A small organic compound having a five-carbon sugar (deoxyribose), nitrogen-containing base, and phosphate group. Nucleotides are the structural units of adenosine phosphates, nucleotide coenzymes, and nucleic acids.

nucleotide coenzyme A protein that transports hydrogen atoms (free protons) and electrons from one reaction site to another in cells.

nucleus (NEW-klee-us) [L. *nucleus*, a kernel] Of atoms, the central core of one or more positively charged protons and (in all but hydrogen) electrically neutral neutrons. In eukaryotic cells, a membranous organelle that physically isolates and organizes the DNA, out of the way of cytoplasmic machinery.

nutrition All those processes by which food is selectively ingested, digested, absorbed, and later converted to the body's own organic compounds.

obesity An excess of fat in the body's adipose tissues, caused by imbalances between caloric intake and energy output.

oligosaccharide A carbohydrate consisting of a short chain of two or more covalently bonded sugar units. One subclass, disaccharides, has two sugar units. *Compare* monosaccharide; polysaccharide.

omnivore [L. *omnis*, all, + *vovare*, to devour] An organism able to obtain energy from more than one source rather than being limited to one trophic level.

oncogene (ON-coe-jeen) Any gene having the potential to induce cancerous transformations in a cell.

oocyte An immature egg.

oogenesis (oo-oh-JEN-uh-sis) Formation of a female gamete, from a germ cell to a mature haploid ovum (egg).

operator A short base sequence between a promoter and the start of a gene; interacts with regulatory proteins.

operon Of transcription, a promoter-operator sequence that services more than a single gene. The lactose operon of *E. coli* is an example.

orbitals Volumes of space around the nucleus of an atom in which electrons are likely to be at any instant.

organ A structure of definite form and function that is composed of more than one tissue.

organ formation Stage of development in which primary tissue layers (germ layers) split into subpopulations of cells, and different lines of cells become unique in structure and function; foundation for growth and tissue specialization, when organs acquire specialized chemical and physical properties.

organ system Two or more organs that interact chemically, physically, or both in performing a common task.

organelle Of cells, an internal, membrane-bounded sac or compartment that has a specific, specialized metabolic function.

organic compound In biology, a compound assembled in cells and having a carbon backbone, often with carbon atoms arranged as a chain or ring structure.

osmosis (oss-MOE-sis) [Gk. *osmos*, act of pushing] Of cells, the tendency of water to move through channel proteins that span a membrane in response to a concentration gradient, fluid pressure, or both. Hydrogen bonds among water molecules prevent water *itself* from becoming more or less concentrated; but a gradient may exist when the water on either side of the membrane has more substances dissolved in it.

ovary (OH-vuh-ree) In female animals, the primary reproductive organ in which eggs form. In seed-bearing plants, the portion of the carpel where eggs develop, fertilization takes place, and seeds mature. A mature ovary (and sometimes other plant parts) is a fruit.

oviduct (OH-vih-dukt) Duct through which eggs travel from the ovary to the uterus. Formerly called Fallopian tube.

ovulation (AHV-you-LAY-shun) During each turn of the menstrual cycle, the release of a secondary oocyte (immature egg) from an ovary.

ovule (OHV-youl) [L. *ovum*, egg] Before fertilization in gymnosperms and angiosperms, a female gametophyte with egg cell, a surrounding tissue, and one or two protective layers (integuments). After fertilization, an ovule matures into a seed (an embryo sporophyte and food reserves encased in a hardened coat).

ovum (OH-vum) A mature female gamete (egg).

oxidation-reduction reaction An electron transfer from one atom or molecule to another. Often hydrogen is transferred along with the electron or electrons.

ozone hole A pronounced seasonal thinning of the ozone layer in the lower stratosphere above Antarctica.

pancreas (PAN-cree-us) Gland that secretes enzymes and bicarbonate into the small intestine during digestion, and that also secretes the hormones insulin and glucagon.

pancreatic islets Any of the two million clusters of endocrine cells in the pancreas, including alpha cells, beta cells, and delta cells.

parasite [Gk. *para*, alongside, + *sitos*, food] An organism that obtains nutrients directly from the tissues of a living host, which it lives on or in and may or may not kill.

parasitism A two-species interaction in which one species directly harms another that serves as its host.

parasitoid An insect larva that grows and develops inside a host organism (usually another insect), eventually consuming the soft tissues and killing it.

parasympathetic nerve Of the autonomic nervous system, any of the nerves carrying signals that tend to slow the body down overall and divert energy to basic tasks; also work continually in opposition with sympathetic nerves to bring about minor adjustments in internal organs.

parathyroid glands (PARE-uh-THY-royd) In vertebrates, endocrine glands embedded in the thyroid gland that secrete parathyroid hormone, which helps restore blood calcium levels.

parenchyma Most abundant of the simple tissues in flowering plant roots, stems, leaves, and other parts. Its cells function in photosynthesis, storage, secretion, and other tasks.

parthenogenesis Development of an embryo from an unfertilized egg.

passive immunity Temporary immunity conferred by deliberately introducing antibodies into the body.

passive transport Diffusion of a solute through a channel or carrier protein that spans the lipid bilayer of a cell membrane. Its passage does not require an energy input; the protein passively allows the solute to follow its concentration gradient.

paternal chromosome One of the chromosomes bearing alleles that are inherited from a male parent.

pathogen (PATH-oh-jen) [Gk. *pathos*, suffering, + *-genēs*, origin]. An infectious, disease-causing agent, such as a virus or bacterium.

pattern formation Of animals, mechanisms responsible for specialization and positioning of tissues during embryonic development.

PCR Polymerase chain reaction. A method used by recombinant DNA technologists to amplify the quantity of specific fragments of DNA.

peat An accumulation of saturated, undecayed remains of plants that have been compressed by sediments.

pedigree A chart of genetic connections among individuals, as constructed according to standardized methods.

pelagic province The entire volume of ocean water; subdivided into neritic zone (relatively shallow waters overlying continental shelves) and oceanic zone (water over ocean basins).

penis A male organ that deposits sperm into a female reproductive tract.

perennial [L. *per-*, throughout, + *annus*, year] A flowering plant that lives for three or more growing seasons.

perforin A type of protein, produced and secreted by cytotoxic cells, that destroys antigen-bearing targets.

pericycle (PARE-ih-sigh-kul) [Gk. *peri-*, around, + *kyklos*, circle] Of a root vascular cylinder, one or more layers just inside the endodermis that gives rise to lateral roots and contributes to secondary growth.

periderm Of vascular plants showing secondary growth, a protective covering that replaces epidermis.

peripheral nervous system (per-IF-ur-uhl) [Gk. *peripherein*, to carry around] Of vertebrates, the nerves leading into and out from the spinal cord and brain and the ganglia along those communication lines.

peristalsis (pare-ih-STAL-sis) A rhythmic contraction of muscles that moves food forward through the animal gut.

peritoneum A lining of the coelom that also covers and helps maintain the position of internal organs.

peritubular capillaries The set of blood capillaries that threads around the tubular parts of a nephron; they function in reabsorption of water and solutes back into the body and in secretion of hydrogen ions and some other substances in the forming urine.

permafrost A permanently frozen, water-impenetrable layer beneath the soil surface in arctic tundra.

peroxisome Enzyme-filled vesicle in which fatty acids and amino acids are digested first into hydrogen peroxide (which is toxic), then to harmless products.

PGA Phosphoglycerate (FOSS-foe-GLISS-er-ate). A key intermediate in glycolysis and in the Calvin-Benson cycle.

PGAL Phosphoglyceraldehyde. A key intermediate in glycolysis and in the Calvin-Benson cycle.

pH scale A scale used to measure the concentration of free hydrogen ions in blood, water, and other solutions; pH 0 is the most acidic, 14 the most basic, and 7, neutral.

phagocyte (FAG-uh-sight) [Gk. *phagein*, to eat, + *kytos*, hollow vessel] A macrophage or certain other white blood cells that engulf and destroy foreign agents.

phagocytosis (FAG-uh-sigh-TOE-sis) [Gk. *phagein*, to eat, + *kytos*, hollow vessel] Engulfment of foreign cells or substances by amoebas and some white blood cells by means of endocytosis.

pharynx (FARE-inks) A muscular tube by which food enters the gut; in land vertebrates, the dual entrance for the tubular part of the digestive tract and windpipe (trachea).

phenotype (FEE-no-type) [Gk. *phainein*, to show, + *typos*, image] Observable trait or traits of an individual; arises from interactions between genes, and between genes and the environment.

pheromone (FARE-oh-moan) [Gk. *phero*, to carry, + *-mone*, as in hormone] A type of signaling molecule secreted by exocrine glands that serves as a communication signal between individuals of the same species. Signaling pheromones elicit an immediate

behavioral response. Priming pheromones elicit a generalized physiological response.

phloem (FLOW-um) Of vascular plants, a tissue with living cells that interconnect and form the tubes through which sugars and other solutes are conducted.

phospholipid A type of lipid that is the main structural component of cell membranes. Each has a hydrophobic tail (of two fatty acids) and a hydrophilic head that incorporates glycerol and a phosphate group.

phosphorus cycle Movement of phosphorus from rock or soil through organisms, then back to soil.

phosphorylation (FOSS-for-ih-LAY-shun) The attachment of unbound (inorganic) phosphate to a molecule; also the transfer of a phosphate group from one molecule to another, as when ATP phosphorylates glucose.

photochemical smog A brown-air smog that develops over large cities when the surrounding land forms a natural basin.

photolysis (foe-TALL-ih-sis) [Gk. *photos*, light, + *-lysis*, breaking apart] A reaction sequence of the noncyclic pathway of photosynthesis, triggered by photon energy, in which water is split into oxygen, hydrogen, and electrons.

photoperiodism A biological response to a change in the relative length of daylight and darkness.

photoreceptor Light-sensitive sensory cell.

photosynthesis The trapping of sunlight energy and its conversion to chemical energy (ATP, NADPH, or both), followed by synthesis of sugar phosphates that become converted to sucrose, cellulose, starch, and other end products. It is the main biosynthetic pathway by which energy and carbon enter the web of life.

photosynthetic autotroph An organism able to synthesize all organic molecules it requires using carbon dioxide as the carbon source and sunlight as the energy source. All plants, some protistans, and a few bacteria are photosynthetic autotrophs.

photosystem One of the clusters of light-trapping pigments embedded in photosynthetic membranes. Photosystem I operates during the cyclic pathway; photosystem II operates during both the cyclic and noncyclic pathways.

phototropism [Gk. *photos*, light, + *trope*, turning, direction] Adjustment in the direction and rate of plant growth in response to light.

photovoltaic cell A device that converts sunlight energy into electricity.

phycobilins (FIE-koe-BY-lins) A class of light-sensitive, accessory pigments that transfer absorbed energy to chlorophylls. They are abundant in red algae and cyanobacteria.

phylogeny Evolutionary relationships among species, starting with most ancestral forms and including the branches leading to their descendants.

phytochrome Light-sensitive pigment molecule, the activation and inactivation of which triggers plant hormone activities governing leaf expansion, stem branching, stem length and often seed germination and flowering.

phytoplankton (FIE-toe-PLANK-tun) [Gk. *phyton*, plant, + *planktos*, wandering] A freshwater or marine community of floating or weakly swimming photosynthetic autotrophs, such as cyanobacteria, diatoms, and green algae.

pigment A light-absorbing molecule.

pineal gland (py-NEEL) A light-sensitive endocrine gland that secretes melatonin, a hormone that influences reproductive cycles and the development of reproductive organs.

pioneer species Typically small plants with short life cycles that are adapted to growing in exposed, often windy areas with intense sunlight, wide swings in air temperature, and soils deficient in nitrogen and other nutrients. By improving conditions in areas they colonize, pioneers invite their own replacement by other species.

pituitary gland Of endocrine systems, a gland that interacts with the hypothalamus to coordinate and control many physiological functions, including the activity of many other endocrine glands. Its posterior lobe stores and secretes hypothalamic hormones; the anterior lobe produces and secretes its own hormones.

placenta (play-SEN-tuh) Of the uterus, an organ composed of maternal tissues and extraembryonic membranes (the chorion especially); it delivers nutrients to the fetus and accepts wastes from it, yet allows the fetal circulatory system to develop separately from the mother's.

plankton [Gk. *planktos*, wandering] Any community of floating or weakly swimming organisms, mostly microscopic, living in freshwater and saltwater environments. *See* phytoplankton; zooplankton.

plant The type of eukaryotic organism, usually multicelled, that is a photosynthetic autotroph—it uses sunlight energy to drive the synthesis of all its required organic compounds from carbon dioxide, water, and mineral ions. Only a few nonphotosynthetic plants obtain nutrients by parasitism and other means.

Plantae The kingdom of plants.

plasma (PLAZ-muh) Liquid component of blood; consists of water, various proteins, ions, sugars, dissolved gases, and other substances.

plasma cell Of immune systems, any of the anitbody-secreting daughter cells of a rapidly dividing population of B cells.

plasma membrane Of cells, the outermost membrane. Its lipid bilayer structure and proteins carry out most functions, including transport across the membrane and reception of extracellular signals.

plasmid Of many bacteria, a small, circular molecule of extra DNA that carries only a few genes and replicates independently of the bacterial chromosome.

plasmodesma (PLAZ-moe-DEZ-muh) Of multicelled plants, a junction between linked walls of adjacent cells through which nutrients and other substances flow.

plasticity Of the human species, the ability to remain flexible and adapt to a wide range of environments.

plate tectonics Arrangement of the earth's outer layer (lithosphere) in slablike plates, all in motion and floating on a hot, plastic layer of the underlying mantle.

platelet (PLAYT-let) Any of the cell fragments in blood that release substances necessary for clot formation.

pleiotropy (PLEE-oh-troe-pee) [Gk. *pleon*, more, + *trope*, direction] A type of gene interaction in which a single gene exerts multiple effects on seemingly unrelated aspects of an individual's phenotype.

polar body Any of three cells that form during the meiotic cell division of an oocyte; the division also forms the mature egg, or ovum.

pollen grain [L. *pollen*, fine dust] Depending on the species, the immature or mature, sperm-bearing male gametophyte of gymnosperms and flowering plants.

pollen sac In anthers of flowers, any of the chambers in which pollen grains develop.

pollen tube A tube formed after a pollen grain germinates; grows down through carpel tissues and carries sperm to the ovule.

pollination Of flowering plants, the arrival of a pollen grain on the landing platform (stigma) of a carpel.

pollutant Any substance with which an ecosystem has had no prior evolutionary experience, in terms of kinds or amounts, and that can accumulate to disruptive or harmful levels. Can be naturally occurring or synthetic.

polymer (POH-lih-mur) [Gk. *polus*, many, + *meris*, part] A molecule composed of three to millions of small subunits that may or may not be identical.

polymerase chain reaction or **PCR** DNA amplification method; DNA containing a gene of interest is split into single strands, which enzymes (polymerases) copy; the enzymes also act on the accumulating copies, multiplying the gene sequence by the millions.

polymorphism (poly-MORE-fizz-um) [Gk. *polus*, many, + *morphe*, form] Of a population, the persistence through the generations of two or more forms of a trait, at a frequency greater than can be maintained by new mutations alone.

polyp (POH-lip) Vase-shaped, sedentary stage of cnidarian life cycles.

polypeptide chain Three or more amino acids joined by peptide bonds.

polyploidy (POL-ee-PLOYD-ee) A change in the chromosome number following inheritance of three or more of each type of chromosome.

polysaccharide [Gk. *polus*, many, + *sakcharon*, sugar] A straight or branched chain of hundreds of thousands of covalently linked sugar units, of the same or different kinds. The most common polysaccharides are cellulose, starch, and glycogen.

polysome Of protein synthesis, several ribosomes all translating the same messenger RNA molecule, one after the other.

population A group of individuals of the same species occupying a given area.

population density The number of individuals of a population that are living in a specified area or volume.

population distribution The general pattern of dispersion of individuals of a population throughout their habitat.

population size The number of individuals that make up the gene pool of a population.

positive feedback mechanism Homeostatic mechanism by which a chain of events is set in motion that intensifies a change from an original condition; after a limited time, the intensification reverses the change.

post-translational controls Of eukaryotes, controls that govern modification of newly formed polypeptide chains into functional enzymes and other proteins.

predation A two-species interaction in which one species (the predator) directly harms the other (its prey).

predator [L. *prehendere*, to grasp, seize] An organism that feeds on and may or may not kill other living organisms (its prey); unlike parasites, predators do not live on or in their prey.

prediction A claim about what you can expect to observe in nature if a theory or hypothesis is correct.

premolar One of the cheek teeth; a tooth having a platform with cusps (surface bumps) that can crush, grind, and shear food.

pressure flow theory Of vascular plants, a theory that organic compounds move through phloem because of gradients in solute concentrations and pressure between source regions (such as photosynthetically active leaves) and sink regions (such as growing plant parts).

primary growth Plant growth originating at root tips and shoot tips.

primary immune response Actions by white blood cells and their products elicited by a first-time encounter with an antigen; includes both antibody-mediated and cell-mediated responses.

primary productivity, gross Of ecosystems, the rate at which the producer organisms capture and store a given amount of energy during a specified interval.

primary productivity, net Of ecosystems, the rate of energy storage in the tissues of producers in excess of their rate of aerobic respiration.

primate The mammalian lineage that includes prosimians, tarsioids, and anthropoids (monkeys, apes, and humans).

probability With respect to any chance event, the most likely number of times it will turn out a certain way, divided by the total number of all possible outcomes.

procambium (pro-KAM-bee-um) Of vascular plants, a primary meristem that gives rise to the primary vascular tissues.

producers, primary Of ecosystems, the organisms that secure energy from the physical environment, as by photosynthesis or chemosynthesis.

progesterone (pro-JESS-tuh-rown) Female sex hormone secreted by the ovaries.

prokaryotic cell (pro-CARRY-oh-tic) [L. *pro*, before, + Gk. *karyon*, kernel] A bacterium; a single-celled organism that has no nucleus or any of the other membrane-bound organelles characteristic of eukaryotic cells.

promoter Of transcription, a base sequence that signals the start of a gene; the site where RNA polymerase initially binds.

prophase Of mitosis, the stage when each duplicated chromosome starts to condense, microtubules form a spindle apparatus, and the nuclear envelope starts to break up.

prophase I Of meiosis, the stage at which the microtubular spindle starts to form, the nuclear envelope starts to break up, and each duplicated chromosome also condenses and pairs with its homologous partner. At this time, their sister chromatids typically undergo crossing over and genetic recombination.

prophase II Of meiosis, a brief stage after interkinesis during which each chromosome still consists of two chromatids.

protein Large organic compound composed of one or more chains of amino acids held together by peptide bonds. Proteins have unique sequences of different kinds of amino acids in their polypeptide chains; such sequences are the basis of a protein's three-dimensional structure and chemical behavior.

Protista The kingdom of protistans.

protistan (pro-TISS-tun) [Gk. *protistos*, primal, very first] Single-celled eukaryote.

proto-oncogene A gene sequence similar to an oncogene but that codes for a protein required in normal cell function; may trigger cancer, generally when specific mutations alter its structure or function.

proton Positively charged particle, one or more of which is present in the atomic nucleus.

protostome (PRO-toe-stome) [Gk. *proto*, first, + *stoma*, mouth] A bilateral animal in which the first indentation in the early embryo develops into the mouth. Includes mollusks, annelids, and anthropods.

protozoan A type of protistan, some predatory and others parasitic; so named because they may resemble the single-celled heterotrophs that presumably gave rise to animals.

proximal tubule Of a nephron, the tubular region that receives water and solutes filtered from the blood.

pulmonary circuit Blood circulation route leading to and from the lungs.

Punnett-square method A way to predict the possible outcome of a mating or an experimental cross in simple diagrammatic form.

purine Nucleotide base having a double ring structure. Adenine and guanine are examples.

pyrimidine (phi-RIM-ih-deen) Nucleotide base having a single ring structure. Cytosine and thymine are examples.

pyruvate (PIE-roo-vate) A compound with a backbone of three carbon atoms. Two pyruvate molecules are the end products of glycolysis.

r Designates net population growth rate; the birth and death rates are assumed to remain constant and so are combined into this one variable for population growth equations.

radial symmetry Body plan having four or more roughly equivalent parts arranged around a central axis.

radioisotope An unstable atom that has dissimilar numbers of protons and neutrons and that spontaneously decays (emits electrons and energy) to a new, stable atom that is not radioactive.

rain shadow A reduction in rainfall on the leeward side of high mountains, resulting in arid or semiarid conditions.

reabsorption Of urine formation, the diffusion or active transport of water and usable solutes out of a nephron and into capillaries leading back to the general circulation; regulated by ADH and aldosterone.

receptor Of cells, a molecule at the surface of the plasma membrane or in the cytoplasm that binds molecules present in the extracellular environment. The binding triggers changes in cellular activities. Of nervous systems, a sensory cell or cell part that may be activated by a specific stimulus.

receptor protein Protein that binds a signaling molecule such as a hormone, then triggers alterations in cell behavior or metabolism.

recessive allele [L. *recedere*, to recede] In heterozygotes, an allele whose expression is fully or partially masked by expression of its partner; fully expressed only in the homozygous recessive condition.

recognition protein Protein at cell surface recognized by cells of like type; helps guide the ordering of cells into tissues during development and functions in cell-to-cell interactions.

recombinant technology Procedures by which DNA (genes) from different species may be isolated, cut, spliced together, and the new recombinant molecules multiplied in quantity in a population of rapidly dividing cells such as bacteria.

red blood cell Erythrocyte; an oxygen-transporting cell in blood.

red marrow A substance in the spongy tissue of many bones that serves as a major site of blood cell formation.

reflex [L. *reflectere*, to bend back] A simple, stereotyped movement elicited directly by sensory stimulation.

reflex pathway [L. *reflectere*, to bend back] Type of neural pathway in which signals from sensory neurons directly stimulate or inhibit motor neurons, without intervention by interneurons.

refractory period Of neurons, the period following an action potential at a given patch of membrane when sodium gates are shut and potassium gates are open, so that the patch is insensitive to stimulation.

regulatory protein A protein that enhances or suppresses the rate at which a gene is transcribed.

releasing hormone A hypothalamic signaling molecule that stimulates or slows down secretion by target cells in the anterior lobe of the pituitary gland.

repressor protein Regulatory protein that provides negative control of gene activity by preventing RNA polymerase from binding to DNA.

reproduction In biology, processes by which a new generation of cells or multicelled individuals is produced. Sexual reproduction requires meiosis, formation of gametes, and fertilization. Asexual reproduction refers to the production of new individuals by any mode that does not involve gametes.

reproduction, sexual Mode of reproduction that begins with meiosis, proceeds through gamete formation, and ends at fertilization.

reproductive isolating mechanism Any aspect of structure, functioning, or behavior that restricts gene flow between two populations.

reproductive isolation An absence of gene flow between populations.

reproductive success The survival and production of the offspring of an individual.

reptile A type of carnivorous vertebrate; its ancestors were the first vertebrates to escape dependency of standing water, largely by means of internal fertilization and amniote eggs. They include turtles, crocodiles, lizards and snakes, and tuataras.

resource partitioning A community pattern in which similar species generally share the same kind of resource in different ways, in different areas, or at different times.

respiration [L. *respirare*, to breathe] In most animals, the overall exchange of oxygen from the environment for carbon dioxide wastes from cells by way of circulating blood. *Compare* aerobic respiration.

respiratory surface The surface, such as a thin epithelial layer, that gases diffuse across to enter and leave the animal body.

respiratory system An organ system that functions in respiration.

resting membrane potential Of neurons and other excitable cells that are not being stimulated, the steady voltage difference across the plasma membrane.

restriction enzymes Class of bacterial enzymes that cut apart foreign DNA injected into them, as by viruses; also used in recombinant DNA technology.

reticular formation Of the vertebrate brainstem, a major network of interneurons that helps govern activity of the whole nervous system.

reverse transcriptase Viral enzyme required for reverse transcription of mRNA into DNA; used in recombinant DNA technology.

reverse transcription Assembly of DNA on a single-stranded mRNA molecule by viral enzymes.

RFLPs Restriction fragment length polymorphisms. Of DNA samples from different individuals, slight but unique differences in the banding pattern of fragments of the DNA that have been cut with restriction enzymes.

Rh blood typing A method of characterizing red blood cells on the basis of a protein that serves as a self-marker at their surface; Rh^+ signifies its presence and Rh^-, its absence.

rhizoid Rootlike absorptive structure of some fungi and nonvascular plants.

ribosomal RNA (rRNA) Type of RNA molecule that combines with proteins to form ribosomes, on which the polypeptide chains of proteins are assembled.

ribosome In all cells, the structure at which amino acids are strung together in specified sequence to form the polypeptide chains of proteins. An intact ribosome consists of two subunits, each composed of ribosomal RNA and protein molecules.

RNA Ribonucleic acid. A category of single-stranded nucleic acids that function in processes by which genetic instructions are used to build proteins.

rod cell A vertebrate photoreceptor sensitive to very dim light and that contributes to coarse perception of movement.

root hair Of vascular plants, an extension of a specialized root epidermal cell; root hairs collectively enhance the surface area available for absorbing water and solutes.

root nodule A localized swelling on the roots of certain legumes and other plants that contain symbiotic, nitrogen-fixing bacteria.

roots Descending parts of a vascular plant that absorb water and nutrients, anchor aboveground parts, and usually store food.

rotifer An invertebrate common in food webs of lakes and ponds.

roundworm A type of parasitic or scavenging invertebrate with a bilateral, cylindrical, cuticle-covered body, usually tapered at both ends. Some cause diseases in humans.

RuBP Ribulose biphosphate. A compound with a backbone of five carbon atoms that is required for carbon fixation in the Calvin-Benson cycle of photosynthesis.

S-shaped curve A curve, obtained when population size is plotted against time, that is characteristic of logistic growth.

sac fungus A type of fungus, usually multicelled, with spores that develop inside the cells of reproductive structures shaped like globes, flasks, or dishes. Yeasts (single-celled) are also in this group.

salination A salt buildup in soil as a result of evaporation, poor drainage, and often the importation of mineral salts in irrigation water.

salivary gland Any of the glands that secrete saliva, a fluid that initially mixes with food in the mouth and starts the breakdown of starch.

salt An ionic compound formed when an acid reacts with a base.

saltatory conduction In myelinated neurons, rapid, node-to-node hopping of action potentials.

saprobe Heterotroph that obtains its nutrients from nonliving organic matter. Most fungi are saprobes.

sarcomere (SAR-koe-meer) Of vertebrate muscles, the basic unit of contraction; a region of myosin and actin filaments organized in parallel between two Z lines of a myofibril inside a muscle cell.

sarcoplasmic reticulum (sar-koe-PLAZ-mik reh-TIK-you-lum) In muscle cells, a membrane system that takes up, stores, and releases the calcium ions required for cross-bridge formation in sarcomeres, hence for contraction.

Schwann cells Specialized neuroglial cells that grow around neuron axons, forming a myelin sheath.

sclerenchyma One of the simple tissues of flowering plants, generally with cells having thick, lignin-impregnated walls. It supports mature plant parts and often protects seeds.

sea squirt An invertebrate chordate with a leathery or jellylike tunic and bilateral, free-swimming larvae that look like tadpoles. Ancient sea squirts may have resembled animals that gave rise to vertebrates.

second law of thermodynamics Law stating that the spontaneous direction of energy flow is from organized (high-quality) to less organized (low-quality) forms. With each conversion, some energy is randomly dispersed in a form, usually heat, that is not as readily available to do work.

second messenger A molecule inside a cell that mediates and generally triggers amplified response to a hormone.

secondary immune response Rapid, prolonged response by white blood cells, memory cells especially, to a previously encountered antigen.

secondary sexual trait A trait that is associated with maleness or femaleness but that does not play a direct role in reproduction.

secretion Generally, the release of a substance for use by the organism producing it. (Not the same as *excretion*, the expulsion of excess or waste material.) Of kidneys, a regulated stage in urine formation, in which ions and other substances move from capillaries into nephrons.

sedimentary cycle A biogeochemical cycle without a gaseous phase; the element moves from land to the seafloor, then returns only through long-term geological uplifting.

seed Of gymnosperms and flowering plants, a fully mature ovule (contains the plant embryo), with its integuments forming the seed coat.

segmentation Of earthworms and many other animals, a series of body units that may be externally similar to or quite different from one another.

segregation, Mendelian principle of [L. *se-*, apart, + *grex*, herd] The principle that diploid organisms inherit a pair of genes for each trait (on a pair of homologous chromosomes) and that the two genes segregate during meiosis and end up in separate gametes.

selective gene expression Of multicelled organisms, activation or suppression of a fraction of the genes in unique ways in different cells, leading to pronounced differences in structure and function among different cell lineages.

selfish behavior A behavior by which an individual protects or increases its own chance of producing offspring, regardless of the consequences of the group to which it belongs.

selfish herd A simple society held together by reproductive self-interest.

semen (SEE-mun) [L. *serere*, to sow] Sperm-bearing fluid expelled from a penis during male orgasm.

semiconservative replication [Gk. *hēmi*, half, + L. *conservare*, to keep] Reproduction of a DNA molecule when a complementary strand forms on each of the unzipping strands of an existing DNA double helix, the outcome being two "half-old, half-new" molecules.

senescence (sen-ESS-cents) [L. *senescere*, to grow old] Sum total of processes leading to the natural death of an organism or some of its parts.

sensation The conscious awareness of a stimulus.

sensory neuron Any of the nerve cells that act as sensory receptors, detecting specific stimuli (such as light energy) and relaying signals to the brain and spinal cord.

sensory system The "front door" of a nervous system; that portion of a nervous system that receives and sends on signals of specific changes in the external and internal environments.

sessile animal Animal that remains attached to a substrate during some stage (often the adult) of its life cycle.

sex chromosome Of most animals and some plants, a chromosome whose presence determines a new individual's gender. *Compare* autosomes.

sexual dimorphism Phenotypic differences between males and females of a species.

sexual reproduction Production of offspring from the union of gametes from two parents, by way of meiosis, gamete formation, and fertilization.

sexual selection A microevolutionary process; natural selection favoring a trait that gives the individual a competitive edge in reproductive success.

shifting cultivation The cutting and burning of trees, followed by tilling of ashes into the soil; once called slash-and-burn agriculture.

shoots The above-ground parts of vascular plants.

sieve tube member Of flowering plants, a cellular component of the interconnecting conducting tubes in phloem.

sink region In a vascular plant, any region using or stockpiling organic compounds for growth and development.

sister chromatids Of a duplicated chromosome, two DNA molecules (and associated proteins) that remain attached at their centromere only during nuclear division. Each ends up in a separate daughter nucleus.

skeletal muscle In vertebrates, an organ that contains hundreds to many thousands of muscle cells, arranged in bundles that are surrounded by connective tissue. The connective tissue extends beyond the muscle (as tendons that attach it to bone).

sliding filament model Model of muscle contraction, in which myosin filaments physically slide along and pull two sets of actin filaments toward the center of the sarcomere, which shortens. The sliding requires ATP energy and cross-bridge formation between the actin and myosin.

slime mold A type of heterotrophic protistan with a life cycle that includes free-living cells that at some point congregate and differentiate into spore-bearing structures.

small intestine Of vertebrates, the portion of the digestive system where digestion is completed and most nutrients absorbed.

smog, industrial Gray-colored air pollution that predominates in industrialized cities with cold, wet winters.

smog, photochemical Form of brown, smelly air pollution occurring in large cities with warm climates.

social behavior Cooperative, interdependent relationships among animals of the same species.

social parasite Animal that depends on the social behavior of another species to gain food, care for young, or some other factor to complete its life cycle.

sodium-potassium pump A transport protein spanning the lipid bilayer of the plasma membrane. When activated by ATP, its shape changes and it selectively transports sodium ions out of the cell and potassium ions in.

solute (SOL-yoot) [L. *solvere*, to loosen] Any substance dissolved in a solution. In water, this means spheres of hydration surround the charged parts of individual ions or molecules and keep them dispersed.

solvent Fluid in which one or more substances is dissolved.

somatic cell (so-MAT-ik) [Gk. *somā*, body] Of animals, any cell that is not a germ cell (which gives rise to gametes).

somatic nervous system Those nerves leading from the central nervous system to skeletal muscles.

sound system Of birds, the brain regions that govern muscles of the vocal organ.

source region Of vascular plants, any of the sites of photosynthesis.

speciation (spee-cee-AY-shun) The evolutionary process by which species originate. One speciation route starts with divergence of two reproductively isolated populations of a species. They become separate species when accumulated differences in allele frequencies prevent them from interbreeding successfully under natural conditions. Speciation also may be instantaneous (by way of polyploidy, especially among self-fertilizing plants).

species (SPEE-sheez) [L. *species*, a kind] Of sexually reproducing organisms, a unit consisting of one or more populations of individuals that can interbreed under natural conditions to produce fertile offspring that are reproductively isolated from other such units.

sperm [Gk. *sperma*, seed] A type of mature male gamete.

spermatogenesis (sperm-AT-oh-JEN-ih-sis) Formation of a mature sperm from a germ cell.

sphere of hydration Through positive or negative interactions, a clustering of water molecules around the individual molecules of a substance placed in water. *Compare* solute.

sphincter (SFINK-tur) Ring of muscle between regions of a tubelike system (as between the stomach and small intestine).

spinal cord Of central nervous systems, the portion threading through a canal inside the vertebral column and providing direct reflex connections between sensory and motor neurons as well as communication lines to and from the brain.

spindle apparatus A type of bipolar structure that forms during mitosis or meiosis and that moves the chromosomes. It consists of two sets of microtubules that extend from the opposite poles and that overlap at the spindle's equator.

spleen One of the lymphoid organs; it is a filtering station for blood, a reservoir of red blood cells, and a reservoir of macrophages.

sponge An invertebrate having a body with no symmetry and no organs; a framework of glassy needles and other structures imparts shape to it. Distinctive for its food-gathering, flagellated collar cells.

sporangium (spore-AN-gee-um), plural **sporangia** [Gk. *spora*, seed] The protective tissue layer that surrounds haploid spores in a sporophyte.

spore Of land plants, a type of resistant cell, often walled, that forms between the time of meiosis and fertilization. It germinates and develops into a gametophyte, the actual gamete-producing body. Of most fungi, a walled, resistant cell or multicelled structure, produced by mitosis or meiosis, that can germinate and give rise to a new mycelium.

sporophyte [Gk. *phyton*, plant] Of plant life cycles, a vegetative body that grows (by mitosis) from a zygote and that produces the spore-bearing structures.

sporozoan One of four categories of protozoans; a parasite that produces sporelike infectious agents called sporozoites. Some cause serious diseases in humans.

spring overturn Of certain lakes, the movement of dissolved oxygen from the surface layer to the depths and movement of nutrients from bottom sediments to the surface.

stabilizing selection Of a population, a persistence over time of the alleles responsible for the most common phenotypes.

stamen (STAY-mun) Of flowering plants, a male reproductive structure; commonly consists of pollen-bearing structures (anthers) on single stalks (filaments).

start codon Of protein synthesis, a base triplet in a strand of mRNA that serves as the start signal for mRNA translation.

stem cell Of animals, one of the unspecialized cells that replace themselves by ongoing mitotic divisions; portions of their daughter cells also divide and differentiate into specialized cells.

steroid (STAIR-oid) A lipid with a backbone of four carbon rings and with no fatty acid tails. Steroids differ in their functional groups. Different types have roles in metabolism, intercellular communication, and (in animals) cell membranes.

steroid hormone A type of lipid-soluble hormone, synthesized from cholesterol, that diffuses directly across the lipid bilayer of a target cell's plasma membrane and that binds with a receptor inside that cell.

stigma Of many flowering plants, the sticky or hairy surface tissue on the upper portion of the ovary that captures pollen grains and favors their germination.

stimulus [L. *stimulus*, goad] A specific change in the environment, such as a variation in light, heat, or mechanical pressure, that the body can detect through sensory receptors.

stoma (STOW-muh), plural **stomata** [Gk. *stoma*, mouth] A controllable gap between two guard cells in stems and leaves; any of the small passageways across the epidermis through which carbon dioxide moves into the plant and water vapor moves out.

stomach A muscular, stretchable sac that receives ingested food; of vertebrates, an organ between the esophagus and intestine in which considerable protein digestion occurs.

stop codon Of protein synthesis, a base triplet in a strand of mRNA that serves as the stop signal for translation, so that no more amino acids are added to the polypeptide chain.

stream A flowing-water ecosystem that starts out as a freshwater spring or seep.

stroma [Gk. *strōma*, bed] Of chloroplasts, the semifluid interior between the thylakoid membrane system and the two outer membranes; the zone where sucrose, starch, and other end products of photosynthesis are assembled.

stromatolite Of shallow seas, layered structures formed from sediments and large mats of the slowly accumulated remains of photosynthetic populations.

substrate A reactant or precursor molecule for a metabolic reaction; a specific molecule or molecules that an enzyme can chemically recognize, bind briefly to itself, and modify in a specific way.

substrate-level phosphorylation The direct, enzyme-mediated transfer of a phosphate group from the substrate of a reaction to another molecule. An example is the transfer of phosphate from an intermediate of glycolysis to ADP, forming ATP.

succession, primary (suk-SESH-un) [L. *succedere*, to follow after] Orderly changes from the time pioneer species colonize a barren habitat through replacements by various species until the climax community, when the composition of species remains steady under prevailing conditions.

succession, secondary Orderly changes in a community or patch of habitat toward the climax state after having been disturbed, as by fire.

surface-to-volume ratio A mathematical relationship in which volume increases with the cube of the diameter, but surface area increases only with the square. Of growing cells, the volume of cytoplasm increases more rapidly than the surface area of the plasma membrane that must service the cytoplasm. Because of this constraint, cells generally remain small or elongated, or have elaborate membrane foldings.

survivorship curve A plot of the age-specific survival of a group of individuals in a given environment, from the time of their birth until the last one dies.

symbiosis (sim-by-OH-sis) [Gk. *sym*, together, + *bios*, life, mode of life] A form of mutualism in which organisms of different species cannot grow and reproduce unless they spend their entire lives together in intimate interdependency. A mycorrhiza is an example.

sympathetic nerve Of the autonomic nervous system, any of the nerves generally concerned with increasing overall body activities during times of heightened awareness, excitement, or danger; also work continually in opposition with parasympathetic nerves to bring about minor adjustments in internal organs.

sympatric speciation [Gk. *sym*, together, + *patria*, native land] Speciation that follows after ecological, behavioral, or genetic barriers arise within the boundaries of a single population. This can happen instantaneously, as when polyploidy arises in a type of flowering plant that can self-fertilize or reproduce asexually.

synaptic integration (sin-AP-tik) The moment-by-moment combining of excitatory and inhibitory signals arriving at a trigger zone of a neuron.

systematics Branch of biology that deals with patterns of diversity among organisms in an evolutionary context; its three approaches include taxonomy, phylogenetic reconstruction, and classification.

systemic circuit (sis-TEM-ik) Circulation route in which oxygen-enriched blood flows from the lungs to the left half of the heart, through the rest of the body (where it gives up oxygen and takes on carbon dioxide), then back to the right side of the heart.

T lymphocyte A white blood cell with roles in immune responses.

tactile signal A physical touching that carries social significance.

taproot system A primary root and its lateral branchings.

target cell Of hormones and other signaling molecules, any cell having receptors to which they can bind.

taxonomy (tax-ON-uh-mee) Approach in biological systematics that involves identifying organisms and assigning names to them.

telophase (TEE-low-faze) Of mitosis, the final stage when chromosomes decondense into threadlike structures and two daughter nuclei form. Of meiosis I, the stage when one of each pair of homologous chromosomes has arrived at one or the other end of the spindle pole. At telophase II, chromosomes decondense and four daughter nuclei form.

telophase II Of meiosis, final stage when four daughter nuclei form.

temperate pathway A viral infection that enters a latent period; the host is not killed outright.

tendon A cord or strap of dense connective tissue that attaches muscle to bones.

territory An area that one or more individuals defend against competitors.

test An attempt to produce actual observations that match predicted or expected observations.

testcross Experimental cross to reveal whether an organism is homozygous dominant or heterozygous for a trait. The organism showing dominance is crossed to an individual known to be homozygous recessive for the same trait.

testis, plural **testes** Male gonad; primary reproductive organ in which male gametes and sex hormones are produced.

testosterone (tess-TOSS-tuh-rown) In male mammals, a major sex hormone that helps control male reproductive functions.

tetanus Of muscles, a large contraction in which repeated stimulation of a motor unit causes muscle twitches to mechanically run together. In a disease by the same name, toxins prevent muscle relaxation.

theory A testable explanation of a broad range of related phenomena. In modern science, only explanations that have been extensively tested and can be relied upon with a very high degree of confidence are accorded the status of theory.

thermal inversion Situation in which a layer of dense, cool air becomes trapped beneath a layer of warm air; can cause air pollutants to accumulate to dangerous levels close to the ground.

thermophile A type of archaebacterium that lives in hot springs, highly acidic soils, and near hydrothermal vents.

thermoreceptor Sensory cell that can detect radiant energy associated with temperature.

thigmotropism (thig-MOTE-ruh-pizm) [Gk. *thigm*, touch] Of vascular plants, growth orientation in response to physical contact with a solid object, as when a vine curls around a fencepost.

threshold Of neurons and other excitable cells, a certain minimum amount by which the voltage difference across the plasma membrane must change to produce an action potential.

thylakoid membrane system Of chloroplasts, an internal membrane system commonly folded into flattened channels and disks (*grana*) and containing light-absorbing pigments and enzymes used in the formation of ATP, NADPH, or both during photosynthesis.

thymine Nitrogen-containing base in some nucleotides.

thymus gland A lymphoid organ with endocrine functions; lymphocytes of the immune system multiply, differentiate, and mature in its tissues, and its hormone secretions affect their functions.

thyroid gland Of endocrine systems, a gland that produces hormones that affect overall metabolic rates, growth, and development.

tissue Of multicelled organisms, a group of cells and intercellular substances that function together in one or more specialized tasks.

tonicity The relative concentrations of solutes in two fluids, such as inside and outside a cell. When solute concentrations are isotonic (equal in both fluids), water shows no net osmotic movement in either direction. When one fluid is hypotonic (has less solutes than the other), the other is hypertonic (has more

solutes) and is the direction in which water tends to move.

tooth Of the mouth of various animals, one of the hardened appendages used to secure or mechanically pummel food; sometimes used in defense.

tracer A radioisotope used to label a substance so that its pathway or destination in a cell, organism, ecosystem, or some other system can be tracked, as by scintillation counters that detect its emissions.

trachea (TRAY-kee-uh), plural **tracheae** An air-conducting tube that functions in respiration; of land vertebrates, the windpipe, which carries air between the larynx and bronchi.

tracheal respiration Of insects, spiders, and some other animals, a respiratory system consisting of finely branching tracheae that extend from openings in the integument and that dead-end in body tissues.

tracheid (TRAY-kid) Of flowering plants, one of two types of cells in xylem that conduct water and dissolved minerals.

transcript-processing controls Of eukaryotic cells, controls that govern modification of new mRNA molecules into mature transcripts before shipment from the nucleus.

transcription [L. *trans*, across, + *scribere*, to write] Of protein synthesis, the assembly of an RNA strand on one of the two strands of a DNA double helix; the base sequence of the resulting transcript is complementary to the DNA region on which it was assembled.

transcriptional controls Of eukaryotic cells, controls influencing when and to what degree a particular gene will be transcribed.

transfer RNA (tRNA) Of protein synthesis, any of the type of RNA molecules that bind and deliver specific amino acids to ribosomes *and* pair with mRNA code words for those amino acids.

translation Of protein synthesis, the conversion of the coded sequence of information in mRNA into a particular sequence of amino acids to form a polypeptide chain; depends on interactions of rRNA, tRNA, and mRNA.

translational controls Of eukaryotic cells, controls governing the rates at which mRNA transcripts that reach the cytoplasm will be translated into polypeptide chains at ribosomes.

translocation Of cells, a change in a chromosome's structure following the insertion of part of a nonhomologous chromosome into it. Of vascular plants, conduction of organic compounds through the plant body by way of the phloem.

transpiration Evaporative water loss from stems and leaves.

transport control Of eukaryotic cells, controls governing when mature mRNA transcripts are shipped from the nucleus into the cytoplasm.

transposable element DNA element that can spontaneously "jump" to new locations in the same DNA molecule or a different one. Such elements often inactivate the genes into which they become inserted and give rise to observable changes in phenotype.

trisomy (TRY-so-mee) Of diploid cells, the abnormal presence of three of one type of chromosome.

trophic level (TROE-fik) [Gk. *trophos*, feeder] All the organisms in an ecosystem that are the same number of transfer steps away from the energy input into the system.

tropical rain forest A type of biome where rainfall is regular and heavy, the annual mean temperature is 25°C, and humidity is 80 percent or more; characterized by great biodiversity.

tropism (TROE-prizm) Of vascular plants, a growth response to an environmental factor, such as growth toward light.

true-breeding Of sexually reproducing organisms, a lineage in which the offspring of successive generations are just like the parents in one or more traits.

tumor A tissue mass composed of cells that are dividing at an abnormally high rate.

turgor pressure (TUR-gore) [L. *turgere*, to swell] Internal pressure applied to a cell wall when water moves by osmosis into the cell.

uniformitarianism The theory that existing geologic features are an outcome of a long history of gradual changes, interrupted now and then by huge earthquakes and other catastrophic events.

upwelling An upward movement of deep, nutrient-rich water along coasts to replace surface waters that winds move away from shore.

uracil (YUR-uh-sill) Nitrogen-containing base found in RNA molecules; can base-pair with adenine.

ureter A tubular channel for urine flow between the kidney and urinary bladder.

urethra A tubular channel for urine flow between the urinary bladder and an opening at the body surface.

urinary bladder A distensible sac in which urine is temporarily stored before being excreted.

urinary excretion A mechanism by which excess water and solutes are removed by way of a urinary system.

urinary system An organ system that adjusts the volume and composition of blood, and so helps maintain extracellular fluid.

urine Fluid formed by filtration, reabsorption, and secretion in kidneys; consists of wastes, excess water, and solutes.

uterus (YOU-tur-us) [L. *uterus*, womb] Chamber in which the developing embryo is contained and nurtured during pregnancy.

vaccine An antigen-containing preparation, swallowed or injected, that increases immunity to certain diseases. It induces formation of huge armies of effector and memory B and T cell populations.

vagina Part of a female reproductive system that receives sperm, forms part of the birth canal, and channels menstrual flow to the exterior.

variable Of a scientific experiment, the only factor that is not exactly the same in the experimental group as it is in the control group.

vascular bundle Of vascular plants, the arrangement of primary xylem and phloem into multistranded, sheathed cords that thread lengthwise through the ground tissue system.

vascular cambium Of vascular plants, a lateral meristem that increases stem or root diameter.

vascular cylinder Of plant roots, the arrangement of vascular tissues as a central cylinder.

vascular plant Plant having tissues that transport water and solutes through well-developed roots, stems, and leaves.

vascular tissue system Xylem and phloem; the conducting tissues that distribute water and solutes through the body of vascular plants.

vein Of the circulatory system, any of the large-diameter vessels that lead back to the heart; of leaves, one of the vascular bundles that thread through photosynthetic tissues.

ventricle (VEN-tri-kuhl) Of the vertebrate heart, one of two chambers from which blood is pumped out. *Compare* atrium.

venule A small blood vessel that accepts blood from capillaries and delivers it to a vein; also overlaps capillaries somewhat in function.

vernalization Of flowering plants, stimulation of flowering by exposure to low temperatures.

vertebra, plural **vertebrae** Of vertebrate animals, one of a series of hard bones arranged with intervertebral disks into a backbone.

vertebrate Animal having a backbone of bony segments, the vertebrae.

vesicle (VESS-ih-kul) [L. *vesicula*, little bladder] Within the cytoplasm of cells, one of a variety of small membrane-bound sacs that function in the transport, storage, or digestion of substances or in some other activity.

vessel member One of the cells of xylem, dead at maturity, the walls of which form the water-conducting pipelines.

villus (VIL-us), plural **villi** Any of several types of absorptive structures projecting from the free surface of an epithelium.

viroid An infectious nucleic acid that has no protein coat; a tiny rod or circle of single-stranded RNA.

virus A noncellular infectious agent, consisting of DNA or RNA and a protein coat; can replicate only after its genetic material enters a host cell and subverts its metabolic machinery.

vision Precise light focusing onto a layer of photoreceptive cells that is dense enough to sample details concerning a given light stimulus, followed by image formation in the brain.

visual signal An observable action or cue that functions as a communication signal.

vitamin Any of more than a dozen organic substances that animals require in small amounts for normal cell metabolism but generally cannot synthesize for themselves.

vocal cord One of the thickened, muscular folds of the larynx that help produce sound waves for speech.

water mold A type of saprobic or parasitic fungus that lives in fresh water or moist soil.

water potential The sum of two opposing forces (osmosis and turgor pressure) that can cause the directional movement of water into or out of a walled cell.

water table The upper limit at which the ground in a specified region is fully saturated with water.

watershed Any specified region in which all precipitation drains into a single stream or river.

wax A type of lipid with long-chain fatty acid tails that help form protective, lubricating, or water-repellent coatings.

white blood cell Leukocyte; of vertebrates, any of the macrophages, eosinophils, neutrophils, and other cells which, together with their products, comprise the immune system.

white matter Of spinal cords, major nerve tracts so named because of the glistening myelin sheaths of their axons.

wild-type allele Of a population, the allele that occurs normally or with greatest frequency at a given gene locus.

wing Of birds, a forelimb of feathers, powerful muscles, and lightweight bones that functions in flight. Of insects, a structure that develops as a lateral fold of the exoskeleton and functions in flight.

X chromosome Of humans, a sex chromosome with genes that cause an embryo to develop into a female, provided that it inherits a pair of these.

X-linked gene Any gene on an X chromosome.

X-linked recessive inheritance Recessive condition in which the responsible, mutated gene occurs on the X chromosome.

xylem (ZYE-lum) [Gk. *xylon*, wood] Of vascular plants, a tissue that transports water and solutes through the plant body.

Y chromosome Of humans, a sex chromosome with genes that cause the embryo that inherited it to develop into a male.

Y-linked gene Any gene on a Y chromosome.

yellow marrow A fatty tissue in the cavities of most mature bones that produces red blood cells when blood loss from the body is severe.

yolk sac Of land vertebrates, one of four extraembryonic membranes. In most shelled eggs, it holds nutritive yolk. In humans, part becomes a site of blood cell formation and some of its cells give rise to the forerunners of gametes.

zero population growth A population for which the number of births is balanced by the number of deaths over a specified period, assuming immigration and emigration also are balanced.

zooplankton A freshwater or marine community of floating or weakly swimming heterotrophs, mostly microscopic, such as rotifers and copepods.

zygospore-forming fungus A type of fungus for which a thick spore wall forms around the zygote; this resting spore germinates and gives rise to stalked, spore-bearing structures.

zygote (ZYE-goat) The first cell of a new individual, formed by the fusion of a sperm nucleus with the nucleus of an egg (fertilization).

CREDITS AND ACKNOWLEDGMENTS

Front Matter

Page i Thomas D. Mangelsen/Images of Nature / **Pages vi–vii** Stock Imagery / **Pages viii–ix** James M. Bell/Photo Researchers / **Pages x–xi** S. Stammers/SPL/Photo Researchers / **Pages xii–xiii** © 1990 Arthur M. Greene / **Pages xiv–xv** Bonnie Rausch/Photo Researchers / **Pages xvi–xvii** Lennart Nilsson from *A Child Is Born*, © 1966, 1977 Dell Publishing Company, Inc. / **Pages xviii–xix** Thomas D. Mangelsen/Images of Nature / **Pages xx–xxi** Jim Doran

Page 1 Tom Van Sant/The GeoSphere Project, Santa Monica, CA

Chapter 1

1.1 Frank Kaczmarek / **Page 3** Art by American Composition & Graphics, Inc. / **1.3** (b) Walt Anderson/Visuals Unlimited; (c) Gregory Dimijian/Photo Researchers; (d) Alan Weaving/Ardea, London / **1.4** Jack deConingh / **1.5** J. A. Bishop and L. M. Cook / **1.6** (a) Tony Brain/SPL/Photo Researchers; (b) M. Abbey/Visuals Unlimited; (c) Dennis Brokaw; (d), (e) Edward S. Ross; (f) Pat & Tom Leeson/Photo Researchers / **1.7** Levi Publishing Company / **Page 13** Photograph Jack deConingh; art by Raychel Ciemma / **Page 15** James M. Bell/Photo Researchers

Chapter 2

2.1 Martin Rogers/FPG / **2.2** Jack Carey / **Page 20** (a) (left) Kingsley R. Stern; (right) Chip Clark / **Page 21** (c) (above) Dr. Harry T. Chugani, M.D., UCLA School of Medicine; (below) Hank Morgan/Rainbow / **2.8** Art by Palay/Beaubois / **2.9** Photograph Richard Riley/FPG / **2.10** Art by Palay/Beaubois; photograph Colin Monteath, Hedgehog House, New Zealand / **2.11** H. Eisenbeiss/Frank Lane Picture Agency / **2.12** Art by Raychel Ciemma / **2.14** Michael Grecco / Picture Group

Chapter 3

3.1 Lewis L. Lainey / **3.5** Peter Steyn/Ardea, London / **3.6** Micrograph Biophoto Associates/Science Source/Photo Researchers / **3.9** (a) David Scharf/Peter Arnold, Inc.; (b) Robert C. Simpson/Nature Stock / **3.10, 3.11** Art by Precision Graphics / **3.12** Clem Haagner/Ardea, London / **3.13** Art by Precision Graphics / **3.14** (a) Kenneth Lorenzen; (b) Larry Lefever/Grant Heilman / **3.15, 3.16** Art by Precision Graphics / **3.19** Art by Palay/Beaubois / **3.20** CNRI/SPL/Photo Researchers; art by Robert Demarest / **3.21, 3.22** Art by Precision Graphics / **3.23** A. Lesk/SPL/Photo Researchers

Chapter 4

4.1 (a) (left) National Library of Medicine; (right) Armed Forces Institute of Pathology; (b) The Bettmann Archive; (d) The Francis A. Countway Library of Medicine / **4.2** Art by Precision Graphics / **Page 55** (d–g) Jeremy Pickett-Heaps, School of Botany, University of Melbourne / **4.5** (a) Micrograph G. Cohen-Bazire; art by Palay/Beaubois; (b) K. G. Murti/Visuals Unlimited / **4.6** (a) Gary Gaard and Arthur Kelman; (b) R. Calentine/Visuals Unlimited / **4.7** Art by Raychel Ciemma / **4.8** Micrograph M. C. Ledbetter, Brookhaven National Laboratory; art by Raychel Ciemma / **4.9** Art by Raychel Ciemma / **4.10** Micrograph G. L. Decker; art by Raychel Ciemma / **4.11** Micrograph Stephen L. Wolfe; art by Raychel Ciemma / **4.12** (a) (left) Don W. Fawcett/Visuals Unlimited; (right) A. C. Faberge, *Cell and Tissue Research*, 151:403–415, 1974 / **4.13** (b) Art by Raychel Ciemma / **4.14** (a), (b) Micrographs Don W. Fawcett/Visuals Unlimited; (below right) art by Robert Demarest / **4.15** (left) Art by Robert Demarest after a model by J. Kephart; micrograph Gary W. Grimes / **4.16** Gary W. Grimes / **4.17** Micrograph Keith R. Porter / **4.18** Micrograph L. K. Shumway; (below) art by Palay/Beaubois / **4.19** (a) Andrew S. Bajer; (b) J. Victor Small and Gottfried Rinnerthaler; (c) art by Precision Graphics / **4.20** (a) C. J. Brokaw; (b) Sidney L. Tamm / **4.21** Art by Precision Graphics after Stephen L. Wolfe, *Molecular and Cellular Biology*, Wadsworth, 1993 / **4.22** (a) H. A. Core, W. A. Coté, and A. C. Day, *Wood Structure and Identification*, Second Editon, Syracuse University Press, 1979; (b) sketch by D. & V. Hennings after P. Raven et al., *Biology of Plants*, Third Edition, Worth Publishers, 1981; (c) P. A. Roelofsen

Chapter 5

5.1 (a) Runk/Schoenberger/Grant Heilman; (b) Inigo Everson/Bruce Coleman Ltd. / **5.3** Art by Raychel Ciemma / **Page 81** Art by Palay/Beaubois; micrograph P. Pinto da Silva and D. Branton, *Journal of Cell Biology*, 45:598, by copyright permission of The Rockefeller University Press / **5.4** Art by Palay/Beaubois / **5.5** Micrograph M. Sheetz, R. Painter, and S. Singer, *Journal of Cell Biology*, 70:193, by copyright permission of The Rockefeller University Press / **5.6** Art by Precision Graphics after Stephen L. Wolfe, *Molecular and Cellular Biology*, Wadsworth, 1993 / **Page 84** (a) Photographs Frank B. Salisbury; (b) Frieder Sauer/Bruce Coleman Ltd. / **5.7** Art by Leonard Morgan; (above) after Alberts et al., *Molecular Biology of the Cell*, Second edition, Garland Publishing Co., 1989 / **5.10** Art by Leonard Morgan / **5.12** M. M. Perry and A. B. Gilbert / **5.13** Art by Palay/Beaubois

Chapter 6

6.1 Gary Head / **6.2** Evan Cerasoli / **6.3** (above) NASA; (below) Manfred Kage/Peter Arnold, Inc. / **6.4** Art by Palay/Beaubois / **6.8** Thomas A. Steitz / **6.9** Art by Palay/Beaubois / **6.11** Photograph Douglas Faulkner/Sally Faulkner Collection / **6.14** Art by Raychel Ciemma and American Composition & Graphics, Inc. after B. Alberts et al., *Molecular Biology of the Cell*, Garland Publishing Co., 1983 / **Page 103** (a), (b) Oxford Scientific Films/Animals Animals; (c) © Raymond Mendez/Animals Animals; (d) Keith V. Wood

Chapter 7

7.1 David R. Frazier/Photo Researchers / **7.2** (a) Hans Reinhard/Bruce Coleman Ltd.; (e), (f) micrographs David Fisher; art by Raychel Ciemma and American Composition & Graphics, Inc. / **7.3** (a) Barker-Blakenship/FPG; (b) art by Precision Graphics after Govindjee; **7.4** Photograph Carolina Biological Supply Company; art by Raychel Ciemma / **7.5** Larry West/FPG / **7.6** Art by Illustrious, Inc. / **7.7** E. R. Degginger / **7.8** Art by Raychel Ciemma and Precision Graphics / **7.9** Art by Precision Graphics / **7.10** Art by Raychel Ciemma and Precision Graphics / **Page 117** NASA / **7.11** Art by Raychel Ciemma / **7.12** Martin Grosnick/Ardea, London

Chapter 8

8.1 Stephen Dalton/Photo Researchers / **8.3** (left) (above) Dennis Brokaw; (below) Janeart/Image Bank; (right) Paolo Fioratti; art by Precision Graphics / **8.4** (right) Art by Palay/Beaubois / **8.5** (a) Micrograph Keith R. Porter; (b) art by L. Calver; (c) art by Raychel Ciemma / **8.11** (b) Adrian Warren/Ardea, London; (c) David M. Phillips/Visuals Unlimited / **8.12** Photograph Gary Head / **Page 137** R. Llewellyn/Superstock, Inc. / **Page 139** © Lennart Nilsson

Chapter 9

9.1 (left and right center) Chris Huss; (right top and bottom) Tony Dawson / **9.2** C. J. Harrison et al., *Cytogenetics and Cell Genetics* 35:21–27, copyright 1983 S. Karger A.G., Basel / **9.3** CNRI/SPL/Photo Researchers / **9.5** Andrew S. Bajer, University of Oregon / **9.6** Micrographs Ed Reschke; art by Raychel Ciemma / **9.9** B. A. Palevitz and E. H. Newcomb, University of Wisconsin/BPS/Tom Stack & Associates / **9.10** Micrographs H. Beams and R. G. Kessel, *American Scientist*, 64:279–290, 1976 / **9.11** (a–c), (e) Lennart Nilsson from *A Child Is Born* © 1966, 1967 Dell Publishing Company, Inc.; (d) Lennart Nilsson from *Behold Man*, © 1974 by Albert Bonniers Forlag and Little, Brown and Company, Boston

Chapter 10

10.1 (a) Jane Burton/Bruce Coleman Ltd.; (b) Dan Kline/Visuals Unlimited / **10.2** Courtesy of Kirk Douglas/The Bryna Company / **10.3** Art by Precision Graphics / **10.4** CNRI/SPL/Photo Researchers / **10.5** Art by Raychel Ciemma / **10.6** Art by Raychel Ciemma and American Composition & Graphics, Inc. / **10.7** Art by Raychel Ciemma / **10.10** Micrograph David M. Phillips/Visuals Unlimited / **10.11** Art by Raychel Ciemma

Chapter 11

11.1 (a) Frank Trapper/Sygma; (b) Focus on Sports; (c) Fabian/Sygma; (d) Moravian Museum, Brno / **11.2** Photograph Jean M. Labat/Ardea, London; art by Jennifer Wardrip / **11.5, 11.7** Art by Hans & Cassady, Inc. / **11.8** Art by Raychel Ciemma / **11.10** Photographs William E. Ferguson / **11.11** Photograph Bill Longcore/Photo Researchers / **11.12** (a), (b) Michael Stuckey/Comstock Inc.; (c) Russ Kinne/Comstock Inc. / **11.13** David Hosking / **11.14** Tedd Somes / **11.15** (top to bottom) Frank Cezus; Frank Cezus; Michael Keller; Ted Beaudin; Stan Sholik/all FPG / **11.16** Dan Fairbanks, Brigham Young University / **11.17** After John G. Torrey, *Development in Flowering Plants*, by permission of Macmillan Publishing Company, copyright © 1967 by John G. Torrey / **11.18** Photograph Jane Burton/Bruce Coleman Ltd.; art by D. & V. Hennings / **Page 184** Evan Cerasoli

Chapter 12

12.1 Eddie Adams/AP Photo / **Page 189** (a) Photograph Omikron/Photo Researchers / **12.3** From Lennart Nilsson, *A Child Is Born*, © 1966, 1977 Dell Publishing Company, Inc. (b) Redrawn by Robert Demarest by permission from page 126 of Michael Cummings, *Human Heredity: Principles and Issues*, Third Edition. Copyright © 1994 by West Publishing Company. All rights reserved.; (c) art by Robert Demarest after Patten, Carlson, and others / **12.4** Photographs Carolina Biological Supply Company / **12.7** Photograph Dr. Victor A. McKusick / **12.11** After Victor A. McKusick, *Human Genetics*, Second edition, copyright 1969. Reprinted by permission of Prentice-Hall, Inc., Englewood Cliffs, NJ; photograph The Bettmann Archive / **12.12** (a, b) Courtesy of G. H. Valentine; (c) C. J. Harrison / **12.13** Art by Raychel Ciemma / **12.14** (a) Cytogenetics Laboratory, University of California, San Francisco; (b) after Collman and Stoller, *American Journal of Public Health*, 52, 1962 / **12.15** (left) Used by permission of Carole Iafrate; (center and right) Courtesy of Peninsula Association for Retarded Children and Adults, San Mateo Special Olympics, Burlingame, CA / **Page 205** (a) Art by Palay/Beaubois; (b) From Fran Heyl Associates, © Jacques Cohen, Computer enhanced by © Pix*Elation / **Page 207** (above) Bonnie Kamin/Stuart Kenter Associates; (below) Carolina Biological Supply Company

Chapter 13

13.1 Photograph A. C. Barrington Brown © 1968 J. D. Watson; model A. Lesk/SPL/Photo Researchers / **13.3** (b) Micrograph Lee D. Simon/Science Source/Photo Researchers / **13.5** Micrograph Biophoto Associates/SPL/Photo Researchers / **13.6, 13.8** Art by Precision Graphics / **13.9** Ken Greer/Visuals Unlimited / **13.10** (a) (left) C. J. Harrison et al., *Cytogenetics and Cell Genetics* 35:21–27, copyright 1983 S. Karger A.G., Basel; (right) U. K. Laemmli from *Cell*, 12:817–828, copyright 1977 by MIT/Cell Press; (b) B. Hamkalo; (c) O. L. Miller, Jr. and Steve L. McKnight; art by Nadine Sokol

Chapter 14

14.1 (left) Dennis Hallinan/FPG; (right) Kevin Magee/Tom Stack & Associates / **Page 220** (a) Gary Head; (b) from Stephen L. Wolfe, *Molecular and Cellular Biology*, Wadsworth, 1993 / **14.3** Art by Hans & Cassady, Inc. / **14.7** (a) Courtesy of Thomas A. Steitz from *Science*, 246:1135–1142, December 1, 1989 / **14.9** Art by Raychel Ciemma and American Composition & Graphics, Inc. / **14.10** Micrograph Dr. John E. Heuser, Washington University School of Medicine, St. Louis, MO; art by Palay/Beaubois / **14.13** Peter Starlinger / **14.14** Art by Raychel Ciemma

Chapter 15

15.1 Lennart Nilsson © Boehringer Ingelheim International GmbH / **15.2** Art by Palay/Beaubois and Hans & Cassady / **15.3** Art by Raychel Ciemma / **15.4** (a) M. Roth and J. Gall / **15.5** Art by Palay/Beaubois / **15.6** (a), (b) Stuart Kenter Associates / **15.7** Jack Carey / **15.8** W. Beerman / **15.9** Frank B. Salisbury / **15.10** Art by Precision Graphics

Chapter 16

16.1 Secchi-Lecague/Roussel-UCLAF/CNRI/SPL/Photo Researchers; (inset) Ted Thai/Time Magazine / **16.2** Dr. Huntington Potter and Dr. David Dressler / **16.5** After Stephen L. Wolfe, *Molecular and Cellular Biology*, Wadsworth, 1993 / **Page 252** (a) Damon Biotech, Inc. / **16.9** Michael Maloney/San Francisco Chronicle / **16.10** (a) W. Merrill; (b) Keith V. Wood / **16.11** (a), (b) Monsanto Company; (c), (d) Calgene, Inc. / **16.12** R. Brinster and R. E. Hammer, School of Veterinary Medicine, University of Pennsylvania / **Page 259** S. Stammers/SPL/Photo Researchers

INDEX